Perspectives on Scientific Argumentation

Myint Swe Khine
Editor

Perspectives on Scientific Argumentation

Theory, Practice and Research

Foreword by Deanna Kuhn

Springer

Editor
Myint Swe Khine
Bahrain Teachers College
University of Bahrain
Manama
Kingdom of Bahrain
mskhine@btc.uob.bh
Dr.MKhine@gmail.com

ISBN 978-94-007-2469-3 e-ISBN 978-94-007-2470-9
DOI 10.1007/978-94-007-2470-9
Springer Dordrecht Heidelberg London New York

Library of Congress Control Number: 2011939071

Printed on acid-free paper

Springer is part of Springer Science+Business Media (www.springer.com)

Foreword

This volume, together with a similar recent collection edited by Erduran and Jiménez-Aleixandre (2007), mark the concentration of attention now being devoted to argumentation as a process central to science and to science education. The idea of science as argument (Kuhn, 1993, 2010) is a key one in that it encompasses both the epistemological and the procedural aspects of doing, teaching, and learning science. Accordingly, it has significant implications for just about all of the concerns of science educators (Duschl, 2008).

The epistemological implications are the most challenging for science educators. It is one thing for us to recognize argument as the core of scientific practice and quite another to get young students, and even older ones, to conceptualize science in this way. In Chapter 2, Nussbaum, Sinatra, and Owens discuss the need to confront students' misconceptions, and perhaps the most pervasive and enduring of these is the misconception of science as an accumulation of facts. If students are to come to believe otherwise, it is essential that educators model scientific practice as something else than this – namely, as a successive revision of theories as new evidence is brought to bear on them. In the absence of such modeling, how can we expect students to replace their naïve epistemologies of science with more robust ones?

Argumentation is of course the heart of scientific practice. With the right kind of modeling, engagement, and practice, students can come to experience this scientific process for themselves. In so doing they stand to transform their conception of the product of science from one of facts to one of arguments and the scientific process to one of coordinating theories and evidence. As Cavagnetto and Hand put it in their chapter (Chapter 3), they come to recognize the distinction between data and evidence. There is as yet little rigorous evidence that engagement in science as argument is sufficient to engender change with respect to this epistemological transformation, but a recent intervention study by Iordanou (2010) contains some promising findings that suggest it can be.

Osborne and Szu, Berland and Hammer, and several other of the authors in this volume recognize and lament the fact that argumentation of any form is seldom present in science classrooms. It is not clear what the most effective path will be toward changing this state of affairs, but one thing that is clear is that the efforts of practitioners and policy makers need to be supported by an expanded knowledge base regarding the nature of argumentive skills and their development. Assessment and analysis are needed at the level of the classroom, as Sampson, Enderle, and Walker have undertaken and described in Chapter 12, at the level of dialogic argumentation, the approach I have taken in my own research (Kuhn, 2010), and at the level of individual argument, which Osborne and Szu address. At each of these levels, although considerable progress has been made, we do not yet know enough about the nature of the competencies involved.

A recurring and key question is whether these competencies are specific to the domain of science or broader in scope, a question that Bricker and Bell, and also Corner, address in Chapters 7 and 10. My own view is that the question should not be conceived as an either/or one. Clearly, and at the very least, there are commonalities in the argumentive strategies people bring to bear regarding scientific matters and matters in other domains. At the same time there are particularities specific to domains that warrant attention in their own right. Again, Iordanou's (2010) research brings important data to bear on this issue in her comparison of the effects of an argumentation intervention in the science domain and one in a social science domain on transfer of skills to the domain not involved in the intervention. She finds transfer, but also asymmetry, with skill development in the science domain showing more transfer to the social domain than vice versa.

A related question, addressed by Cavagnetto and Hand, is the pedagogical one of whether to foster the development of argument skills in a relatively stand-alone or deeply content-embedded manner. My own view, again, is that the question is not an either/or one. All instruction of course needs to be conducted in the context of meaningful content. But there is no reason that the balance between skill goals and content goals need be constant across instructional units. Skill goals may be most prominent in some cases, and content goals in another. Most important is that we be able to *identify* with the greatest possible clarity, the skills (as well as content) that we wish students to master. And here again, Osborne and Szu remind us that work remains to be done.

The surge of interest in argumentation in science education has led to a wide range of pedagogical efforts to support development of students' skills, and these warrant careful analysis and comparison, as Simon, Richardson, and Amos note in Chapter 6. And they of course depend on rigorous analysis of the skills and subskills at stake, as Bulgren and Ellis describe in Chapter 8. Technologically supported scaffolding of argumentation is promising, as Clark et al. and Wu and Tsai describe in Chapters 9 and 11. My own studies rely on technology to support (as well as to assess) students' argumentation (Kuhn, 2010; Goldstein, Crowell, & Kuhn, 2009). Yet we need to be careful that we do not jump in with so much or such highly structured support as to obscure observation of students' own reasoning and the kinds of strategies they bring to bear in engaging in direct, authentic debate with their peers. These are the phenomena we seek to understand and to support, but we cannot hope to design the most effective supports without the requisite understanding. Appearance of the current volume suggests a level of interest and engagement on the part of both researchers and practitioners that stands to move us forward in this regard.

Teachers College, Columbia University
New York, USA
Deanna Kuhn

References

Duschl, R. (2008). Science education in three-part harmony: Balancing conceptual, epistemic, and social learning goals. *Review of Research in Education, 32*, 268–291.

Erduran, S., & Jiménez-Aleixandre, M. P. (Eds.). (2007). *Argumentation in science education: Perspectives from classroom-based research*. New York: Springer.

Goldstein, M., Crowell, A., & Kuhn, D. (2009). What constitutes skilled argumentation and how does it develop? *Informal Logic, 29*, 379–395.

Iordanou, K. (2010). Developing argument skills across scientific and social domains. *Journal of Cognition and Development, 11*, 293–327.

Kuhn, D. (1993). Science as argument: Implications for teaching and learning scientific thinking. Science Education, *77*(3), 319–337.

Kuhn, D. (2010). Teaching and learning science as argument. *Science Education, 94*(5), 810–824.

Contents

Contributors

Ruth Amos Institute of Education, University of London, London, UK, r.amos@ioe.ac.uk

Philip Bell University of Washington, Seattle, WA, USA, pbell@u.washington.edu

Leema Kuhn Berland University of Texas, Austin, TX, USA, leema.berland@mail.utexas.edu

Leah A. Bricker University of Washington, Seattle, WA, USA, lbricker@u.washington.edu

Janis A. Bulgren University of Kansas, Lawrence, KS, USA, jbulgren@ku.edu

Andy Cavagnetto Binghamton University, State University of New York, Binghamton, NY, USA, acavagne@binghamton.edu

Hsin-Yi Chang National Kaohsiung Normal University, Kaohsiung City, Taiwan, hsinyichang@nknucc.nknu.edu.tw

Douglas B. Clark Vanderbilt University, Nashville, TN, USA, doug.clark@vanderbilt.edu

Adam Corner School of Psychology, Cardiff University, Cardiff, Wales, UK, corneraj@cardiff.ac.uk

James D. Ellis University of Kansas, Lawrence, KS, USA, jdellis@ku.edu

Patrick J. Enderle Florida State University, Tallahassee, FL, USA, patrick.enderleadi@gmail.com

David Hammer Tufts University, Medford, MA, USA, david.hammer@tufts.edu

Brian Hand University of Iowa, Iowa City, IA, USA, Brian-Hand@uiowa.edu

Gregory J. Kelly The Pennsylvania State University, University Park, PA, USA, gkelly@psu.edu

Myint Swe Khine Bahrain Teachers College, University of Bahrain, Manama, Kingdom of Bahrain, mskhine@btc.uob.bh; Dr.MKhine@gmail.com

Deanna Kuhn Teacher College, Columbia University, New York, NY, USA, Dk100@columbia.edu

Anna MacPherson Stanford University, Stanford, CA, USA, annamac@stanford.edu

Scott P. McDonald The Pennsylvania State University, University Park, PA, USA, smcdonald@psu.edu

E. Michael Nussbaum University of Nevada, Las Vegas, NV, USA, Nussbaum@unlv.nevada.edu

Jonathan Osborne Stanford University, Stanford, CA, USA, osbornej@stanford.edu

Marissa C. Owens University of Nevada, Las Vegas, NV, USA, owensm17@unlv.nevada.edu

Alexis Patterson Stanford University, Stanford, CA, USA, alexisdp@stanford.edu

Katherine Richardson Institute of Education, University of London, London, UK, katherine.richardson@teachfirst.org.uk

Victor Sampson Florida State University, Tallahassee, FL, USA, vsampson@fsu.edu

Beat Schwendimann University of California, Berkeley, CA, USA, Beat.Schwendimann@gmail.com

Shirley Simon Institute of Education, University of London, London, UK, s.simon@ioe.ac.uk

Gale M. Sinatra University of Nevada, Las Vegas, NV, USA, gsinatra@usc.edu

Evan Szu Stanford University, Stanford, CA, USA, eszu@stanford.edu

Erika D. Tate University of California, Berkeley, CA, USA, erikadtate@gmail.com

Chin-Chung Tsai National Taiwan University of Science and Technology, Taipei, Taiwan, cctsai@mail.ntust.edu.tw

Joi Phelps Walker Tallahassee Community College, Tallahassee, FL, USA, WALKERJ@tcc.fl.edu

Ying-Tien Wu Graduate Institute of Network Learning Technology, National Central University, JhongLi, Taiwan, ytwu@mail.ntust.edu.tw

Helen Zhang University of California, Berkeley, CA, USA, zhang.zhihui@gmail.com

Part I
Theoretical Premises of the Study of Argumentation

Chapter 1
Introduction

Jonathan Osborne, Anna MacPherson, Alexis Patterson, and Evan Szu

Why another book on argumentation? Is this just yet another passing enthusiasm amongst the research community? Such questions deserve a serious response. First, we should point out that argumentation is not just a focus of work in science education but also a focus of research in mathematics education (Resnick, Michaels, & O'Connor, in press; Yackel & Cobb, 1996), language arts (Andrews, 1995; Applebee, Langer, Nystrand, & Gamoran, 2003) and history teaching (Wineburg, 2001). More fundamentally, the nature of human reasoning has been a substantial concern of psychologists with major contributions from a 30-year programme of work by Deanna Kuhn and her co-workers (Kuhn, 1991, 1999, 2009) – the author of the Foreword to this book. For psychologists, the focus of interest has been providing descriptive accounts of how humans reason and developing theoretical models. Zimmerman (1999, 2007) offers two major reviews of the development of scientific thinking skills arguing for the significance of studying reasoning in informing our understanding of the development of human cognition. The nature of rationality and reasoning has also been an enduring feature of study for philosophers (Hempel, 1962; Nercessian, 2008; Siegel, 1989; Thagard, 2008) who offer normative rather than descriptive accounts.

Why then this interest in argumentation? Much of the motivation is driven by a view expressed by Billig that 'learning to argue is learning to think' – in essence that argument and dialogue are the processes by which we learn (Andriessen, 2006). Some of the interest comes too from the higher expectations society has of its educational system. The past two decades have seen a rise in the belief that education is too important to be left solely to the educators. The contemporary economic environment is cast as a competition between countries and as a race to the top – the eponymous title of the current US policy initiative in education and of a UK report reviewing the needs of the science and technology base (Lord Sainsbury of Turville, 2007). The common thread to these reports is that societies that educate people to be hewers of wood and drawers of water are doomed to lose this competition as

J. Osborne (✉)
Stanford University, Stanford, CA, USA
e-mail: osbornej@stanford.edu

M.S. Khine (ed.), *Perspectives on Scientific Argumentation*,
DOI 10.1007/978-94-007-2470-9_1, © Springer Science+Business Media B.V. 2012

increasing automation of more and more elements of life means that such forms of employment are diminishing. Rather, countries need individuals increasingly who are capable of analytical and creative thinking. In short, higher order thinking.

Gilbert (2005), for instance, makes the telling point that 'in a world where there is an oversupply of information, the ability to make sense of information is now the scarce resource' (p. 197). Consequently, what our education systems must seek to develop is the capacity to think and learn – a capacity which is seen as highly malleable. The function of education then is to foster this capacity and not, as has been the historical case, act as a sieve to identify those already possessing such ability.

Likewise Hill (2008), in a reflective essay on the demands of future advanced societies, argues that we are moving to a 'post-scientific' society where

> the demands on innovators are very great. They must have not only a core understanding of scientific and technical principles but an equally strong preparation in business principles, communications skills, multicultural understanding, a foreign language or two, human psychology, and one or more of the creative arts. Their education must emphasize making connections among ideas, people, organizations, and cultures, often across boundaries that no one has thought to try to cross before.

Evidence that there is some substance to Hill's arguments come from employers who complain that there is a mismatch between the kinds of skilled workers they seek and the ranks of the unemployed (Hanushek, Peterson, & Woessmann, 2010; Schäfer, 2010; Weitzman, 2010). However, it must be pointed out that this rhetoric has been a pervasive argument for science on the curriculum throughout the years. The report of the committee appointed by the prime minister to enquire into the position of natural science in the educational system of Great Britain in 1918 argued that 'a nation thoroughly trained in scientific method and stirred with enthusiasm for penetrating and understanding the secrets of nature, would no doubt reap a rich material harvest of comfort and prosperity' (Committee to Enquire into the Position of Natural Science in the Educational System of Great Britain, 1918, p. 7).

The challenge is how does science education contribute to developing the kind of higher order thinking skills which contemporary society values? To date, science education has dealt in well-established, consensually agreed knowledge. The educational metaphor that dominates its thinking is one of transmission (Lyons, 2006). Not surprisingly, students perceive science as something which is monolithic, fixed and finished – a body of knowledge to be learnt – rather than something which is organic, developing and a way of knowing that has freed us from the shackles of received wisdom. As Ravetz (2002) points out:

> The inherited institution of science education is one of the last surviving authoritarian social-intellectual systems in Europe. Its teaching style is dogmatic, and it is designed around the social function of training and selecting future scientific research workers. By example and exclusion, students absorb the lesson that every real scientific problem has only one simple, correct answer. This mindset can be seriously disabling for all who eventually deal with science-related policy problems.

Horton too came to similar conclusions when examining the differences between African traditional thought and the epistemic basis of belief in Western societies by pointing to the fact that the grounds for accepting the explanatory models proposed by the science teacher were often no different for the grounds for the beliefs taught by elders in African villages to their youth (Horton, 1971). In both cases belief was accepted because it was presented by 'the accredited agents of tradition'. To paraphrase an infamous statement by the German theologian and existential philosopher, Paul Tillich, the passion for reason is extinguished by answers which bear the imprint of undisputed authority.

The deep irony is that more than anything science is *the* domain of human inquiry which has been responsible for the transformation in Western thinking to the use of evidence rather than tradition as the basis of belief. Yet the system it has devised for formal education in science does so little to promote the critical qualities which are the hallmark of the scientist. As Rogers (1948) points out, 'we should not assume that mere contact with science, which is so critical, will make students think critically' (p. 7).

The contradiction is made more puzzling by the fact that, ever since its inception, one of the major justifications for science has been in its ability to develop the 'general faculties of the mind' (Layton, 1973). This work became more systematic with the insights provided by psychologists such as Robert Gagné and Jean Piaget emphasizing the importance of reasoning and its development. The work of the former led to the production of the well-known course – *Science a Process Approach* (American Association for the Advancement of Science, 1964) whilst Piaget was highly influential on a number of research and curriculum development projects – notably *Thinking Science* (Adey, Shayer, & Yates, 1989). The 1960s too were a time when Joseph Schwab was to work on the new Biology Curriculum Studies Committee biology course framing the whole course as an inquiry into inquiry (Schwab, 1962). Suffice to say that none of these has ever taken root within the common practice of most teaching of science.

Why might that be? Our suggestion is that, to date, developing an account of scientific reasoning, how it functions and why it matters has been an incomplete project. The approach of psychologists, whilst illuminating specific acts conducted in the pursuit of science such as observing, classifying and using logico-mathematical reasoning, was only partial. Their descriptive accounts of scientific reasoning ignored or neglected the insights of the normative accounts of philosophers who pointed to the criteria that are used for evaluation (Popper, 1963) and seeing science as a community governed by social norms (Kuhn, 1962). Both of these lacked the perceptions provided by ethnographic studies of laboratory life that have brought new insights presenting science as a set of social practices where argumentation is a central feature (Latour, 1985; Pickering, 1995; Traweek, 1988).

The major flaw with the psychological approach to scientific reasoning was identified by Millar and Driver (1987) who argued that there were two inherent problems to the idea that processes could be taught. First there were no models of progression. What, for instance, was 'easy classifying' as opposed to 'difficult classifying'? The point they sought to make was Rumelahrt and Norman's (1981) contention that we reason in a context and that our ability to reason is highly dependent on the prior conceptual knowledge that we bring to that context. In short, there can be no process which is independent of content and that the choice of content matters. Their second point is that the processes were pedagogic means – devices the teachers used to engage their students – rather than educational goals in themselves.

Nevertheless, the original and basic imperative for teaching scientific reasoning still remained. In the United States and elsewhere, the normative description of a scientific method emerging from the work of the logical positivists had offered a simple algorithm of the procedural steps science used to derive new knowledge – the scientific method. A concept which became enshrined through its pervasive articulation in US textbooks appearing commonly in a single chapter never to be referred to again. More sophisticated descriptions of the epistemic nature of science or arguments that there was no single dominant method failed to gain much traction. However, given the importance of the scientific mode of thought as a rationale for teaching science, attempts to articulate what should be taught and how it should be taught persisted. In the United Kingdom, the requirement to teach a form of scientific reasoning was enshrined in the component of the National Curriculum as a component entitled Sc1 – Scientific Enquiry (Department for Education and Science, 1991). In the United States, the National Standards for Science Education recommended that science should be taught as a process of inquiry (National Academy of Science, 1995). Inquiry was defined as

> a multifaceted activity that involves making observations; posing questions; examining books and other sources of information to see what is already known; planning investigations; reviewing what is already known in light of experimental evidence; using tools to gather, analyze, and interpret data; proposing answers, explanations, and predictions; and communicating the results. Inquiry requires identification of assumptions, use of critical and logical thinking, and consideration of alternative explanations. (p. 23)

However, whilst this may define what kind of student behaviours are expected with sufficient clarity, it failed to offer any sense of what the end product of such activity should be. Students taught through a process of inquiry may acquire the skill required to perform scientific investigation (knowing how), but what specific knowledge and understanding would such activity develop? Engaging in inquiry was simply an end in itself. Moreover, as most classroom activities meet at least some aspect of this definition, how is the observer to distinguish inquiry-orientated lessons from other lessons and what distinguishes good inquiry-based lessons from those that are weaker.

In the UK, although the goals of inquiry were more well defined, a specific problem was that the nature of student performances was operationalized through a formal, teacher assessment that was part of students' final grade on a high stakes assessment. The consequence was a narrow, algorithmic interpretation of the methods of science, dominated by hypothetico-deductive processes, which was simplified by teachers to ensure their student's success (Donnelly, Buchan, Jenkins, Laws, & Welford, 1996). The ultimate outcome was the reduction of any attempt to foster the skills of scientific reasoning to one homogeneous procedure that was applied mechanistically.

Changing Perceptions

In the 1990s, emerging from the work of the historians of science, philosophers, cognitive scientists and sociologists was a body of work that viewed science as a set of material practices. This scholarly analysis illuminated how science is actually done, both in the short term (e.g. studies of activity in a particular laboratory or a programme of study) and historically (e.g. studies of laboratory notebooks, published texts, eyewitness accounts) (Brown, Collins, & Duiguid, 1989; Latour, 1999; Latour & Woolgar, 1986; Lave & Wenger, 1991; Rogoff & Lave, 1984). This body of work saw reasoning, in particular argumentation, as a component of a larger ensemble of social practices that included networks of participants and institutions (Latour, 1999; Longino, 2002) and specialized ways of talking and writing (Bazerman, 1988). Driver, Newton, and Osborne (2000) were the first to identify the epistemic importance of this work for school science. These authors noted that, while argument was a central feature of the discursive practices of science, only scant attention had been paid by formal science education to this practice. Drawing on the normative account of the role of argument of philosophers such as Toulmin and the ethnographic account provided by sociologists such as Latour, they argued that opportunities for students to engage in argumentation were essential to challenge the erroneous impression given by the lingering grasp of logical positivism about the nature of science. More fundamentally, engaging in argumentation provided an opportunity to demonstrate science as a way of knowing where evidence was fundamental to the basis of belief. Moreover, it would offer a window into the 'epistemology, the practices and methods of science, and its nature as a social practice through studies of science-in-the-making, whether historical or in contemporary practice' (p. 297). Having made the case for argumentation, they then outlined a programme of research that was needed to establish the extent to which engaging in argumentation developed student's conceptual understanding, their capability to engage in inquiry and their understanding of scientific epistemology.

That these arguments resonated with the community comes from our search of the ERIC database which shows how the field has expanded over the past two decades (Fig. 1.1).

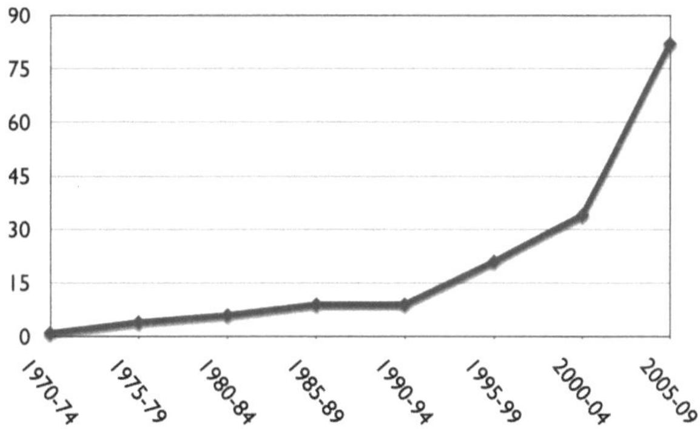

Fig. 1.1 Number of articles found using the search terms 'argument', 'science' and 'education' per 5-year period from 1970 to 2009 in the ERIC database

Likewise, in a review of research trends between 2003 and 2007, Lee, Wu, and Tsai (2009) show that the three most highly cited papers between 1998 and 2002 were on argumentation and five out of ten between 2003 and 2007 were on argumentation or informal reasoning.

Granted then that the work on argumentation as a vehicle for promoting scientific reasoning has gained considerable traction within the research community. Whether it can attain the same status in the everyday practice of school science is yet to be determined. The current national curriculum for England and Wales requires that pupils should be able to 'present information, develop and draw a conclusion'. The practice of argumentation is a distinct feature of the National Academies Report – Taking Science to School (National Research Council, 2007) where 'science is a set of practices ... where what counts as evidence is contingent on making careful observations and building arguments' (chapter 6, p. 1). Likewise the new framework for the common standards in science education incorporates constructing and evaluating argument as one of eight specific scientific practices that should be taught to students.

Argumentation and Critique

A key feature missed by Driver et al. (2000) in their paper published in *Science Education* was the role of criticism in the construction of knowledge. Rather, it was left to Ford (2008) to identify that the construction of knowledge depends on 'a dialectic between construction and critique' and cannot be solely reliant on activities which explore construction. Given that the core project of science learners is the construction of knowledge, the minimal opportunity to engage in critique in school science – to identify why the wrong idea is wrong – would, therefore, seem to be a serious omission. Critique and argumentation serve as a vital epistemic function, enabling beliefs to be tested against counter-arguments. As Kuhn (1992) argues, 'Only by considering alternatives — by seeking to identify what is not — can one begin to achieve any certainty about what is.' Given that many of the ideas that science has to present about the material world defy common sense and appear unnatural (Cromer, 1993; Wolpert, 1992), establishing their plausibility depends as much on demonstrating why common sense misleads as it does on demonstrating the veracity of the canonical scientific explanation. For instance, in their seminal paper on the structure of DNA, Watson and Crick (1953) begin not by arguing for their structure, but instead, arguing why the three-strand model of Pauling and Corey must be wrong. Only then do they advance an argument for their structure. Further evidence that the ability to engage in critique and evaluation is important emerges from the work of Sandoval (2003) in a study of 69 high school students asked to construct explanations for the natural selection of finches in a set of data drawn from the Galápagos islands. He found that students were unable to critique each other's work because they did not see evidence as central to the justification of an explanation nor recognize its absence in the explanations provided by others.

The major message for science education is that if knowledge construction for scientists is dependent on the twin processes of construction and critique, then likewise, knowledge construction for the learner requires the opportunity to engage in critique and the higher order thinking skills of evaluation and synthesis. Yet, science education is notable by the absence of critique. Where, for instance, are students asked to explain why an explanation might be wrong or why the interpretation of a set of data is flawed? Changing this state of affairs requires nothing more than a fundamental reconception of the project of science education. Whilst educators have a duty to develop a knowledge and understanding of the major explanatory ideas that have emerged from study of the natural world, they also need to demonstrate that it is not a monolith of knowledge to be remembered but a set of ideas which are to be tested against the evidence that the material world provides. And that, as a consequence, the identification of error is an essential part of its practice.

Assessing Scientific Reasoning

However, none of this transformation will happen unless we can (a) define the outcomes we seek and (b) devise reliable and valid measures of students' facility with argumentation and reasoning. To date, much of the effort has been devoted to using some form of modified version of the Toulmin framework (Toulmin, 1958) (e.g. Kelly & Takao, 2002; Osborne, Erduran, & Simon, 2004; Zohar & Nemet, 2002). As Sampson and Clark point out though – what makes the Toulmin framework so appealing – its domain generality and relative simplicity – is also its weakness in that the lack of specificity of constructs makes it difficult to resolve the linguistic function served by the specific elements of an argument. In particular, the difficulty of resolving data from warrants (Sampson & Clark, 2008). One common solution is to collapse data, warrants and backings into one single category of justifications (Zohar & Nemet, 2002).

Moreover, only one tentative model of a learning progression for argumentation exists (Berland & McNeill, 2010), which resolves progression into three components of instructional context, argumentative product and argumentative process. However, this has not been empirically tested. An additional element of complexity arises from Sandoval's finding that there are both conceptual and epistemic elements to an argument (Sandoval & Millwood, 2005) inviting the question of whether it is possible, or even desirable, to assess argument independent of any content. We think not. Reasoning always takes place in a context and must be examined in the context of its use. Knowing why an answer is wrong relies on the ability to identify erroneous conceptual understanding (content knowledge) or flaws in the methods that have been deployed (procedural knowledge). The real challenge for the field is to develop a repertoire of generic frameworks for assessing scientific reasoning that can be adapted to different domains. One model that holds promise is offered by Garrat, Overton, and Threlfall (1999). In an attempt to break the mould of the standard form of assessment that dominates undergraduate chemistry, this group asked a series of question using common stems such as which one of the following is the best statement of the flaw in this argument? Which one of the following is the most sensible inference to draw from the above observations? Or, which one of the following conclusions can you draw from this information? These enabled questions to be produced which tested student ability both to construct an argument and to understand and evaluate other arguments in the context of chemistry.

Changing Teachers' Pedagogy

If the assessment is problematic, so too is developing teachers' pedagogy to make greater use of a discursive and dialogic approach to teaching science. The essential question is whether the use of argumentation will gain any traction in the practices of science teachers? Here too little work has been done to provide a definitive answer. The difficulty for the teacher, whose rhetorical task is to persuade students to believe the accepted scientific explanation, is that presenting or even discussing common misconceptions is perceived to run the risk of reinforcing erroneous ideas (Osborne, 2001). As few science teachers have actually been scientists, few have acquired the insight developed by engaging in the doing of science that reveal critique is essential for advancing knowledge and understanding. Thus rather than adapting their instructional practice that reflect a model of science-as-it-is, teachers of science tend to present a model of science-as-it-was-taught-to-them to their students simply sustaining its extant problems (Gallagher, 1991). If the defining feature of scientific reasoning is a commitment to evidence as the basis of belief, then it should not be unreasonable to expect teachers of science to hold up to the light of day the claims we wish our students to believe and ask the question 'is there evidence to support this claim or is it flawed?' And to ask, furthermore, 'what are the limits of the evidence and is the interpretation offered justified?' For, if we never ask students to engage in critical thought in science, what hope is there that they will develop the critical disposition that is the hallmark of the scientist?

In an interview study of 30 teachers (Sampson, 2009), teachers reported that teachers of science valued argumentation as a way to improve the teaching and learning of science but they had a number of reservations about its use. The most common reservation being that they lacked the ability to teach argumentation. The second most stated reservation was that average- and low-performing students do not have the desire or ability to engage in argumentation – a finding which is supported by the work of Zohar (2004).

Over the past decade, research has focused on integrating argumentation into the classroom in several ways:

- the development of new curricula (Krajcik, McNeill, & Reiser, 2008; Stewart, Cartier, & Passmore, 2005),
- technology-enhanced learning environments (Clark & Sampson, 2008; Linn, Davis, & Bell, 2003; Sandoval & Reiser, 2004; Toth, Suthers, & Lesgold, 2002),
- instructional strategies (Kuhn & Reiser, 2006; Osborne, Erduran, & Simon, 2004)

All of these have been developed such that science teachers can promote and support the use of argumentation as an instructional practice. Despite this body of work, research (McNeill, Lizotte, Krajcik, & Marx, 2006; Simon, Erduran, & Osborne, 2006) has found that teachers have difficulty with using argumentation as an instructional practice. These findings would suggest that the implementation of argumentation will require teachers who are knowledgeable and understand argumentation and its role in learning such that they can utilize argumentation for learning successfully. Research, however, demonstrates that changing teachers' instructional practices is a significant challenge (Borko, 2004; Garet, Porter, Desimone, Birman, & Yoon, 2001) – a finding confirmed by Simon et al. (2006) who found that only 6 out of the 12 teachers they worked with in their study adopted the use of argumentation as an instructional practice with any degree of competence.

Kuhn and Reiser (2006) argue that integrating argumentation into the classroom is hard for both teachers and students suggesting that 'scientific argumentation requires that students and teachers take on new and different roles in the classroom' and that 'typical classroom practices often inhibit student participation in this practice' (p. 2). Because knowledge is presented as a set of stable and isolated facts, there is no necessity to question the information provided by the teacher or the textbook. Moreover, as students have no sense of how scientific knowledge is constructed, they do not see the value critique and argumentation might have in establishing a secure basis for belief. When disagreements around concepts do arise, Kuhn and Reiser find that there are no established processes or means for students to resolve conflicts other than to turn to the teacher for resolution as the ultimate arbiter of knowledge. A further challenge to developing a disposition to argue, identified by these researchers, is that when students write – because they are writing for a knowledgeable audience, the teacher – they do not see the need for foregrounding the evidence in their explanations as there is no need to justify to the teacher what he or she already knows.

The nature of the challenge for teachers of science is also shown by a detailed case study of one teacher developing and adapting her practice (Martin & Hand, 2009) found that nearly 2 years was required for the teacher to make a substantive change in their practice leading the authors to conclude that

> the shift in pedagogical orientation is not easily achieved and requires time to occur. Experienced teachers have developed successful pedagogical strategies and these have become entrenched as automatic ways of operating. Even with constant interaction with a professional development person, there was an 18 month period before significant shifts in the teacher's practices were observed. (p. 35)

This is a finding that is supported by other work on teacher professional development (Borko, 2004; Sisk-Hilton, 2009). To date, then, our knowledge of how to support teachers in this form of discursive practice is still limited. How much time does it take to adapt and what mechanisms of support are needed? What practices are readily adopted and which cause difficulty? Hence much remains to be done to improve our understanding of how argumentation can be enacted and supported as a common instructional practice.

Summary

The answer then to those who wonder why this book is necessary is that there is still much that we do not know about the role and value of argumentation in the teaching of science. There is a set of small-scale interventions conducted by researchers with limited sample sizes to suggest that engaging in argumentation does lead to enhanced conceptual understanding in students (Mercer, Dawes, Wegerif, & Sams, 2004; Venville & Dawson, 2010; Zohar & Nemet, 2002). However, how argumentation functions, how student facility with argumentation might be assessed and how the use of argumentation as a common instructional practice might become more widely valued and used by teachers of science are all questions where our knowledge is limited. The contributions in this book are to be valued, therefore, as they offer a body of work about how we might find better answers to such questions. Or to paraphrase Yeats – if not light at least a half-light that will guide our way forward.

References

Adey, P., Shayer, M., & Yates, C. (1989). *Thinking science: The curriculum materials of the CASE project*. London: Thomas Nelson and Sons.

American Association for the Advancement of Science. (1964). *Science: A process approach*. Washington, DC: American Association for the Advancement of Science.

Andrews, R. (1995). *Teaching and learning argument*. London: Cassell.

Andriessen, J. (2006). Arguing to learn. In K. Sawyer (Ed.), *Handbook for the learning sciences* (pp. 443–460). Cambridge: Cambridge University Press.

Applebee, A., Langer, J., Nystrand, M., & Gamoran, A. (2003). Discussion-based approaches to developing understanding: Classroom instruction and student performance in middle and high school English. *American Educational Research Journal, 40*(3), 685.

Bazerman, C. (1988). *Shaping written knowledge*. Madison, WI: The University of Wisconsin Press.

Berland, L. K., & McNeill, K. L. (2010). A learning progression for scientific argumentation: Understanding student work and designing supportive instructional contexts. *Science Education, 94*(5), 765–793.

Borko, H. (2004). Professional development and teacher learning: Mapping the terrain. *Educational Researcher, 33*(8), 3–15.

Brown, J., Collins, A., & Duiguid, P. (1989). Situated cognition and the culture of learning. *Educational Researcher, 18*(1), 32–42.

Clark, D. B., & Sampson, V. (2008). Assessing dialogic argumentation in online environments to relate structure, grounds, and conceptual quality. *Journal of Research in Science Teaching, 45*(3), 293–321.

Committee to Enquire into the Position of Natural Science in the Educational System of Great Britain. (1918). *Report of the committee appointed by the prime minister to enquire into the position of natural science in the educational system of Great Britain*. London: Her Majesty's Stationary Office.

Cromer, A. (1993). *Uncommon sense: The heretical nature of science*. New York: Oxford University Press.

Department for Education and Science. (1991). *Science in the national curriculum*. London: HMSO.

Donnelly, J., Buchan, A., Jenkins, E., Laws, P., & Welford, G. (1996). *Investigations by order: Policy, curriculum and science teachers' work under the Education Reform Act*. Nafferton: Studies in Science Education.

Driver, R., Newton, P., & Osborne, J. F. (2000). Establishing the norms of scientific argumentation in classrooms. *Science Education, 84*(3), 287–312.

Ford, M. (2008). Disciplinary authority and accountability in scientific practice and learning. *Science Education, 92*(3), 404–423.

Gallagher, J. J. (1991). Prospective and practicing secondary school science teachers' knowledge and beliefs about the philosophy of science. *Science Education, 75*(1), 121–133.

Garet, M. S., Porter, A. C., Desimone, L., Birman, B., & Yoon, K. (2001). What makes professional development effective? *American Educational Research Journal, 38*(4), 915–945.

Garratt, J., Overton, T., & Threlfall, T. (1999). *A question of chemistry: Creative problems for critical thinkers*. Harlow: Pearson.

Gilbert, J. (2005). *Catching the knowledge wave? The knowledge society and the future of education*. Wellington, New Zealand: NZCER Press.

Hanushek, E. A., Peterson, P. E., & Woessmann, L. (2010). *U.S. math performance in global perspective: How well does each state do at producing high-achieving students?* Cambridge, MA: Program on Education Policy and Governance, Harvard University.

Hempel, C. G. (1962). Deductive-nomological vs. statistical explanation. *Minnesota Studies in the Philosophy of Science, 3*, 98–169.

Hill, C. (2008). The post-scientific society. *Issues in Science and Technology on Line, 24*(1), 78–84.

Horton, R. (1971). African traditional thought and Western science. In M. D. Young (Ed.), *Knowledge and control* (pp. 208–266). London: Colin-MacMillan.

Kelly, G., & Takao, A. (2002). Epistemic levels in argument: An analysis of university oceanography students' use of evidence in writing. *Science Education, 86*, 314–342.

Krajcik, J., McNeill, K. L., & Reiser, B. J. (2008). Learning goals driven design model: Developing curriculum materials that align with national standards and incorporate project based pedagogy. *Science Education, 92*(1), 1–32.

Kuhn, D. (1991). *The skills of argument*. Cambridge: Cambridge University Press.

Kuhn, D. (1992). Thinking as argument. *Harvard Educational Review, 62*(2), 155–178.

Kuhn, D. (1999). A developmental model of critical thinking. *Educational Researcher, 28*(2), 16–46.

Kuhn, D. (2009). Do students need to be taught how to reason? *Educational Research Review, 4*(1), 1–6.

Kuhn, L., & Reiser, B. (2006). *Structuring activities to foster argumentative discourse*. Paper presented at the Annual meeting of the American Educational Research Association.

Kuhn, T. E. (1962). *The structure of scientific revolutions*. Chicago: University of Chicago Press.

Latour, B. (1985). Science in action: How to follow scientists and engineers through society. Cambridge, MA: Harvard University Press.

Latour, B. (1999). *Pandora's hope: Essays on the reality of science studies*. Cambridge, MA: Harvard University Press.

Latour, B., & Woolgar, S. (1986). *Laboratory life: The construction of scientific facts* (2nd ed.). Princeton, NJ: Princeton University Press.

Lave, J., & Wenger, E. (1991). *Situated learning: Legitimate peripheral participation*. Cambridge: Cambridge University Press.

Layton, D. (1973). *Science for the people: The origins of the school science curriculum in England*. London: Allen and Unwin.

Lee, M.-H., Wu, Y.-T., & Tsai, C.-C. (2009). Research trends in science education from 2003 to 2007: A content analysis of publications in selected journals. *International Journal of Science Education, 31*(15), 1999–2020.

Linn, M. C., Davis, E. A., & Bell, P. (2003). *Internet environments for science education*. Mahwah, NJ: Lawrence Erlbaum.

Longino, H. E. (2002). *The fate of knowledge*. Princeton, NJ: Princeton University Press.

Lord Sainsbury of Turville. (2007). *The race to the top: A review of government's science and innovation policies*. London: HM Treasury.

Lyons, T. (2006). Different countries, same science classes: Students' experience of school science classes in their own words. *International Journal of Science Education, 28*(6), 591–613.

Martin, A., & Hand, B. (2009). Factors affecting the implementation of argument in the elementary science classroom. A longitudinal case study. *Research in Science Education, 39*(1), 17–38.

McNeill, K., Lizotte, D., Krajcik, J., & Marx, R. (2006). Supporting students' construction of scientific explanations by fading scaffolds in instructional materials. *Journal of the Learning Sciences, 15*(2), 153–191.

Mercer, N., Dawes, L., Wegerif, R., & Sams, C. (2004). Reasoning as a scientist: ways of helping children to use language to learn science. *British Education Research Journal, 30*(3), 359–377.

Millar, R., & Driver, R. (1987). Beyond processes. *Studies in Science Education, 14*, 33–62.

National Academy of Science. (1995). *National science education standards*. Washington, DC: National Academy Press.

National Research Council. (2007). *Taking science to school: Learning and teaching in grades K-8*. Washington DC: National Research Council.

Nercessian, N. (2008). Model-based reasoning in scientific practice. In R. A. Duschl & R. E. Grandy (Eds.), *Teaching scientific inquiry: Recommendations for research and implementation* (pp. 57–79). Rotterdam, Netherlands: Sense.

Osborne, J. F. (2001). Promoting argument in the science classroom: A rhetorical perspective. *Canadian Journal of Science, Mathematics and Technology Education, 1*(3), 271–290.

Osborne, J. F., Erduran, S., & Simon, S. (2004). Enhancing the quality of argument in school science. *Journal of Research in Science Teaching, 41*(10), 994–1020.

Pickering, A. (1995). *The mangle of practice: Time, agency & science*. Chicago: University of Chicago Press.

Popper, K. (1963). *Conjectures and refutations: The growth of scientific knowledge*. London: Routledge and Kegan Paul.

Ravetz, J. (2002). *Reflections on the new tasks for science education*. Unpublished Evidence submitted to the House of Commons Committee for Science and Technology.

Resnick, L., Michaels, S., & O'Connor, C. (in press). How (well-structured) talk builds the mind. In R. Sternberg & D. Preiss (Eds.), *From genes to context: New discoveries about learning from educational research and their applications*. New York: Springer.

Rogers, E. M. (1948). Science in general education. In E. J. McGrath (Ed.), *Science in general education* (pp. 1–23). Dubuque, IA: Wm. C. Brown Co.

Rogoff, B., & Lave, J. (1984). *Everyday cognition: Its development in social context*. Cambridge, MA: Harvard University Press.

Rumelahrt, D. E., & Norman, D. A. (1981). Analogical processes in learning. In J. R. Anderson (Ed.), *Cognitive skills and their acquisition* (pp. 335–360). Hillsdale, NJ: Lawrence Erlbaum.

Sampson, V. (2009). *Science teachers and scientific argumentation: Trends in practice and beliefs*. Paper presented at the Annual Conference of the National Association of Research in Science Teaching (NARST). Retrieved.

Sampson, V., & Clark, D. B. (2008). Assessment of the ways students generate arguments in science education: Current perspectives and recommendations for future directions. *Science Education, 92*(3), 447–472.

Sandoval, W. A. (2003). Conceptual and epistemic aspects of students' scientific explanations. *Journal of the Learning Sciences, 12*(1), 5–51.

Sandoval, W. A., & Millwood, K. (2005). The quality of students's use of evidence in written scientific explanations. *Cognition and Instruction, 23*(1), 23–55.

Sandoval, W. A., & Reiser, B. J. (2004). Explanation driven inquiry: Integrating conceptual and epistemic scaffolds for scientific inquiry. *Science Education, 88*(3), 345–372.

Schäfer, D. (2010). Skills shortage threatens German engineering. *Financial Times*.

Schwab, J. J. (1962). *The teaching of science as enquiry*. Cambridge, MA: Harvard University Press.

Siegel, H. (1989). The rationality of science, critical thinking and science education. *Synthese, 80*(1), 9–42.

Simon, S., Erduran, S., & Osborne, J. F. (2006). Learning to teach argumentation: Research and development in the science classroom. *International Journal of Science Education, 28*(2–3), 235–260.

Sisk-Hilton, S. (2009). *Teaching and learning in public*. New York: Teachers College Press.

Stewart, J., Cartier, J. L., & Passmore, C. M. (2005). Developing understanding through model-based inquiry. In J. D. Bransford (Ed.), *How students learn science in the classroom* (pp. 515–565). Washington, DC: National Research Council.

Thagard, P. (2008). Explanatory coherence. In J. E. Adler & L. J. Rips (Eds.), *Reasoning* (pp. 471–513). Cambridge: Cambridge University Press.

Toth, E. E., Suthers, D. D., & Lesgold, A. M. (2002). ìMapping to knowî: The effects of representational guidance and reflective assessment on scientific inquiry. *Science Education, 86*(2), 264–286.

Toulmin, S. (1958). *The uses of argument*. Cambridge: Cambridge University Press.

Traweek, S. (1988). *Beamtimes and lifetimes: The world of high energy physicists*. Cambridge, MA: Harvard University Press.

Venville, G. J., & Dawson, V. M. (2010). The impact of a classroom intervention on grade 10 students' argumentation skills, informal reasoning, and conceptual understanding of science. *Journal of Research in Science Teaching, 47*(8), 952–977.

Watson, J. D., & Crick, F. H. C. (1953). A structure for deoxyribose nucleic acid. *Nature, 171*, 737–738.

Weitzman, H. (2010). US manufacturers face skills shortage. *Financial Times*.

Wineburg, S. (2001). *Historical thinking and other unnatural acts*. Philadelphia, PA: Temple University Press.

Wolpert, L. (1992). *The unnatural nature of science*. London: Faber and Faber.

Yackel, E., & Cobb, P. (1996). Sociomathematical norms, argumentation, and autonomy in mathematics. *Journal for Research in Mathematics Education, 27*(4), 458–477.

Zimmerman, C. (1999). The development of scientific reasoning skills. *Developmental Review, 20*(1), 99–148.

Zimmerman, C. (2007). The development of scientific thinking skills in elementary and middle school. *Developmental Review, 27*(2), 172–223.

Zohar, A. (2004). *Higher order thinking in science classrooms: Students' learning and teachers professional development*. Dordrecht: Kluwer.

Zohar, A., & Nemet, F. (2002). Fostering students' knowledge and argumentation skills through dilemmas in human genetics. *Journal of Research in Science Teaching, 39*(1), 35–62.

Chapter 2
The Two Faces of Scientific Argumentation: Applications to Global Climate Change

E. Michael Nussbaum, Gale M. Sinatra, and Marissa C. Owens

Introduction

The title, *The two faces of scientific argumentation*, is meant to convey that, in science, there are at least two types of discourse: theoretical discourse, pertaining to what theories of the world best fit the data, and practical, deliberative discourse, regarding how to apply those theories to reach practical goals. It certainly could be argued that the latter type of discourse is not, technically speaking, "scientific." Nevertheless, we argue that in the science classroom, it is important for students to understand both faces of scientific argumentation.

Students need to understand "classic" scientific argumentation because science itself can be described as "one long argument" (Mayr, 1991). Students need to understand how scientists build a case for a theory as being equivalent to constructing a scientific argument, by working from a set of premises, gathering and testing data, and using the results of empirical research and reasoned logic as backing and warrants to support (or refute) a theory.

Students also need to understand the "other face," deliberative, practical argumentation, for several reasons. First, it is important to understand how a purely scientific argument may differ from practical and other types of arguments. Second, it is also important for students to understand the role of science in public policy deliberations and in addressing the socio-scientific challenges facing modern society (Zeidler, Sadler, Applebaum, & Callahan, 2009). The authors, as well as many of our readers, live in technologically sophisticated, industrial democracies, where too often citizens, and the policy makers they elect, make decisions based on pseudoscience or a lack of evidence (e.g., banning genetically modified foods).

An overarching issue facing industrial and industrializing nations is what to do about climate change. Climate change threatens to destabilize nations by changing the global distribution of precipitation (thereby potentially causing droughts, famine, and more severe storms), increasing sea levels, fomenting forest fires,

---

E.M. Nussbaum (✉)
University of Nevada, Las Vegas, NV, USA
e-mail: Nussbaum@unlv.nevada.edu

M.S. Khine (ed.), *Perspectives on Scientific Argumentation*,
DOI 10.1007/978-94-007-2470-9_2, © Springer Science+Business Media B.V. 2012

and causing other ecological disruptions (National Science Foundation, 2009). Yet while 90–97% of earth scientists agree that human activity is contributing to climate change (Doran & Kendall Zimmerman, 2009), there is markedly less certainty among the general public (although the amount of skepticism varies, depending on the time and place of polling) (Leiserowitz, Maibach, Light, 2009; Pew Research Center, 2009; Yale Project on Climate Change, 2010). It is imperative that the public understand the scientific argument for *anthropogenic* (human-caused) climate change to be able to engage in productive argumentative discourse as to how we (collectively and individually) should respond to it. These twin imperatives require individuals to understand both the theoretical and the practical sides of scientific issues, and more generally the nature of both scientific and practical argumentation.

The goal of this chapter is to introduce readers to the research that we have been conducting on the two "faces" of scientific argumentation in relationship to global climate change (GCC). This chapter is organized as follows. We first discuss the nature of argumentation. We then present a case study of practical argumentation: a discussion about the merits of a carbon tax that was conducted in a seventh-grade classroom. We contrast this research with a more current project designed to teach students about the scientific basis of climate change and how it may be contributing to drought in the southwest part of the United States. This project involves conveying scientific arguments to students rather than having them debate the arguments among themselves. In both cases, however, the theoretical and practical sides of scientific argumentation become intermixed. This is not a new perspective; advocates of the study of socio-scientific issues have long recognized this interdependence (Zeidler, Osborne, Erduran, Simon, & Monk, 2006). We do show, however, how analytic concepts from Walton's (1989) dialogue theory, a contemporary philosophic theory of argumentation, can be used to help students better understand the nature of scientific argumentation.

What Is Argumentation?

During the last several decades, researchers and reformers in science educators have increasingly emphasized the importance of teaching students about scientific argumentation (Chin & Osborne, 2010; Duschl, 2008; Jiménez-Aleixandre, & Erduran, 2008; Kuhn, 2005). The importance of students learning scientific argumentation is also stressed in state and national science standards (Sunal, 2006). This raises the questions of what exactly is argumentation (generally) and scientific argumentation (specifically). Angell (1964) defines an argument as a conclusion supported by at least one reason. Govier (1987) distinguishes argument from explanation in holding that in an argument the conclusion should also be uncertain. Now we may ask what, then, is a *scientific* argument? Certain qualities come to mind; for example, the argument should be supported by empirical evidence, or at least capable of being verified, falsified, or weakened by such evidence (Erduran, 2008). The purpose of such arguments should be to explain some phenomenon in the natural or social world. There should also be adherence to some standards appropriate for the field of inquiry; for example, using random sampling in survey research rather than personal anecdotes for evidence.

We have defined *argument* as a series of propositions. *Argumentation*, on the other hand, is the social process where two or more individuals construct and critique arguments (Kuhn & Udell, 2003; Nussbaum, 2002). The process of *scientific argumentation* should involve the construction and critique of scientific arguments, one that involves the consideration of alternative hypotheses (Lawson, 2003). Aristotle and Descarte (cited in Walton, 1998) conceived of scientific argumentation as a form of inquiry, in which the truth of conclusions is established cumulatively, based on other conclusions that are well established with evidence. Aristotle called a scientific argument a *demonstration*, a syllogism with scientific knowledge as a conclusion and premises that are immediate, better known, and causative of the conclusion. However, studies of the sociology of science suggest that scientific argumentation is often more like a debate or what Walton (1998) calls a *persuasive dialogue*, in which different scientists try to win people over to their points of view and to weaken opposing views with evidence and rational argument (see Broad & Wade, 1982; Keith & Rehg, 2008; Kitcher, 1993). However, this ignores the fact that scientific argumentation is not always like a debate between opposing parties; there is also the general scientific community within a field that makes judgments in a somewhat cumulative fashion, although the amount of consensus can wax and wane over time (Solomon, 2008).

We find this, for example, in the discourse on climate change. There was initially some debate as to the role of human activity in climate change, but as more evidence became available, the general scientific view among climate scientists shifted toward endorsing the reality of anthropogenic change. In our view, scientific argumentation involves both elements of persuasion and inquiry, and often times a shift between these forms of discourse. The ultimate goals of science are those of inquiry (to obtain insight into the truth of things), but persuasive inquiry is often embedded in this process and can further the goals of inquiry to the extent that scientific controversies and debates help the scientific community explore and evaluate alternative hypotheses (Solomon, 2008).

There is another, very important type of argumentative discourse that scientists often engage in, what Walton (1998) calls a *deliberation*. In a deliberation, the parties discuss what course of action is best. For example, regarding climate change, parties might deliberate on how best to decrease greenhouse gas emissions. Climate scientists engage in these deliberative discussions along with social scientists (Antholis & Talbert, 2010), engineers (Hoffert et al., 2002), policy analysts (Nemet & Kammen, 2007), advocates (Krupp & Horn, 2008), politicians ("Obama Announces Climate Change Deal," 2009), media commentators (Will, 2010), and the general public. Technically, a deliberative argument is not scientific, as it involves inferences as to what *ought* to be done and not what is the case. Deliberative arguments involve a value component, and it is not possible to establish the truth of values scientifically.

This raises the question of whether deliberative, policy-oriented arguments should be part of the science curriculum. Certainly studying the evidence for or against anthropogenic climate change could make students concerned about this problem and anxious to discuss solutions, be it in science classes or other subject areas, such as history, economics, humanities, or mathematics. Some take the position that although deliberation should be allowed in the science classroom to make science more meaningful and engaging to students, deliberation should not be portrayed as science (Environmental Literacy Council, 2007) or, in our terms, as scientific argumentation.

We take a slightly different stance. Deliberative arguments, in our view, can have a scientific component. They can be presented to students in this manner, while at the same time having students recognize that there is a value component. Despite different points of view, there are often common core values (e.g., saving the planet), and value-based arguments can often be justified by reference to this common ground (Clark & Brennan, 1991; Aristotle, cited in Walton, 1998). For example, regarding climate change, most people would agree that avoiding large-scale droughts or floods is a good thing. On the other hand, there is disagreement as to what countries should make the biggest sacrifices to diminish global warming (Wintour, 2009). Students should understand the reasons for these differences in perspectives. The amount of time given to the nature of "values" and "deliberation" can range along an entire spectrum from a brief 5-minute lecture (to clarify what a scientific argument is or is not) to an integrated curriculum addressing socio-scientific issues through the in-depth study of science, technology, ethics, and other disciplines. There is precedence for the latter in the movement to introduce the impact of technology and other socio-scientific issues into the science curriculum (Sismondo, 2008; Zeidler et al., 2009).

In addition to the potential for using socio-scientific issues to make science more engaging to students, another rationale for teaching science students about deliberative arguments is because such arguments are part of the social practices engaged in by "applied scientists" such as engineers, doctors, policy analysts, and behavioral and learning scientists. In addition, scientists make deliberative arguments, often to funding agencies, regarding what to study and how to study it (Keith & Rehg, 2008). These practices reflect how science is used in the real world and reflect the "other face" of scientific argumentation reflected in the title of this chapter.

Toward Higher Levels of Argumentation

There are two major goals for introducing students to argumentation. First, students who master complex argumentation skills are in a very real sense developing critical thinking and judgment. This is often cited as one reason for teaching students about the so-called scientific method (De Avila & Torres, 2010), scientific inquiry (Dewey; 1910, National Research Council (NRC), 2000, 2001), and other process skills. The second goal of teaching argumentation is to help foster conceptual change (Nussbaum & Sinatra, 2003). Argumentation can promote socio-cognitive conflict (Doisy, Mugny, & Perrett-Clermont, 1976), which, when combined with hypothesis testing (Schwarz, Neuman, & Biezuner, 2000) or knowledge building (Chan,

Burtis, & Bereiter, 1997), can promote conceptual change. A few studies support this viewpoint (Asterhan & Schwarz, 2007; Nussbaum, Sinatra, & Poliquin, 2008; Zohar & Nemet, 2002) although more such studies are needed (see Nussbaum, 2008a). However, conceptual change can be either weak or strong (Dole & Sinatra, 1998). For strong conceptual change to occur, students need to engage deeply with the content. We contend that students who make complex, high-quality arguments are more likely to engage deeply and thus experience conceptual change. However, in order to do this, students need to learn about the qualities of good arguments.

Nussbaum et al. (2008) taught undergraduate students standards for a good scientific argument (describe a causal mechanism, account for all facts, consider alternative hypotheses, etc.). A simple tutorial improved students' argumentative reasoning on a physics problem, conducted in dyads. Specifically students tended to consider more factors and constraints in their reasoning and produced better developed arguments and explanations. Chin and Osborne (2010) used question prompts (King, 1990) to help high-school students generate questions regarding puzzling aspects of a scientific experiment (and about missing information and applicable conditions). The authors presented qualitative evidence that this questioning produced richer and more productive argumentation. Chin and Osborne therefore showed that it is productive to combine the research traditions on questioning with that of argumentation. Along these lines, Nussbaum, Winsor, Aqui, and Poliquin (2007) proposed that the teaching of evaluative standards could be promoted by having students ask *critical questions* when evaluating both sides of an argument. The authors found that doing so resulted in undergraduate students generating arguments that included and integrated more factors and constraints; this also led to attitude change on a social science topic.

The notion of critical questions comes from Walton's model of argumentation (Walton, Reed, & Macagno, 2008; see also Nussbaum, 2011), although the genesis of the idea can be traced backed to Hastings (1963). Critical questions are evaluative questions that should be asked about specific types of arguments (Walton calls these *argument schemes*). For example, consider the argument scheme, correlation to cause. This is the type of argument that is often made by climate scientists in support of the greenhouse effect: There is a strong correlation between levels of greenhouse gases in the atmosphere and mean global temperatures, with data going back hundreds of thousands of years. Critical questions that should be asked about this type of argument scheme include: "Is there good evidence that the causal relationship goes from A to B rather than B to A?" and "Can it be ruled out that the correlation is caused by some third factor?" (Walton et al., 2008, p. 329). In contrast, critical questions about deliberative-type arguments should be based on consequences (One should do X because it will lead to good consequence or the avoidance of bad ones). For example, "How strong is the likelihood that the claimed consequences may occur?" "How strong is the evidence supporting these claims?" and "Are there opposite consequences (bad as opposed to good, for example) that should be taken into account?" (Walton et al., 2008, p. 102). In Walton's framework, arguments where these types of questions can be answered satisfactorily are judged as strong, whereas when a critical question cannot be answered satisfactorily, this weakens the argument. It therefore becomes important to teach students to ask critical questions about various types of arguments. One might predict, based on previous research, that doing so would strengthen the quality of students' arguments by encouraging students to think "deeper" and address more factors and constraints in their arguments. The following section demonstrates this through an analysis of a discussion among middle schoolers on the topic of climate change.

Discussion of Climate Change in One Seventh-Grade Classroom

The topic of climate change is rich in possibilities for argumentation, both regarding whether anthropogenic climate change is occurring, the time and probability of likely effects, and possible solutions. In this section, we analyze a classroom discussion on climate change that was conducted in a seventh-grade classroom. The discussion was part of a larger study on teaching students to ask *critical questions.* The study was conducted in three different sections of a social studies class, all taught by the same teacher. A case study of one of these sections was presented in Nussbaum and Edwards (2011); we focus here on one of the other class sections. Our purpose is to analyze the discussion in terms of the effect of asking critical questions on the quality and depth of students' deliberative argumentation.

The topic of the discussion was whether the federal government should adopt a carbon tax to reduce greenhouse gas emissions. The big question was specifically phrased as "To slow global warming, should the federal government raise taxes on gasoline so people will drive less?" About 11 students participated in the discussion. The school where the study was conducted was an ethnically diverse public school located in a large city in the Southwest (Las Vegas); of the participating students, 36% were Latino, 33% were Caucasian, 20% were African American, and 10% were Asian/Pacific Islander. Regarding gender, 27% were males and 73% were females.

The discussion on climate change lasted two class periods. The two periods were spaced one week apart. The participating teacher allowed the first author, MN, to conduct the class once per week in order to discuss current events and, in the process, to collect data for the study.

The first part of the discussion was more teacher-controlled so as to help build students' background knowledge and to introduce the topic. The facilitator (MN) asked if anyone knew what global warming was. Students' prior understanding was hazy. One student, Robert (all student names are pseudonyms), noted that global warming is "when the emissions of cars get into the atmosphere." MN clarified that pollution does cause global warming, but that it is specifically associated with particular emission such as carbon dioxide. He proceeded to explain the greenhouse effect and how humans have contributed to the effect since the industrial revolution, as well as the analogy with an actual greenhouse.

Students then took turns reading out loud a *Newsweek* article on the greenhouse effect (Begley, 2007/2008). The article discussed the effects of prolonged global warming, such as melting ice caps and sea level rise, and how coastal cities all over the world could become flooded. In the arid Southwest, however, the likely effect would be drought. Students were provided a picture of Lake Mead (see Fig. 2.1), which is the lake, fed by the Colorado River, where Las Vegas obtains 90% of its water. Lake Mead is the largest man-made reservoir in the United States and is also a major source of recreation. The picture provided to students showed how water levels in the lake have been declining over the last decade, leaving a visible white ring of mineral deposits. This was dramatic evidence of the effects of the drought and how it is endangering the lake. The facilitator closed this section of the discussion by noting "there used to be debate on whether or not global warming is occurring, but now scientists are in agreement." This discourse move was intended to shift the discussion from information seeking and building to a deliberative discussion of solutions, specifically a carbon tax.

MN first conducted a straw poll of students to find out how many of them thought raising taxes on gasoline was a good idea—not a single student initially supported the idea. However, one student, Robert, asked if "this would be a temporary solution until global warning stopped?" MN responded that "we live in a democracy and so let the people decide." He asked the students to imagine that they were members of Congress and had to make all these decisions. This took the focus off of MN as the

The Southwestern U.S Drought - Some Hydrological Impacts-
Lake Mead Elevation at Hoover Dam Continues to Fall

Dr. Ken Dewey,
High Plains Regional Climate Center and National Drought Mitigation Center

October 31, 2006 Report

Image © Ken Dewey, HPRCC. Lake Mead at 51% of capacity (October 31, 2006).

The white "bathtub ring" is the result of exposing rocks that were at one time under the water and collecting mineral deposits.A clear glass, for example, dipped in water and then allowed to dry will have mineral deposit "spots" on the glass.

Fig. 2.1 Handout on Lake Mead used in the seventh-grade discussion on climate change. Global warming is thought to contribute to the drought, which in turn is causing a dramatic decline in water levels. Downloaded from http://www.nebraskaweatherphotos.org/Lake-Mead-2006.html. Copyright 2006 by Dr. Ken Dewey, School of Natural Resources, University of Nebraska-Lincoln. Adapted with permission

expert and shifted greater authority to the students. Robert at this point indicated that it would be a good idea and that, as a society, we should use more wind power. He was immediately challenged by Valerie, who noted that wind power would not be a good idea because wind towers would be a hazard to planes flying over. The lights on the towers would also be a distraction to the people living nearby, and Valerie noted how she was distracted by such lights on a recent road trip to Utah. Another student, Gary, suggested that some kind of "blocker" be put up over the lights as a shield, but Robert noted that "if we block the lights, how would people see them?"

As an example of deliberative argumentation, up to this point the discussion had mixed results. On the positive side, some of the students were challenging one another and making arguments. The discourse was not, however, directly focused on the pros and cons of a carbon tax. It was a type of deliberation known as *practical reasoning* (Walton et al., 2008) in which students make arguments about how best to achieve a goal (in this case, how to safely implement wind power). To refocus the discussion, MN at this point made two discourse moves. First, he took another straw poll, first on whether students thought the red lights on wind towers would bother drivers. No one voted yes. Second, he asked "What are some of the critical questions that we could ask?" One major critical question that MN had been asking students in prior discussions to consider is whether one value is more important than another. As an example, he asked what is more important: the fact that windmill lights might bother some drivers or that we have an alternative source of energy? Valerie responded that "pollution kills people," implying that finding alternative sources of energy was more important. This allowed MN to shift the discussion to a different topic, particularly the burning of coal to generate electricity. Robert attempted to design a solution to the pollution generated from burning coal; he suggested that instead of sending the pollution into the air, "maybe redirect it and put it in some form of container." Valerie noted a constraint, however, asking "what happens when the containers fill up?"

A pattern had started to form in which Valerie played the role of the "critic." She criticized the idea that people should take the bus more because "a bus is bigger than a car and would probably end up costing more in fuel." She also criticized the idea of employee carpools, noting that "her mother has tried it and had to drive all over town to pick up everybody you'll be carpooling with." All these points were refutable, but the next speaker (Daniela), rather than refuting them, changed the subject by noting that "there's a guy on the news that uses vegetable oil in his car." Valerie—continuing in her critical role—noted that "vegetable oil could wreck the car." MN used this as an opportunity to underscore another critical question that he encouraged students to ask when a design-oriented argument is made, "Is it practical?" Daniela proposed that "we should buy a special engine that uses the vegetable oil." Then Robert proposed "solar-powered cars," but Valerie countered that it would be a waste of money to buy new engines, and that you can't drive a solar-powered car at night. Furthermore, a solar-powered car could not be recharged when parked inside a parking garage, and if cars were parked outside, the car could get stolen. Another student countered, however, that cars could be stolen even when parked inside a parking garage.

Although the discussion was not, at this point, ostensibly about a carbon tax, there was strong resistance among the students against raising taxes, and so we infer that they were therefore searching for alternatives to the carbon-tax proposal, specifically by exploring alternative forms of transportation. They likely did not understand that a carbon tax would encourage more people to buy alternative types of vehicles and that, this in turn, would make these forms of transportation more affordable. They were, however, discovering that there were constraints and costs involved in these alternative forms, and that more technological research on them is needed.

During the second day of discussion, MN tried to have students take a broader view of the topic by having them complete an argumentation vee diagram (AVD). An AVD (see Nussbaum, 2008b) is a graphic organizer that lists pro and con arguments and then has students complete an overall integrative argument. Students do not necessarily have to choose a side but—as suggested by the vee—can take a middle position. Vee diagrams were originally used by Novak and Gowan (1984) for an entirely different purpose (teaching the scientific method) and were adapted by Nussbaum (2008b) to support pro/con argumentation. The actual arguments written on the vee in Fig. 2.2 were prepared by MN as a summary of various arguments and counterarguments made by students in all three sections during the first day of discussion. In this study, students were given 15 minutes to complete their AVDs and then these were used as a springboard for further discussion.

MN began the second discussion by having one student, Estella, read her integrating paragraph. Estella argued that people should walk or take the bus more. Robert counterargued that if everyone took the bus, you would need more buses, and buses pollute, just like cars. This argument appeared to have been appropriated from Valerie, who made it during the first day of discussion. The argument reflected a misconception that "bigger" means more (in this case, more pollution), ignoring the amount of pollution produced per person. However, later in the discussion, another student reiterated the argument and Robert countered "buses are good, because more people fit on a bus," so he appeared to have filled in the missing piece of his understanding. Daniella noted, however, that buses take longer than cars "because they are always stopping." MN then summarized the issue here by saying "The bus would use less gas (because there are more people on it), but [taking the bus] would take longer." Estella proposed that there could be shuttle system with some express buses, and Robert also endorsed the idea, proposing that there should be express buses just to get people to work. When asked by MN, several students indicated that this would "save the bus option."

As can be seen in Fig. 2.2, the AVD encouraged students to engage in creative designing but at the same time to ask themselves the critical question, "Is it practical?" Students attempted to identify constraints and to search for solutions that satisfied multiple constraints. Students seemed to have a natural proclivity to engage in design reasoning, but the critical questions asked on the AVD, and reinforced by the facilitator during the discussion, may have encouraged students to generate deeper arguments. (There is supporting evidence for this point, as students in those class sections that had critical questions placed on their AVDs

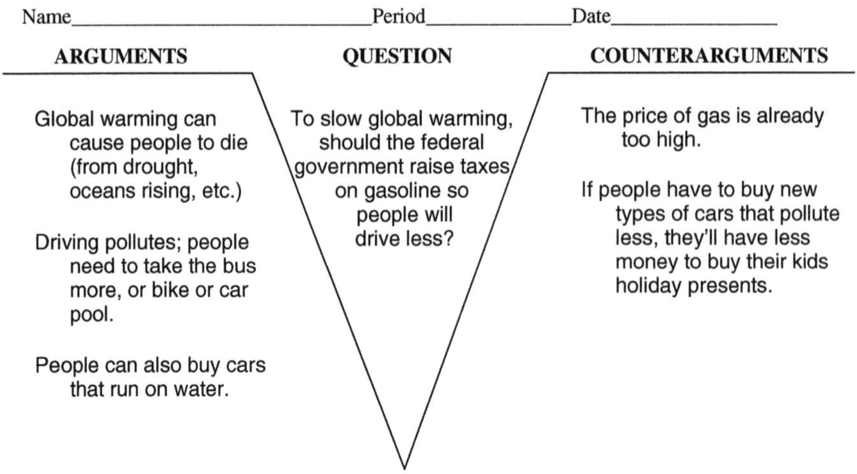

CRITICAL QUESTIONS

Look at all the above arguments, and answer each of the following questions

	Circle Yes or No	Which Argument?
Are any of the arguments **not as important** as others?	Yes No	
Are any of the arguments **unlikely?**	Yes No	
Is there a **creative solution** to any problems raised?	Yes No	
Is the creative solution **practical?** (Consider costs.)	Yes No	
For any argument, can you think of any **other likely explanations?** Or **evidence to the contrary?**	Yes No	

INTEGRATE

Using your answers, explain why one side may be stronger and the other side weaker? Or is there an "in-between" solution?

Fig. 2.2 Argumentation Vee Diagram (AVD) used in the seventh-grade discussion on climate change. The arguments and counterarguments are summaries of arguments made by the students in the preceding discussion

generated significantly more *practical* design arguments than in the control section; see Nussbaum and Edwards (2011)).

As noted previously, we also emphasized the critical question, "Is one value more important than another?" So in response to the bus issue, MN asked the students which was more important, saving the planet or taking longer to get to work? Robert, not giving the answer that MN expected, responded that "You have to work in order to get money to live," but Estella countered "If the Earth is dying, then you'll eventually be dying." Jim added that "buses are fun" and MN added that one can do work while riding on the bus. Finally, there was some discussion of whether some students could save gas by being home-schooled and ways that they could still make friends.

At the end of the discussion, MN took a final straw poll on how many students supported a carbon tax, and almost all now supported it. The argument that students could take the bus more seemed to resonate with these students the most. In contrast, in the control class, where the critical questions were not introduced, there was some attitude change but it was less strong. (Only 4 students changed their minds to support a carbon tax, as opposed to 10 in the experimental condition. The difference was statistically significant, $p < 0.05$, Fisher's Exact Test.) In the control group's discussion, the misconception regarding buses being more gas-intensive surfaced but was never resolved, and the discussion instead came to focus on whether the government should give subsidies for individuals to purchase alternative types of vehicles. We speculate that this discussion may have been less effective in inducing attitude change because (a) the solution discussed was more government oriented (and the students may have had some aversion to "big government"), and/or (b) students had a harder time sorting through the issues due to the absence of critical questions.

Students appeared to have fun with these discussions. We were initially concerned that the topic of climate change might cause fear and anxiety among the students, but the general tone was playful and design oriented. Furthermore, the facilitator's asking of critical questions did in fact appear to encourage students to consider more constraints and affordances in their reasoning (Greeno & van de Sande, 2007), where affordances are the opportunities for action provided by things (such as buses), and in particular that buses afford less per capita fuel consumption than a car. Critical questions also helped students weigh values. We elsewhere present evidence that some of these students appropriated the critical questions and began asking them on their own (see Nussbaum & Edwards, 2011).

"Losing the Lake": Understanding the Effect of Climate Change on Water Resources in the Las Vegas Valley

As the focus of the media and the scientific community debates climate change, the issue has been brought to the general public's attention. In the Las Vegas Valley specifically, this topic has been entangled in a misleadingly more pressing topic: drought. The Las Vegas Valley has experienced over 10 years of drought, with no clear end in sight (Brean, 2009). It is not surprising that recent newspaper articles have appeared in the area with titles such as "Portending doom," "Water and our economy," and "Drought requires community to pursue all of its options" (Kelly, 2009; Lindzen, 2009; Sisolak, 2009).

Although these topics are cause for concern, perhaps even more alarming is the persistence of the misconceptions individuals have surrounding climate change, drought, and their connection. The effect of climate change on snowpack in the Rocky Mountains, and as a result, water flow into Lake Mead, is not always clear. This makes the situation that much more difficult for students and citizens of the Las Vegas Valley to understand. These misconceptions negatively influence people's decision making regarding climate change and drought, and they contribute to a lack of understanding of the issues involved.

An interdisciplinary team, funded by a grant from a larger NSF EPSCoR grant on climate change, has been developing an educational computer simulation game designed to promote sustainability awareness of the water supply in the Las Vegas Valley from Lake Mead. The ultimate goal of this simulation game is to educate residents and students of the Las Vegas Valley on water resources, water conservation, water usage, and how all of these are connected to climate change. We call the research and development project and the game "Losing the Lake." By its nature (designing a simulation game), the project involves presenting scientific and practical arguments to students, rather than having them discuss these issues among themselves.

Understanding Student Preconceptions

Before designing the game, we first conducted a qualitative study to investigate some of the preconceptions that students may have about both water resources and climate change. Although the simulation game is being designed for middle- and high-school students, we used college undergraduates for the qualitative study due to convenience. The sample was small ($N = 10$; 8 females and 2 males) and not intended to be representative of a general population, but we wanted to find out what misconceptions (or lack of conceptions) some local students might have about these issues. Participants were recruited through a department subject pool. The subject pool consists of undergraduate and graduate students who are enrolled in Educational Psychology classes. As compensation for participating in the pilot study, all students received research credit applied toward a course requirement.

Participants were questioned by the third author in semi-structured, individual interviews. Each interview lasted 60 minutes and was audio-recorded. Participants were asked questions about the region's water resources and climate change. These questions were designed to determine participants' understanding of climate change and how that affects water supply in the Las Vegas Valley. The interview consisted of 20 questions, including: *Where does the water in Nevada go once we are finished using it? Where does the Colorado River receive its water from? Why are Lake Mead water levels dropping?* and *What is causing climate change?* Table 2.1 shows the complete list of questions asked during the interview and some examples of participants' answers.

For purposes of analysis, participants' answers were compiled in a table to easily identify common themes. One hundred percent of the participants mentioned that Southern Nevada receives its water from Lake Mead and 90% of the participants knew that Lake Mead receives its water from the Colorado River. In contrast, there were several topics where participants demonstrated a lack of knowledge. These topics were less familiar to participants: the relationship between water use and energy use, the nature of the greenhouse effect, and the water allocation of each state in the Southwest from the Colorado River. Seventy percent of the participants could not explain the greenhouse effect. Fifty percent of participants could not identify

Table 2.1 Losing the Lake preconception study results

Interview question	Percent with misconception or lack of knowledge
Where does the water in Nevada go once we are finished using it?	50
Where does Lake Mead receive its water from?	10
Where does the Colorado River receive its water from?	70
What is happening with Lake Mead's water levels?	0
Why are Lake Mead water levels dropping?[a]	60
How long have water levels been dropping?	60
What are the two or three most important factors that affect lake levels?	50[b]
Do you think Lake Mead will ever run out of water?	90
Which states receive their water from the Colorado River?	90[c]
What state uses the most amount of water?[d]	80
Which state uses the least amount of water?	80
What can you tell us about climate change? Can you provide an example?	60
How does weather differ from climate?	60
Do you think the climate is changing?[e]	80
By what percent do you think climate change will reduce the net inflow of water into Lake Mead over the next 50 years?	40
What is causing climate change?	90
Can you explain the greenhouse effect?	70
What is a scientific model?	100
If all Las Vegans conserve water, how do you think that will affect Lake Mead's water levels?	40
Do you see any relationship between water use and energy use?	50

[a] Representative quote: "I guess it's dropping because the Las Vegas Valley has grown so much and it needs that much more water."
[b] Figure reflects citing one incorrect factor.
[c] Reflects ability to name at least six of the seven states.
[d] Representative quote regarding which state receives the most amount of water: "[Nevada does] cuz Colorado has like snow and probably other lakes and stuff and Utah has a lot and we have like no water. We are in the middle of a desert." Nevada actually received the least amount of water, California receives the most.
[e] Representative quote regarding whether climate is changing: "I don't know.…The only thing that I know about global warming is that it is controversial and my family argues about it.…The Republicans don't believe in it and the Democrats in my family do. I just try to stay neutral."

any relationship between energy use and water use. No participants mentioned allocations or could identify all the states that receive water from Lake Mead.

In addition to identifying knowledge and knowledge gaps, misconceptions were also identified. These misconceptions included: (a) Lake Mead's water levels are dropping because of population growth or excessive water use and (b) weather and climate are interchangeable concepts.

Fifty percent of participants believed that Lake Mead's water levels are dropping because of population growth or excessive water use: Specifically, 30% mentioned excessive water use and 20% mentioned population growth. For example, one participant stated, "I guess it's dropping because the Las Vegas Valley has grown so much and it needs that much more water." In fact, Lake Mead water levels are dropping primarily due to drought. Less snowfall in the Rocky Mountains has reduced the water flow in the Colorado River and therefore the amount of water flowing into Lake Mead. Furthermore, while some of the upper Colorado River states, such as Utah, are using more of their water allocation due to population growth, that is not the case with Nevada; Nevada's water allocation from the Colorado River is small (300 thousand acre feet), smaller than any other state, and Nevada has been using almost its full water allocation for some time. The association of declining lake levels and decreased snowpack in the Rocky Mountains was weak or nonexistent for many students. The mountains are remote and not within the realm of many students' personal experience. (Even if they have been to the Rockies, they may not have seen how much the snowpack has declined or made the link to Lake Mead water levels.)

One cannot make the argument that the extended drought is being caused by climate change. Climate involves a long-term average of temperature and precipitations levels (usually over 30 years or more; see Environmental Literacy Council (2007)), and so it is not possible to argue that any single weather event is being caused by climate change rather than natural variability in the weather. (An accumulation of such events over the long-term is, however, evidence of climate change.) Scientists do know that climate change will have a future impact on Colorado River flow: Climate models predict that climate change will cause the water flow in the Colorado River to decline 10–20% (or more) during the next several decades than would otherwise be the case. It is highly plausible that climate change is contributing to the drought. So it is important for students to understand the mechanisms of climate change. Only a couple of the students that we interviewed could, however, even explain the greenhouse effect.

Consistent with findings of other researchers, participants demonstrated misconceptions about the relationship between weather and climate. Sixty percent of our participants believed that these terms are somewhat interchangeable. One participant stated:

> [Climate change is] just a shift in the normal trend of what the climate has been in the past...Um, anything to do with weather and temperature and...Um, weather is, wow, um, different...different types of, um, I wanna say climate, but I already defined climate, um...different types of environmental trends. I think weather is more like wind and rain and climate is like desert ....

In fact, weather involves short duration events (hours, days, months, and years) at a particular location, and climate is weather conditions averaged over long-term periods of at least three decades and over wide areas (National Climatic Data Center, 2008). One participant stated that the climate is changing because "winter is coming on." It is this confusion between weather and climate that allows some in the media (e.g., "Snowpocolypse," 2010) to argue against "global warming" because of increased snowfall in the Northeast (ignoring the fact that increased solar energy is needed to move more water into the atmosphere, and that what is at issue is long-term spatial and temporal averages and extremes, not a single weather event).

Designing the Game

In designing the computer game, our intent is to make the arguments that (a) anthropogenic climate change is occurring, (b) that climate change will contribute to reduced snowfall in the Rocky Mountains, and (c) this in turn is causing Lake Mead water levels to decline. This is the theoretical side of our argument. On the practical side, we wanted to make the argument that students should take actions to conserve water (and which actions are most promising). Water conservation (e.g., taking shorter showers or replacing lawns with desert landscaping) relates to concrete activities that are meaningful to many students, whereas discussion of climate and hydrological systems are more complex and abstract. Furthermore, research in climate change education has shown that when instruction appeals to fear and catastrophe, individuals often "shut down" and reduce their level of cognitive processing, whereas when educated about things that can be controlled, both cognitive processing and self-efficacy increase (Moser & Dilling, 2007).

We initially thought it would make sense to first introduce users to a model of water flow, then to the nature of the problem (why lake levels are declining), and then to possible solutions (related to conservation or increasing supply). However, given the complexity of these issues, especially for eighth graders, we thought it best to start with something tangible and within students' range of experience (water conservation in residences), and then to proceed to increasingly more complex realms (the community, the region, the earth) and then to end with a simulation of Lake Mead water levels. This approach also solved another problem: Because Nevada's Colorado River water allocation is so small (and fixed), water conservation by Southern Nevadans will have a negligible impact on lake levels. Yet we did not want to give users the impression that water conservation is unimportant. Rather, it is very important to retain an adequate supply of water, especially if Nevada's water allocation is cut. This shows how "the making of an argument" (or more precisely, a line of argument) came to frame our design of the game and the sequence of activities.

The simulation game consists of five activities. Activity 1 is designed to build the user's background knowledge through multiple-choice questions pertaining to water resources and climate change. The user will be asked a series of questions, such as: *Do you know what's the most important factor determining the amount of water flowing in the Colorado River*? *Of the seven states that get their water from the Colorado River, which state gets the least*? and *Do you know what happens to most of the water that is used by Southern Nevadans after its treated*? After each question is answered, the user will have the opportunity to view the correct answer and receive additional information about the question. They will then be directed to Activity 2.

Activity 2 is designed to educate the user on what citizens' can do in their homes to conserve water. The user will go through a model of a house and pick the three most important options to ensure that household makes some changes to increase their sustainable water usage. Some options for the user to pick include: fix leaky toilets; remove lawns; cover swimming pool; and replace showerheads. Activity 3 is similar to Activity 2 in that the user is asked to pick options for water conservation but the options are related to the Las Vegas Valley community as a whole, rather than individual households. Some of the options in Activity 3 for the user to choose from are turn off casino fountains (a weak option, as the amount of water is negligible and most of it recycled); stop watering golf courses; and raise water prices. Users earn points in these two activities for picking the best options for water conservation.

Activity 4 is designed to develop the user's understanding of climate change and its potential impact on Southern Nevada's water supply. The user will have the opportunity to view images of receding glaciers, investigate how even small temperature changes can affect snow levels in the Rocky Mountains, and view a graph depicting the effect CO_2 levels have on average global temperature. After these four background knowledge building activities, users are ready for Activity 5 which focuses on lowering water levels in Lake Mead. Here the user makes a prediction as to when water levels in Lake Mead may fall to a level (1000 feet above sea level) where Las Vegas is no longer able to pump water from the lake because lake levels are below the lowest available intake pipe. The user then runs one or more computer simulations of declining water levels to assess the accuracy of their prediction. The computer simulation will reflect the results of an actual simulation of a GCC model under one of three different global economic growth scenarios (each with a different level of greenhouse gas emissions). The user can replay this part of the game again under a different scenario. A more advanced level of the game (for tenth graders and adults) will introduce users to the fact that there are multiple GCC models, and that the prediction of these models diverge, but only within a certain range.

Argumentation in "Losing the Lake"

In summary, the line of argument that the game will develop is (a) we are in a drought, (b) water conservation is important (and possible), (c) climate change will cause the drought to become worse, (d) we only have so much time before Las Vegas loses its water supply, and so (e) we need to act now! This is ultimately a practical and ethical argument, because it pertains to what should be the case (and not just to how the world works). At the same time, there is a whole lot of science involved in this chain of argument. The qualitative study showed that students' mental models of both global warming and regional water flow are likely to be impoverished, and so a considerable number of activities are being included in the game on these topics. On top of this, we also need to make the theoretical case that human activities are contributing to climate change. All GCC models recognize this fact; what is uncertain is the severity and timing of the effects. Nevertheless, students may need to be persuaded that the assumptions being made by the models are reasonable. One of the most compelling pieces of evidence that we plan to use is the close association between levels of greenhouse gases in the atmosphere and average global temperatures over the last 350,000 years. In Walton's argumentation framework, this argument employs a "correlation to cause" argument scheme (Walton et al.,

2008). As discussed previously, this is a type of argument that backs up a causal claim using correlational data.

Now it is frequently argued among social scientists that it is fallacious to use correlational data to establish causation. Is the argument for a greenhouse effect therefore based on a logical fallacy? Not at all. The idea of a "fallacious argument" is an outdated one (Walton, 1989), because Walton (1995) and others have shown that specific modes of argument that may be considered fallacious in some situations are perfectly sound in others. Instead, Walton uses the terminology of "argument schemes" and being able to appropriately answer the critical questions involved in those schemes. For the correlation to cause scheme, a key question is whether there is a third "variable" that is causing a correlation between the other two variables. No one has yet identified such a third variable and developed a GCC model around it, whereas there are well-elaborated causal mechanisms for how greenhouse gases affect global climate, and these are reflected in the GCC models. This shows that, while correlation by itself does not establish causation, it can as part of a larger argument be embedded within scientific inquiry of the critical questions. One has to understand the nature of the argument scheme.

In conclusion, the "Losing the Lake" game will employ both a practical reasoning argumentation scheme and a "correlation to cause" argumentation scheme. Other argumentation schemes that will be used are "argument from value" (saving water is good), "argument from expert opinion" (scientists who are in a position to know believe X), "argument from analogy" (analogy with a greenhouse), "argument from cause to effect" (less snowfall in the Rockies will cause water levels to drop), and to some degree, argument from abduction (because GCC models have made successful predictions, this is evidence for their underlying assumptions). One final argumentation scheme is "Negative Reasoning from Normal Expectations" (Walton et al., 2008): If the current situation was normal (i.e., with no anthropogenic climate change), one would expect lower global atmospheric and oceanic temperatures than has been observed, so the situation is not normal (and the only plausible explanation for the difference is anthropogenic climate change).

Conclusion

In both the case of the seventh-grade discussion and the Losing the Lake game, asking critical questions appropriate to the argument schemes being used is key to achieving our pedagogical goals. In the first case, however, the first author taught students to ask critical questions, whereas in the second case, we asked the critical questions ourselves in framing arguments that formed the basis of the game design. Related to this, the first intervention was highly dialogical, with students interacting with one another, whereas the second was more monological. We hope at some point to add a more dialogical component to the computer game; for example, if used as part of a middle- or high-school science curriculum, students could argue with one another over a discussion board over the merits of building a pipeline to transport ground water from Eastern to Southern Nevada. We did not include this as an "option" in our initial game because it is a highly controversial proposal and the environmental effects and financial costs are still uncertain. The issue could, however, be fodder for a classroom debate or collaborative argumentation discussion.

We started this chapter with some discussion of the two faces of scientific argumentation (theoretical and practical), and how these two sides are interwoven. In fact, there are probably more than just "two sides" because a number of different argumentation schemes may be involved. Some of these schemes are more purely empirical than others, but all of the ones discussed here contain some sort of scientific dimension. Whether an argument is "scientific" may therefore be more a matter of degree than a categorical attribute.

It is important, in our view, for students to understand a number of different argumentation schemes and the critical questions that should be asked about them. They need a more nuanced view of argumentation than is involved in just teaching about "standard components" of a scientific argument or the scientific method. Even with these, students have a hard time learning to go beyond simply describing the results of an experiment and engaging in model-based reasoning (Windschitl, 2004) or relation-based reasoning (where an outcome is predicted by a "correlation or linear causal sequence") (Sunal, 2006). Introducing students to the language of argumentation, in general, and those of some argumentation schemes in particular (such as argument from correlation to cause, from cause to effect, or from value), can help convey to students the type of reasoning that is valued, provided that the nature of the schemes—and the associated critical questions—are adequately understood. Discussing argumentation terminology can also provide a context for exploring the strength and limitations of the different argument schemes. In a sense, the different schemes provide different perspectives on the world and our attempts to change or adapt to it, which is extremely important is this era of rapid ecological change.

Scientific argumentation, like science itself, is a rich, multifaceted construct. Like science itself, it does not take one form, follow one pattern, or hold to one set of rules. Scientists use multiple methods to achieve their goals of defining and describing the natural and physical world. So too can an argument take many forms and still be "scientific." From our perspective, the more students learn about the nature of argumentation, the more they can critically think and reason about science and the socio-scientific challenges facing society.

References

Angell, R. B. (1964). *Reasoning and logic.* New York: Appleton-Century-Crofts.

Antholis, W., & Talbott, S. (2010). *Fast forward: Ethics and politics in the age of global warming.* Washington, DC: Brookings Institution Press.

Asterhan, C. S. C., & Schwarz, B. B. (2007). The effects of monological and dialogical argumentation on concept learning in evolutionary theory. *Journal of Educational Psychology, 99,* 626–639.

Begley, S. (2007, December 31/2008, January 7). Learning to love climate 'adaptation.' *Newsweek, 151*(1), 84.

Brean, H. (2009, April 14). Drought lingers: Lake sinking near 1965 level. *Las Vegas Review-Journal*. Retrieved from http://www.lvrj.com/news/42949442.html

Broad, W., & Wade, N. (1982). *Betrayers of the truth*. New York: Simon & Schuster.

Chan, C., Burtis, J., & Bereiter, C. (1997). Knowledge building as a mediator of conflict in conceptual change. *Cognition & Instruction, 15*, 1–40.

Chin, C., & Osborne, J. (2010). Supporting argumentation through students' questions: Case studies in science classrooms. *The Journal of the Learning Sciences, 19*, 230–284.

Clark, H. H., & Brennan, S. E. (1991). Grounding in communication. In L. Resnick, J. M. Levine, & S. D. Teasley (Eds.), *Perspectives on socially shared cognition* (pp. 127–149). Washington, D C: American Psychological Association.

De Avila, P. Jr., & Torres, B. B. (2010). Introducing undergraduate students to science. *Biochemistry and Molecular Biology Education, 38*(2), 70–78.

Dewey, J. P. (1910). *How we think*. Lexington, MA: D.C. Heath.

Doise, W., Mugny, G., & Perrett-Clermont, A. (1976). Social interaction and cognitive development: Further evidence. *European Journal of Social Psychology, 56*, 245–247.

Dole, J. A., & Sinatra, G. M. (1998). Reconceptualizing change in the cognitive construction of knowledge. *Educational Psychologist, 33*(2/3), 109–128.

Doran, P. T., & Kendall Zimmerman, M. (2009). Examining the scientific consensus on climate change. *Eos Trans. AGU, 90*(3). doi:10.1029/2009EO030002.

Duschl, R. A. (2008). Quality argumentation and epistemic criteria. In S. Erduran & M. P. Jimenez-Aleixandre (Eds.), *Argumentation in science education: Perspectives from classroom-based research* (pp. 159–175). Dordrecht, The Netherlands: Springer.

Environmental Literacy Council and the National Science Teachers Association. (2007). *Global climate change: Resources for environmental literacy*. Arlington, VA: NSTA Press.

Erduran, S. (2008). Methodological foundations in the study of argumentation in science classrooms. In S. Southerland (Series Ed.), S. Erduran & M. P. Jiménez-Aleixandre (Volume Eds.), *Argumentation in science education: Perspectives from classroom-based research* (pp. 47–69). Dordrecht, The Netherlands: Springer.

Govier, T. (1987). *Problems in argument analysis and evaluation*. Providence, RI: Foris.

Greeno, J. G., & van de Sande, C. (2007). Perspectival understanding of conceptions and conceptual growth in interaction. *Educational Psychologist, 42*, 9–23.

Hastings, A. C. (1963). *A reformulation of the modes of reasoning in argumentation*. Unpublished doctoral dissertation, Northwestern University, Evanston, IL.

Hoffert, M. I., Caldeira, K., Benford, G., Criswell, D. R., Green, C., Herzog, H., et al. (2002). Advanced technology paths to global climate stability: Energy for a greenhouse planet. *Science, 28*, 981–987. doi: 10.1126/science.1072357

Jiménez-Aleixandre, M. P., & Erduran, S. (2008). Argumentation in science education: An overview. In S. Southerland (Series Ed.), S. Erduran & M. P. Jiménez-Aleixandre (Volume Eds.), *Argumentation in science education: Perspectives from classroom-based research* (pp. 3–27). Dordrecht, The Netherlands: Springer.

Keith, W., & Rehg, W. (2008). Argumentation in science: The cross-fertilization of argumentation theory and science studies. In E. J. Hackett, O. Amsterdamska, M. Lynch, & J. Wajcman (Eds.), *The handbook of science and technology studies* (3rd ed., pp. 211–239). Cambridge, MA: MIT Press.

Kelly, K. (2009, August 16). Water and our economy. If we don't develop the resources, we're at risk. *Las Vegas Review-Journal*. Retrieved from http://www.lvrj.com/opinion/53343277.html

King, A. (1990). Enhancing peer interaction and learning through guided student-generated questioning. *Educational Psychologist, 27*, 111–126.

Kitcher, P. (1993). *The advancement of science*. New York: Oxford University Press.

Krupp, F., & Horn, M. (2008). *Earth: The sequel: The race to reinvent energy and stop global warming*. New York: W. W. Norton.

Kuhn, D. (2005). *Education for thinking*. Cambridge, MA: Harvard University Press.

Kuhn, D., & Udell, W. (2003). The development of argument skill. *Child Development, 74*, 1245–1260.

Lawson, A. E. (2003). The nature and development of hypothetico-predictive argumentation with implications for science teaching. *International Journal of Science Education, 25*, 1387–1408.

Leiserowitz, A., Maibach, E., & Light, A. (2009). *Global warming's six Americas: An audience segmentation analysis.* Center for America Progress. Retrieved from http://www.americanprogress.org/issues/2009/05/6americas.html

Lindzen, R. S. (2009, December 6). Portending doom. *Las Vegas Review Journal.* (Reprinted from "The climate science isn't settled," *The Wall Street Journal.*) Retrieved from http://online.wsj.com/article/SB10001424052748703939404574567423917025400.html

Mayr, E. (1991). *One long argument: Charles Darwin and the genesis of modern evolutionary thought.* Cambridge, MA: Harvard University Press.

Moser, S. C., & Dilling, L. (2007). Toward the tipping point. In S. C. Moser & L. Dilling (Eds.), *Creating a climate for change: Communicating climate change and facilitating social change* (pp. 491–516). New York: Cambridge University Press.

National Climatic Data Center (2008). *Weather/climate events.* Retrieved from http://www.ncdc.noaa.gov/oa/climateresearch.html

National Research Council. (2000). *Inquiry and the national science education standards.* Washington, DC: National Academy Press.

National Research Council. (2001). *How students learn: History, mathematics, and science in the classroom.* Washington, DC: National Academy Press.

National Science Foundation. (2009). *Solving the puzzle: Researching the impacts of climate change around the world.* Retrieved from http://www.nsf.gov/news/nsf09202/nsf09202.pdf

Nemet, G. F., and Kammen, D. M. (2007). U.S. energy R&D: Declining investment, increasing need, and the feasibility of expansion. *Energy Policy, 35*, 746–755.

Novak, J. D., & Gowan, D. B. (1984). *Learning how to learn.* New York: Cambridge University Press.

Nussbaum, E. M. (2002). How introverts versus extroverts approach classroom argumentative discussions. *The Elementary School Journal, 102*, 183–197.

Nussbaum, E. M. (2008a). Collaborative discourse, argumentation, and learning: Preface and literature review. *Contemporary Educational Psychology, 33*, 345–359.

Nussbaum, E. M. (2008b). Using argumentation vee diagrams (AVDs) for promoting argument/counterargument integration in reflective writing. *Journal of Educational Psychology, 100*, 549–565.

Nussbaum, E. M. (2011). Argumentation, dialogue theory, and probability modeling: Alternative frameworks for argumentation research in education. *Educational Psychologist, 46*, 84–106.

Nussbaum, E. M., & Edwards, O. V. (2011). Critical questions and argument stratagems: A framework for enhancing and analyzing students' reasoning practices. *The Journal of the Learning Sciences, 20*, 443–488.

Nussbaum, E. M., & Sinatra, G. M. (2003). Argument and conceptual engagement. *Contemporary Educational Psychology, 28*, 384–395.

Nussbaum, E. M., Sinatra, G. M., & Poliquin, A. M. (2008). The role of epistemic beliefs and scientific argumentation in science learning. *International Journal of Science Education, 30*, 1977–1999.

Nussbaum, E. M., Winsor, D. L., Aqui, Y. M., & Poliquin, A. M. (2007). Putting the pieces together: Online argumentation vee diagrams enhance thinking during discussions. *International Journal of Computer-Supported Collaborative Learning, 2*, 479–500.

Obama announces climate change deal with China, other nations. (2009, December 18). *CNN Politics*. Retrieved from http://www.cnn.com/2009/POLITICS/12/18/obama.copenhagen/index.html

Pew Research Center for the People and the Press. (2009). *Fewer Americans see solid evidence of global warming. Modest support for "cap and trade" policy.* Retrieved from http://people-press.org/report/556/global-warming

Schwarz, B., Neuman, Y., & Biezuner, S. (2000). Two wrongs may make a right ... If they argue together! *Cognition & Instruction, 18*, 461–494.

Sismondo, S. (2008). Science and technology studies and an engaged program. In E. J. Hackett, O. Amsterdamska, M. Lynch, & J. Wajcman (Eds.), *The handbook of science and technology studies* (3rd ed., pp. 13–31). Cambridge, MA: MIT Press.

Sisolak, S. (2009, August 19). Drought requires community to pursue all of its options. *Las Vegas Sun.* Retrieved from http://www.lasvegassun.com/news/2009/aug/19/drought-requires-community-pursue-all-its-options/

Snowpocolypse vs. global warming. (2010, February 10). *True/Slant.* Retrieved from http://politifi.com/news/Snowpocalypse-Vs-Global-Warming-272306.html

Solomon, M. (2008). STS and social epistemology of science. In E. J. Hackett, O. Amsterdamska, M. Lynch, & J. Wajcman (Eds.), *The handbook of science and technology studies* (3rd ed., pp. 241–258). Cambridge, MA: MIT Press.

Sunal, C. S. (2006). Argumentation and the science standards: The intersection of scientific and historical reasoning and inquiry. In D. W. Sunal & E. L. Wright (Eds.), *The impact of state and national standards on K-12 science teaching* (pp. 257–300). Greenwich, CN: IAP.

Teixeira, R. (2010, June 14). *Public opinion snapshot: The public believes global warming is happening and is ready for action.* Center for American Progress. Retrieved from http://www.americanprogress.org/issues/2010/06/public_and_warming.html

Walton, D. N. (1989). Dialogue theory for critical thinking. *Argumentation, 3*, 169–184.

Walton, D. N. (1995). *A pragmatic theory of fallacy.* Tuscaloosa, AL: University of Alabama Press.

Walton, D. N. (1998). *The new dialectic.* University Park, PA: Pennsylvania State University Press.

Walton, D., Reed, C., & Macagno, F. (2008). *Argumentation schemes.* New York: Cambridge University Press.

Will, G. F. (2010, February 21). Climate science tantrums. *Jewish World Review.* Retrieved from http://www.jewishworldreview.com/cols/will022110.php3

Windschitl, M. (2004). Folk theories of "inquiry:" How preservice teachers reproduce the discourse and practices of an atheoretical scientific method. *Journal of Research in Science Teaching, 41*, 481–512.

Wintour, P. (2009, August 8). UN climate change deal needs more sacrifices by West, John Prescott warns. *The Guardian.* Retrieved from http://www.guardian.co.uk/environment/2009/aug/08/copenhagen-kyoto-climate-change-talks

Yale Project on Climate Change/George Mason University Center for Climate Change Communication. (2010). *Climate change in the American mind: Americans climate change beliefs, attitudes, policy preferences, and actions.* Retrieved from http://www.climatechangecommunication.org/images/files/Climate_Change_in_the_American_Mind.pdf

Zeidler, D. L., Osborne, J., Erduran, S., Simon, S., & Monk, M. (2006). The role of argument during discourse about socioscientific issues. In D. L. Zeidler (Ed.), *The role of moral reasoning on socioscientific issues and discourse in science education* (pp. 97–116). Dordrecht, The Netherlands: Kluwer Academic.

Zeidler, D. L., Sadler, T. D., Applebaum, S., & Callahan, B. E. (2009). Advancing reflective judgment through socioscientific issues. *Journal of Research in Science Teaching, 46*, 74–101.

Zohar, A., & Nemet, F. (2002). Fostering students' knowledge and argumentation skills through dilemmas in human genetics. *Journal of Research in Science Teaching, 39*, 35–62.

Chapter 3
The Importance of Embedding Argument Within Science Classrooms

Andy Cavagnetto and Brian Hand

Introduction

Evolving understandings of the nature of science and the role of language in student learning has led to an increase in the emphasis placed on argument and argumentation in education, particularly science education, contexts. The wave of argument-based interventions comes with a great deal of diversity in the characteristic nature of the interventions. Not only are the interventions diverse, but the methods by which the interventions are studied are as well. The diversity of interventions and methods for assessment is positive, but also has the potential to create a nebulous sea of argument. That is, much like the notion of inquiry, argument has the potential for becoming an instructional strategy that is undefined and therefore underutilized, yet always considered a positive technique to improve student learning. Like inquiry, the effectiveness of argument is dependent on the goal of instruction. In order to prevent the translation from research to practice from becoming an ill-defined instructional strategy, researchers need to move toward understanding the particular aspects that are critical to moving students toward specific goals. This process seems to be underway at the level of intervention, but this likely is not enough. After all, there are particular interventions for inquiry that have been shown to move students toward defined goals, yet the translation to classroom application has still broken down. One method for identifying the critical components for argument would be to deconstruct interventions to isolate particular areas that prove to be effective. This method is possible in a controlled setting but becomes increasingly more difficult in most classroom contexts. An alternative would be to continue to define aspects of argument and begin comparing key characteristics of the various argument-based interventions. This chapter will compare the use of language in argument-based interventions in an effort to stimulate further discussion about the variations in the argument in science education literature.

A. Cavagnetto (✉)
Binghamton University, State University of New York, Binghamton, NY, USA
e-mail: acavagne@binghamton.edu

M.S. Khine (ed.), *Perspectives on Scientific Argumentation*,
DOI 10.1007/978-94-007-2470-9_3,

Language in Science

The focus on language, as viewed in terms of the concept of science literacy, has shifted many times over the last century. Much of the early focus on science literacy was ensuring that a learner could read the science textbook and use the words of science correctly. However, as suggested by a number of international standards and publications (Australian Curriculum, Assessment and Reporting Authority, 2009; National Research Council (NRC), 2007; Organisation for Economic Co-Operation and Development, 2003; The United Kingdom Department for Children, Schools, and Families, n.d.) the emphasis has changed from replication of terminology to a focus on the ability to use the language to build understanding of the topic and the ability to communicate to a broad audience the science knowledge gained from studying the topic. This shift in emphasis has meant that much more attention is now focused on the relationship between language and science. Students are now expected to do much more than simply remember lists of facts, be able to spell words correctly, or recognize and define the bold words in the science text. These international documents appropriately, although not explicitly, situate language as central to science with their emphasis on the epistemic nature of science.

Norris and Phillips (2003) are much more explicit in the role of language in scientific literacy when they defined the two essential senses of literacy that frame science. The first is the derived sense of literacy in which "reading and writing do not stand only in a functional relationship with respect to science, as simply tools for the storage and transmission of science. Rather, the relationship is a constitutive one, wherein reading and writing are constitutive parts of science" (p. 226). For Norris and Phillips this is critical because these constituents are the "essential elements of the whole" (p. 226), that is, remove these language elements and there is no science. Science is not something that can be done without language. For the authors, Norris' and Phillips' derived sense of science literacy should also include the different modes of representation. While this is implicit within reading and writing, there is a need to understand that different modes of science are integral to the concept of reading and writing. Science is more than just text as evidenced by other modes used by scientists to construct understanding (e.g., graphs, equations, tables, diagrams, and models).

The second essential sense of literacy is the fundamental sense of science literacy. For Norris and Phillips the fundamental sense involves the "reasoning required to comprehend, interpret, analyse, and criticise any text" (p. 237). Importantly, they argue that science has to move past oracy and the oral traditions because "without text, the social practices that make science possible could not be engaged with" (p. 233). The important recording, presentation, and re-presentation of ideas, and debates and arguments that constitute the nature of the discipline are not possible without text. These two essential senses of literacy are critical to the development of scientific literacy. Simply viewing acquiring of science content knowledge (the derived sense) as success denies the importance of being able to apply the reasoning structures of science (the fundamental sense) required for reading and writing about science. This important emphasis on language is critical when discussing science argument and learning of science. The vehicle for advancing science knowledge is argument, and argument is only advanced through the use of language. Without language science cannot advance.

Orientations Toward Learning Argument

A recent review of argument-based interventions by the first author suggests different positions have been adopted for learning science argument (Cavagnetto, 2010). The review of 54 articles that reported on argument interventions categorized argument-based interventions with regard to (a) when argument is used in the intervention, (b) what the interventions are designed to stimulate argument about, and (c) what aspects of science are present in the interventions. The author found three orientations toward argument were present in the research literature. The orientations are: (a) immersion in science for learning scientific argument (immersion), (b) learning the structure of argument to learn and apply scientific argument (structure), and (c) experiencing the interaction between science and society to learn scientific argument (socioscientific).

As suggested in Cavagnetto (2010), a number of the interventions are guided by the notion that it is best to learn scientific argument by embedding argument within investigative contexts. From this immersion orientation, argument serves as a tool for the construction of understanding of both the epistemic practices of science and scientific concepts. These interventions accomplish this through the use of scaffolding prompts and cognitive conflict. One such example is the ExplanationConstructor, a computer program that scaffolds students' understanding of the relationship between investigation and explanation. A second example, which will be elaborated on in more detail later in the chapter, is the Science Writing Heuristic (SWH) approach (Hand & Keys, 1999). The SWH utilizes questions to prompt student construction and critique of arguments. Other interventions such as the use of personally seeded discussions (Clark and Sampson, 2007) and concept cartoons (Keogh & Naylor, 1999) attempt to establish cognitive conflict as a way to foster argument. Personally seeded discussions use a computer program to match opposing student explanations of natural phenomena. The students then engage in dialogue to determine which explanation captures the phenomena best. Concept cartoons use cartoons that depict scenarios centered on common misconceptions as a means to generate dialogue and interest in understanding the science principle. While argument is important in these interventions, it is in pursuit of understanding science content.

Another way to facilitate argument competence is by explicitly teaching a structure for argument and subsequently asking students to apply the structure. The structure orientation has been advanced primarily by the work published as part of the IDEAS project (Erduran, Simon, & Osborne, 2004; Osborne, Erduran, & Simon, 2004; Simon, Erduran, & Osborne, 2006; and von Aufschnaiter , Erduran, Osborne, & Simon, 2008) and the Claims, Evidence, and Reasoning structure identified in McNeill (2009); McNeill, Lizotte, Krajcik, and Marx (2006); and McNeill and Krajcik (2008). In the IDEAS project, students are taught Toulmin's (1958) argument structure and then gain experience with its application across nine argument topics. In the published studies to date, students were most often asked to generate explanations by evaluating evidence for competing mechanisms for a phenomenon. Similarly, the claims, evidence, and reasoning structure reported by McNeill et al. were developed as a more digestible version of Toulmin's structure. These interventions emphasize the structure of argument as a scaffold to critical thinking and a product of inquiry.

A third orientation toward learning scientific argument emphasizes the interaction between science and society, including moral, ethical, and political influences on decision making in scientific contexts. In these interventions, socioscientific and science, technology, society issues-based interventions are used as contexts for engaging students in argument. Argument then serves as a vehicle for students to gain an understanding of the social and cultural elements that influence science. This orientation has been advanced by Zeidler, Sadler, and colleagues. For example, Walker and Zeidler (2007) report on a policy-making debate about genetically modified foods. Students participated in explorations using web-based investigative software and applied their conceptual understandings in the policy-making debate. A similar intervention occurred in Sadler, Chambers, and Zeidler (2004). These interventions have primarily focused on realizing moral, ethical, and political considerations associated with the application of science knowledge rather than argument for constructing an understanding of scientific principles. However, Zeidler and Sadler now appear to be moving toward using socioscientific issues as a curricular context for courses of study (Fowler, Zeidler, & Sadler, 2009).

When looking across these three orientations the diversity among argument is evident. The diversity of the characteristics and goals of the interventions certainly illustrate a clear movement by the science education community to broaden the instructional goals from those historically found in school curricula or emphasized in textbooks. In addition to highlighting the diversity of these interventions, and for the authors more importantly, Cavagnetto's characterization of the three orientations allows researchers to gain a clearer understanding of the perspective of language in which these orientations are founded.

There has been an ongoing debate about the best approach to introduce language instruction within classrooms, particularly in relation to science classrooms. The work of Halliday and Martin (1993) clearly emphasized the need for students to have to engage with the structure of the genres of science as a precursor to doing science. This position adopts the view that there is a need to learn to use the language prior to learning the science. For example, students need to learn the structure of the laboratory report prior to using the format to engage with laboratory activities. Gee (2004) has argued for the opposite position, contending that language must be embedded in the learning experience in order for students to gain a rich understanding of the disciplinary language. From Gee's stance, language is viewed as a learning tool. That is, learning how to use the language is not separated from learning science.

While there is much debate about the relative merits at the extremes of these positions, Hand and Prain (2006) have argued for the need for some convergence of these positions. They believe there is a continuum of positions such that while there is a requirement for students to engage with the language of the discipline as a learning tool in order to learn the content, they also believe students need to understand the structure of the genres used within science. Klein (2006), in discussing the relative importance of first- and second-order cognitive science with respect to science literacy, also argues that there is no one position in terms of language that should be adopted. He suggests that in

> the middle of the spectrum are practices that integrate expressive features of human thought and language with denotative features of authentic science text, such as concept mapping, graphing and the SWH. The result is that contemporary reforms in science literacy education accommodate students' cognition and language, while preparing them to participate in disciplinary knowledge construction. Furthermore, the central hypothesis overarching these interpretations is that enhanced science literacy in the fundamental sense will result in improved understanding of the big ideas of science and fuller participation in the public debate about science, technology, society, and environment issues—the derived sense of science literacy. (p. 171)

The importance of recognizing the need to have some middle ground also applies to the concept of science argument. Much of the structure orientation (e.g., work done by Osborne et al. (2004)) is based on the work of Halliday and Martin. The work has focused on the dealing with promoting argument as a structure to be learnt prior to using argument within class. As they suggest "argument is a discourse that needs to be explicitly taught, through the provision of suitable activity, support, and modeling" (Simon et al., 2006, p. 237).

Conversely, the emphasis of the immersion orientation is to embed science argument within the context of doing inquiry, that is, students need to use science argument as a critical component of building understanding of the content. By using the scaffolds as a guide for completing the science inquiry, students are required to both construct understanding and build their understanding around an argument framework. Building on the writing to learn framework (Prain & Hand, 1996), there is a need for students to engage with the language of the science, the language of argument and all the negotiation of meaning that is required in moving between the various elements of the argument structure. Having students use science argument within the context of the topics that they have to build understanding of means that students are not separating the concept of argument from how knowledge is constructed in science.

Ford recognized this consistency among building understanding of science concepts and understanding of the epistemology of science. In his work related to the basis of authority in science, Ford (2008) argues for moving students toward a grasp of scientific practice because it facilitates the learning of scientific principles. As Ford contends, "a grasp of practice is necessary for scientists to participate in the creation of new knowledge because it provides an overview of its architecture and how to navigate it. A grasp of practice motivates and guides a search within this architecture for the informational content, indeed the meaning, of canonical scientific knowledge" (p. 406). For Ford, the epistemological nature of science serves as a framework for negotiating content.

This continuum of positions for introducing language has implications not only for broad characteristics of argument-based instructional interventions, but also for the way in which terms associated with science and argument are used within the argument-based interventions. One area that appears to illustrate these implications is the relationship between data and evidence. One example of the turbidity of the relationship between data and evidence is found in the National Research Council's *Ready Set Science* (p. 133). The book highlights a claim, evidence, and reasoning structure and provides the following explanation:

- Claim: What happened, and why did it happen?
- Evidence: What information or data support the claim?
- Reasoning: What justification shows why the data count as evidence to support the claim?

The NRC follows these points by suggesting that students who utilize this framework (and the curriculum of which this framework is a fundamental component) "make sense of the phenomena under study (claim), articulate that understanding (evidence), and defend that understanding to their peers" (p. 133). While we recognize that the claim, evidence, and reasoning structure is more clearly defined in McNeill et al. (2006), we would suggest that as characterized by the NRC, reasoning is undervalued. That is, the characterization suggests that reasoning occurs only at a defined point of inquiry rather than throughout as a critical aspect of the entire process. While we suspect that McNeill and colleagues may not have intended for such an interpretation, practitioners are most likely to work from the NRC compilation rather than directly excavating ideas from the research literature. Therefore, this claim, evidence, and reasoning framework warrants discussion. Does this oversight stem from a mechanistic view of language? From a mechanistic perspective, the argument structure must be mastered and is therefore the focus. Argument appears to be a product of inquiry rather than a means of inquiry. To instruct students in the structure of argument, they do not need to interpret or analyze data. While we acknowledge that some of the structure-oriented interventions do include data analysis, many do not. In many instances, evidence is provided and students need to make a claim based on the evidence. This is found not only in some of the Investigating and Questioning our World through Science and Technology (IQWST) materials but also in the commonly cited Ideas, Evidence, and Argument in Science (IDEAS) materials. Importantly, we are not suggesting that these materials do not hold value, rather we are contending that the structures that drive these interventions blur the lines between data and evidence.

Data and Evidence

What is the relationship between data and evidence? In framing this discussion we recognize that there are two orientations that are generally used within classrooms. The first is prepared arguments that are given to students, that is, examples that scientists have been involved with or textbook examples that students are required to use. The second is the argument-based inquiry activities that students undertake as part of the work in which they have to generate an argument from the activity. Our discussions are focused on the second orientation. In other words, we are focused on how students move from an inquiry activity to frame an argument as a consequence of the investigation. As such we need to address the use of such words as data and evidence.

Before moving to discuss the use of the words, we would state that for us the concept of immersion within the process of argument is a critical component of the learning process. Using language as a learning tool as the conceptual frame for this work means that using argument as a learning tool becomes a subset of this concept given that arguments are language based. That is, arguments are based in language where language is represented across all the modes necessary to frame the argument (e.g., mathematical, symbolic, textual, etc.). Arguments do not exist until we create them. The production of any argument involves a learner/participant in negotiating publically and personally across a variety of settings. These include observations, written text (reading), oral text (debate), and in the construction of written text (written report).

Data

What constitutes data? When doing an inquiry what is considered data? For most of the work we do in school science inquiries, and in science generally, data is taken as being the observations completed for the investigation. What is seen and recorded is generally taken as being the data obtained from the inquiry. We constantly encourage children to write down everything they observe—colors, smells, changes in position, and so forth. We try where possible to get them to be as diligent as they can be and to record these data in some systematic manner. This recording can be in a table format or in the form of accurate notes. Thus at the end of the investigation and before the write up takes place, we are anticipating that students will have a rich source of data to use as they move forward to complete the write up.

However, are data to be gained only from the hands-on activities where observation skills dominate? Does information from sources other than hands-on activities provide data? That is, when reading to see what others say about the investigation, whether the authors are scientists or other students, is this serving as evidence or data? When collecting information from these sources the learner is trying to build support for his/her argument. The information acts as data—the learner uses the information as part of the support for a particular line of argument.

The immediate question that needs to be asked is how does one use this data? We are constantly saying or are constantly told data does not speak! That is, there is no neon sign from nature explaining its thinking. So if data does not speak the question goes—how do we move from data to evidence?

Evidence

If data does not speak, does not tell us what is useful or not useful, does not tell us which points are related to each other, how do we generate evidence? What is necessary to move from data to evidence? Who makes decisions about what is good evidence? Sufficient evidence? Appropriate evidence? While there are a number of different teaching/learning strategies that can be discussed, theoretically we have to deal with the cognitive perspective.

Shifting from data to evidence requires us to engage in cognitive work. A student has to analyze and synthesize the data points into some coherent series. There are critical decisions that need to be made such as what to keep, what to discard, and how well the data points are connected. That is, data does not speak and so the learner has to apply some critical thinking and reasoning to be able to make decisions to produce the required evidence he/she needs to make an argument. Thus, data plus reasoning result in evidence.

This distinction is important because we know that students have difficulty in moving from a series of observation points to constructing a logical line of reasoning that allows him/her to connect the data points in some coherent manner. The learner has to negotiate between their prior ideas and beliefs and with what they have collected as observations, that is, their data. Evidence is not separate from reasoning. The problem with a structure that highlights a claim, evidence, and reasoning approach is that evidence construction appears to occur separate from reasoning.

This concept of reasoning is further highlighted when we ask students to move to determine what scientists may say about the ideas they are exploring. Students have to negotiate with the text they are reading to understand what is being described, and then negotiate with themselves in terms of what knowledge they can associate with that being described in text. We need to be constantly asking them why they use particular bits of information (data) from text in support of the claim they wish to make.

Students do not choose which information or data point to use and then reason about it. They choose the point because they have to think critically and reason in the process of choosing. While this may be taken as a given, a structure that appears to separate reasoning from evidence has the great potential for teachers and learners to think that providing an answer for each is a separate process. First you provide evidence and then you supply the reasoning. The question is how can they be separate.

Argument-Based Inquiry: An Immersion Experience

After a chance meeting at the NARST conference in Chicago in 1997 one of the authors (Hand) collaborated with Carolyn Keys to explore the idea of building a framework that would link inquiry, argumentation, and an emphasis on language. The result was the development of the SWH approach. The SWH approach consists of a framework to guide activities as well as a metacognitive support to prompt student reasoning about data. Similar to Gowin's vee heuristic (1981, p. 157), the SWH provides learners with a heuristic template to guide science activity and reasoning in writing. Further, the SWH provides teachers with a template of suggested strategies to enhance learning from laboratory activities (see Table 3.1). As a whole, the activities and metacognitive scaffolds seek to provide authentic meaning-making opportunities for learners. As suggested in the teacher template, the negotiation of meaning occurs across multiple formats for discussion and writing. The approach is conceptualized as a bridge between informal, expressive writing modes that foster personally constructed science understandings, and more formal, public modes that focus on canonical forms of reasoning in science. In this way the heuristic scaffolds learners in both understanding their own lab activity and connecting this knowledge to other science ideas. The template for student thinking (see Table 3.1) prompts learners to generate questions, claims, and evidence for claims. It also prompts them to compare their laboratory findings with others, including their peers and information in the textbook, Internet, or other sources. The template for student thinking also prompts learners to reflect on how their own ideas have changed during the experience of the laboratory activity. The SWH can be understood as an alternative format for laboratory reports, as well as an enhancement of learning possibilities of this science genre. Instead of responding to the five traditional sections, purpose, methods, observations, results, and conclusions, students are expected to respond to

Table 3.1 The two templates for the SWH: the teacher template and the student template

The science writing heuristic

A template for teacher-designed activities to promote laboratory understanding	A template to scaffold students
1. Exploration of pre-instruction understanding through individual or group concept mapping	1. Beginning ideas—What are my questions?
2. Pre-laboratory activities, including informal writing, making observations, brainstorming, and posing questions	2. Tests—What did I do?
3. Participation in laboratory activity	3. Observations—What did I see?
4. Negotiation phase I – writing personal meanings for laboratory activity (e.g., writing journals)	4. Claims—What can I claim?
5. Negotiation phase II – sharing and comparing data interpretations in small groups (e.g., making group charts)	5. Evidence—How do I know? Why am I making these claims?
6. Negotiation phase III—comparing science ideas to textbooks or other printed resources	6. Reading—How do my ideas compare with other ideas?
7. Negotiation phase IV—individual reflection and writing (e.g., creating a presentation such as a poster or report for a larger audience)	7. Reflection—How have my ideas changed?
8. Exploration of post-instruction understanding through concept mapping	

prompts eliciting questioning, knowledge claims, evidence, description of data and observations, methods, and to reflect on changes to their own thinking.

While the SWH recognizes the need for students to conduct laboratory investigations that develop their understanding of scientific practice, the teachers' template also seeks to provide a stronger pedagogical focus for this learning. In other words, the SWH is based on the assumption that science writing genres in school should reflect some of the characteristics of scientists' writing, but also be shaped as pedagogical tools to encourage students to "unpack" scientific meaning and reasoning. The SWH is intended to promote both scientific thinking and reasoning in the laboratory, as well as metacognition, where learners become aware of the basis of their knowledge and are able to monitor more explicitly their learning. Because the SWH focuses on canonical forms of scientific thinking, such as the development of links between claims and evidence, it also has the potential to build learners' understandings of the nature of science, strengthen conceptual understandings, and engage them in authentic argumentation processes of science.

The SWH emphasizes the collaborative nature of scientific activity, that is, scientific argumentation, where learners are expected to engage in a continuous cycle of negotiating and clarifying meanings and explanations with their peers and teacher. In other words, the SWH is designed to promote classroom discussion where students' personal explanations and observations are tested against the perceptions and contributions of the broader group. Learners are encouraged to make explicit and defensible connections between questions, observations, data, claims, and evidence. When students state a claim for an investigation, they are expected to describe a pattern, make a generalization, state a relationship, or construct an explanation.

The SWH promotes students participation in setting their own investigative agenda for laboratory work, framing questions, proposing methods to address these questions, and carrying out appropriate investigations. Such an approach to laboratory work is advocated in many national science curriculum documents on the grounds that this freedom of choice will promote greater student engagement and motivation with topics. However, in practice much laboratory work follows a narrow teacher agenda that does not allow for broader questioning or more diverse data interpretation. When procedures are uniform for all students, where data are similar and where claims match expected outcomes, then the reporting of results and conclusions often lacks opportunities for deeper student learning about the topic or for developing scientific reasoning skills. To address these issues the SWH is designed to provide scaffolding for purposeful thinking about the relationships between questions, evidence, and claims.

Research on the SWH Approach

A number of quasi-experimental studies have been conducted to test the efficacy and impact of the SWH approach. These include the following:

A comparison between traditional teaching approaches and the SWH approach. Traditional teaching refers to the approaches that the teachers were using at the time of the study. In the first study this involved didactic teaching and some laboratory activities, while in the second study this involved using student recipe-type laboratory activities and report formats.

- A study by Akkus, Gunel, and Hand (2007) examined if there was a difference in student test performance between high levels of traditional science teaching and high-quality SWH implementation, with seven teachers. The results from the teacher-generated tests were very interesting in that the study looked at the difference between high achievers and low achievers in each group. The difference between high achievers in each group was not significant—they were essentially equal. However, when comparing the effect size difference between high and low achievers the following results were obtained: for the high traditional teaching the effect size difference was 1.23 between high and low achievers, while for the high SWH teaching the effect size difference was only 0.13. These results are very encouraging and indicate that the SWH approach is effective for all learners in the classroom.
- A study by Rudd, Greenbowe, and Hand (2007) was carried out to determine the effectiveness of using the SWH approach in university freshman general chemistry laboratory activities compared to the traditional formats used, with particular focus on the topic of equilibrium. To determine whether students in the SWH or standard sections exhibited better understanding of the concept of chemical equilibrium, the student explanations on the lecture exam problem were analyzed using mean explanation scores. Using the baseline knowledge score as a covariate, the ANCOVA results ($F = 4.913$; $df = 1.49$; $p = 0.031$) indicated a statistically significant association between higher explanation scores and the SWH format. The SWH sections demonstrated a greater ability than standard sections to identify the equilibrium condition and to explain aspects of equilibrium despite these sections having a lower baseline knowledge score.
- A study by Hohenshell and Hand (2006) examined the difference in student performance with Year 10 biology students who completed the laboratory activities using traditional approaches versus the SWH approach. The study examined students' test performance immediately after completing all the laboratory activities and then after completing a written summary report of the unit of study. Results from the first round of testing indicated that there were no significant differences on recall and conceptual question scores between the control (traditional laboratory approaches and report format) and the treatment group (SWH approach). However, on the second round of test, after completing the summary writing activities the SWH students scored significantly better on conceptual questions than the control students ($F(1.43) = 5.53$, $p = 0.023$, partial $\eta^2 = 0.114$).

Studies examining the impact of the quality of implementation of the SWH approach on student success on examinations. The purpose of these studies was to begin the process of determining the importance of adopting the particular strategies required when using the SWH approach. Rather than compare the SWH approach to traditional approaches, these studies compared student performance resulting from different levels of implementation of the SWH approach.

- A National Science Foundation-funded project to adapt the SWH approach to freshman general chemistry for science and engineering majors' laboratory activities demonstrated that the quality of implementation impacts performance on standardized tests and positively impacts the performance of females and low-achieving students—two groups that are viewed as disadvantaged in science classrooms. When comparing the difference between low and high implementation of the SWH approach, the following results were obtained of students' scores on American Chemical Society (ACS) standardized tests. On the pretest (ACS California diagnostic test) the difference, measured as Cohen d statistic, between students with high implementing teaching assistants (TAs) and low implementing TAs was 0.07. At the end of the semester the difference between the two groups on the ACS end of semester 1 test was 0.45 (medium effect size difference). The gap between males and females decreased from 0.45 (medium effect) on the pretest to 0.04 (no effect) on the posttest; while the gap between high and low achieving students decreased from 2.7 (large effect) to 0.7 (large effect) (Poock, Burke, Greenbowe & Hand, 2007).
- In a study by Mohammad (2007) of a one-semester freshman chemistry course for the "soft" sciences (agricultural, food science, etc.) students at the same university similar results were obtained, particularly with benefits to females in high implementation use of the SWH approach.
- In a study of six middle/secondary school science teachers, Gunel (2006) tracked the impact of implementation of the SWH approach on students' performance on the Iowa Test of Basic Skills (ITBS)/Iowa Test of Educational Development science tests across a 3-year period. His results show that for teachers who remained low across the period, the magnitude of effect size change in students' scores ranged from 0 to 0.4. For the teacher who shifted from traditional instruction to high-level implementation, there was an effect size change in his students' scores of 1.0 across the 3-year time period.

Studying the impact at the elementary level. Results from work done at the elementary level reflect those obtained from the other studies at middle/secondary and college level.

Results from a 3-year State of Iowa funded project involving 32 K 6 teachers who implemented the SWH approach for teaching science again supported the previous results when examining the quality of implementation. The results reported are for the first 2 years of the project. Teachers were rated as low, medium, or high in their implementation and the students performance on the ITBS science test were tracked (no teacher was rated as high). Grade equivalency growth scores were calculated and used in order to use teacher ranking as the dependent variable given that the analysis collapsed teachers into the low or medium level regardless of grade level. The results show that there were significant differences in ITBS science scores between students whose teachers are low and medium implementers of the SWH approach. This gap increased from an effect size of 0.073 in year 1 to an effect size of 0.268 in year 2. The gap between low SES students in low and medium implementation in year 1 (effect size 0.291) was almost the same as in year 2 (effect size 0.284). The gap between IEP students in low and medium implementation grew from year 1 (effect size 0.158) to year 2 (effect size 0.229).

The importance of such studies is that they begin to provide evidence for the claim that argument-based inquiry practices are valuable in promoting learning opportunities within science classrooms. These studies described above provide evidence that the SWH is an argument-based inquiry approach that can be used across a broad range of ages. While recognition is given to the argument that the sophistication of constructed arguments varies across the grade levels, the critical point for the authors is that we are able to involve children in the formation of arguments at an early age. This is important—we can involve all children in building science arguments by introducing a question, claims, and evidence approach to doing science. Students can be involved in constructing and critiquing arguments, differentiating between data and evidence, and as a consequence improve their understanding of science concepts.

Conclusion

For the authors, argument should be central to school science primarily because it is a vehicle for students to develop their understandings of scientific principles. We find this important, as we feel the current move toward argument has become one with an end goal that focuses on argument structure. While we acknowledge that rhetorical structures have importance in participation in science we believe that the structural level offered by models such as Toulmin's (1958) is simply too defined to be practical scaffolds for learning science. We align with Ford (2008) who suggests that understanding the practice of science is important because it allows students to understand scientific principles as scientists understand principles.

As suggested by the previously cited work on the SWH approach, public negotiations require students to engage with science principles at a level not achieved by traditional instruction. Subsequently, argument has the potential to increase student achievement on standardized metrics such as national and state exams. While there is some evidence of student achievement on standardized metrics with the use of the SWH approach, most argument interventions choose not to focus on this politically critical aspect of achievement. As such, it is difficult to make broad empirical claims of the benefits of argument on learning science content. This is a critical point for science educators because this lack of recognition of argument as a vehicle for learning content reduces the potential policy impacts of argument in science. While argument has been recognized as an important goal for science education in documents like *Ready, Set, Science* it is not recognized as an important goal by school administrators and teachers who are judged on students' abilities to pass content-based standardized exams. Being armed with numbers indicating increased student performance on such measures allows for a greater case to be made for inclusion. Using standardized exams as metrics for measuring argument-based interventions would also allow the research community to benefit more from individual contributions. Currently, it is difficult to conduct any meta-analysis to capture current findings as each group of researchers tends to use different outcome measures. While we recognize that research has specific questions that require unique methods of analysis, we feel it is important to highlight that content matters. That is, the level of sophistication of argument is linked not only to rhetorical structure but also to student content understanding.

References

Akkus, R., Gunel, M., & Hand, B. (2007). Comparing an inquiry based approach known as the Science Writing Heuristic to traditional science teaching practices: Are there differences? *International Journal of Science Education,* 29, 1745–1765.

Australian Curriculum, Assessment and Reporting Authority. (2009, May). *Shape of the Australian curriculum: Science*. Retrieved December 28, 2009, from http://www.acara.edu.au

Cavagnetto, A. R. (2010). Argument to foster scientific literacy: A review of argument interventions in K-12 science contexts. *Review of Educational Research, 80*(3), 336–371.

Clark, D. B., & Sampson, V. D. (2007). Personally-seeded discussions to scaffold online argumentation. *International Journal of Science Education, 29*(3), 253–277.

Erduran, S., Simon, S., & Osborne, J. (2004). TAPping into argumentation: Developments in the application of Toulmin's argument pattern for studying science discourse. *Science Education, 88*(6), 915–933.

Ford, M. J. (2008). Disciplinary authority and accountability in scientific practice and learning. *Science Education, 92*(3), 404–423.

Fowler, S., Zeidler, D., & Sadler, T. (2009). Moral sensitivity in the context of socioscientific issues in high school science students. *International Journal of Science Education, 31*(2), 279–296.

Gee, J. P. (2004). Language in the science classroom: Academic social languages as the heart of school-based literacy. In E. W. Saul (Ed.), *Crossing borders in literacy and science instruction: Perspectives on theory and practice* (pp. 13–32). Arlington, VA: NSTA Press.

Gowin, D. B. (1981). *Educating*. Ithaca, NY: Cornell University Press.

Gunel, M. (2006). *Investigating the impact of teachers' implementation practices on academic achievement in science during a long-term professional development program on the Science Writing Heuristic*. Unpublished PhD thesis, Iowa State University.

Halliday, M. A. K., & Martin, J. R. (1993). *Writing science: Literacy and discursive power.* London: Falmer Press.

Hand, B., & Prain, V. (2006). Moving from crossing borders to convergence of understandings in promoting science literacy. *International Journal of Science Education, 28*, 101–107.

Hand, B., & Keys, C. (1999). Inquiry investigation. *The Science Teacher, 66*(4), 27–29.

Hohenshell, L., & Hand, B. (2006). Writing-to-learn strategies in secondary school cell biology. *International Journal of Science Education, 28*, 261–289.

Keogh, B., & Naylor, S. (1999). Concept cartoons, teaching and learning in science: An evaluation. *International Journal of Science Education, 21*(4), 431–446.

Klein, P. D. (2006). The challenges of scientific literacy: From the viewpoint of second-generation cognitive science. *International Journal of Science Education, 28*, 143–178.

McNeill, K. L. (2009). Teachers' use of curriculum to support students in writing scientific arguments to explain phenomena. *Science Education, 93*(2; 2), 233–268.

McNeill, K. L., & Krajcik, J. (2008). Scientific explanations: Characterizing and evaluating the effects of teachers' instructional practices on student learning. *Journal of Research in Science Teaching, 45*(1), 53–78.

McNeill, K., Lizotte, D., Krajcik, J., & Marx, R. (2006). Supporting students' construction of scientific explanations by fading scaffolds in instructional materials. *Journal of the Learning Sciences, 15*(2), 153–191.

Mohammad, E. G. (2007). Using the science writing heuristic approach as a tool for assessing and promoting students' conceptual understanding and perceptions in the general chemistry laboratory ProQuest Dissertations And Theses; Thesis (Ph.D.)–Iowa State University, 2007.; Publication Number: AAI3274832; ISBN: 9780549148166; Source: Dissertation Abstracts International, Volume: 68-07, Section: A, page: 2792.; 143 p.

National Research Council. (2007). *Taking science to school: Learning and teaching science in grades K-8*. Washington, DC: The National Academies Press.

Norris, S. P., & Phillips, L. M. (2003). How literacy in its fundamental sense is central to scientific literacy. *Science Education, 87*(2), 224–240.

Organisation for Economic Co-Operation and Development. (2003). *The PISA 2003 assessment framework*. Retrieved from December 28, 2009, from http://www.pisa.oecd.org

Osborne, J., Erduran, S., & Simon, S. (2004). Enhancing the quality of argumentation in school science. *Journal of Research in Science Teaching, 41*(10), 994–1020.

Poock, J. A., Burke, K. A., Greenbowe, T. J., & Hand, B. M. (2007) Using the Science Writing Heuristic to improve students' academic performance. *Journal of Chemical Education, 84*, 1371–1379.

Prain, V., & Hand, B. (1996). Writing for learning in secondary science: Rethinking practices. *Teaching and Teacher Education, 12*(6), 609–626.

Rudd, II, J. A., Greenbowe, T. J., & Hand, B. M. (2007) Using the Science Writing Heuristic to improve students' understanding of general equilibrium. *Journal of Chemical Education, 84*, 2007–2011.

Sadler, T., Chambers, F., & Zeidler, D. (2004). Student conceptualizations of the nature of science in response to a socioscientific issue. *International Journal of Science Education, 26*(4), 387–409.

Simon, S., Erduran, S., & Osborne, J. (2006). Learning to teach argumentation: Research and development in the science classroom. *International Journal of Science Education, 28*(2–3), 235–260.

Toulmin, S. (1958). *The uses of argument*. Cambridge: Cambridge University Press.

United Kingdom Department for Children, Schools, and Families. (n.d.). *The national strategies*. Retrieved December 28, 2009, from http://nationalstrategies.standards.dcsf.gov.uk/

von Aufschnaiter, C., Erduran, S., Osborne, J., & Simon, S. (2008). Arguing to learn and learning to argue: Case studies of how students' argumentation relates to their scientific knowledge. *Journal of Research in Science Teaching, 45*(1), 101–131.

Walker, K. A., & Zeidler, D. L. (2007). Promoting discourse about socioscientific issues through scaffolded inquiry. *International Journal of Science Education, 29*(11), 1387–1410.

Chapter 4
Scientific Reasoning and Argumentation from a Bayesian Perspective

Evan Szu and Jonathan Osborne

Introduction

In this chapter, we seek to develop a specific account of scientific reasoning, its role, and its value in science education. One of the defining characteristics of science (and scientists) is the critical spirit that is central to science as a practice. Such reasoning is essential for the construction of claims to knowledge which are based on data and warrants which are then used to justify a claim. Typically, arguments may be either deductions about the world from a set of a priori premises such as those used in the development of kinetic theory; inductive generalizations about what patterns may exist typified by laws such as the law of conservation of energy; or inferences to the best explanation such as those used by Darwin in developing his argument for evolutionary theory. As important as the use of reasoning for the construction of knowledge is its use for critical review and evaluation for, as Ford (2008) argues, it is "critique which motivates authentic construction of scientific knowledge." Claims must be defended against arguments that question either the validity or reliability of the data, the warrant that justifies the significance of the data to the claim, or the background theoretical assumptions. The formal embodiment of this process is peer review and it is through this practice of discourse and argument that science maintains its objectivity (Longino, 1990).

However, whilst all might concur that such discursive practices are characterized by the use of reasoning, what are the salient features that distinguish scientific reasoning? Some conceptual clarity is needed if we are to distinguish good reasoning from that which is weak, wanting, or simply erroneous. In this chapter, therefore, we seek to explore briefly what are the some common characterizations of scientific reasoning. Our goal here is to suggest that all of these fail to capture an account of scientific reasoning which captures how individuals really reason. Instead, our main argument is that it is a form of Bayesian reasoning that offers the most comprehensive articulation of reasoning in a scientific context. As we will show, not only

J. Osborne (✉)
Stanford University, Stanford, CA, USA
e-mail: osbornej@stanford.edu

M.S. Khine (ed.), *Perspectives on Scientific Argumentation*,
DOI 10.1007/978-94-007-2470-9_4, © Springer Science+Business Media B.V. 2012

does it explain existing controversies that exist within the body of empirical research but it also offers an explanatory account of why critique is an essential element of scientific practice and effective pedagogy in science.

Scientific Reasoning

Historically, there have been three fundamental perspectives on the nature of scientific reasoning—the psychological, the philosophical, and the sociological. The psychological perspective is probably most strongly associated with the work of Jean Piaget (Inhelder & Piaget, 1958; Piaget, 1929, 1953) who saw reasoning in science as a practice dominated by a set of logico-mathematical operations such as compensation, seriation, classification, and logical reasoning typified by hypothetico-deductive arguments of the form "if...then...therefore." Such reasoning required the ability to identify and control variables and manipulate abstract representations. Children were seen as progressing through a set of stages of mental development, attaining the highest level, formal reasoning, in adolescence. The apotheosis of the influence of this perspective on the classroom was Shayer & Adey's curriculum intervention for middle school science—*Thinking Science* (Michael Shayer, 1999; Michael Shayer & Adey, 1992). This was a two-year course consisting of interventions once every two weeks that were designed to cognitively accelerate children's ability to undertake these operations. Much of the research has focused specifically on developing children's capability to identify and control variables as this is seen as a cognitive operation which is core to the process of inquiry. Zimmerman offers a good summary of much of this work (Zimmerman, 1999, 2007). Clearly, this form of reasoning is an essential feature of experimental design as experiments where all the relevant variables are not identified, or where there is more than one dependent variable produce results which are confounded and cannot make claims to knowledge. The use of this reasoning strategy is very much at the core of double-blind trials of new pharmaceutical products.

There are many well-known objections to the Piagetian account—most notably those summarized by Metz (1995). However, the substance of the critique is that while such reasoning is required by science, the common interpretation of an implied deterministic developmental pathway is simply flawed and not supported by the evidence. Children, it is argued, are much more capable than the Piagetian account would suggest. Our critique, however, is somewhat different. Essentially, a focus on a specific set of logico-mathematical operations as the principal forms of reasoning in science offers only a narrow and incomplete vision of scientific reasoning. In short, reasoning is always situated in a context and only makes sense when judged within that context. Judgments about what constitute good data depend critically on well-established "concepts of evidence" (Gott & Duggan, 1996) such as whether the data are valid, are subject to random or systematic error, how reliable they are, and what the degree of error might be. Further, reasoning is also context dependent in that judgments about the validity of any scientific argument are reliant on the construction of meaning from scientific texts or discourse (Norris & Phillips, 2003). Only an individual who has an appropriate level of scientific knowledge is able to construct the meaning necessary to reason with. Scientific reasoning, therefore, does not take place in some contextual vacuum. This is the essence of the critique mounted by Koslowski and her co-workers that individual performance varies significantly when subjects have credible theoretical justifications for why two variables might co-vary (Koslowski, 1996; Koslowski, Marasia, Chelenza, & Dublin, 2008). Finally, research in psychology has focused overwhelmingly on student's capability to achieve an agreed performance. Little of the work has examined student's capability to detect erroneous reasoning and justify why it is flawed. Given that the ability to engage in critique is a major element of scientific reasoning, this omission is surprising. Thus, our view of this perspective on scientific reasoning is not so much that it is wrong or flawed but rather it offers a partial or incomplete view of the edifice.

A somewhat different perspective is offered by philosophical accounts of scientific reasoning. These have ranged through Baconian descriptions of science as a process of generalizations emerging from empirical enquiry; Popperian notions of science as a process of conjecture and refutation; Kuhn's view that science was a community of practice governed by internal norms that framed the paradigm in which scientists work; and to the more radical views of Feyerabend that there was no common, identifiable method that could characterize science (Chalmers, 1999). All of these have attempted to describe the normative criteria used by science in its search for knowledge which would help distinguish science from other forms of cultural activity. To date, most would agree that this has been a failed project. Rather, each of these descriptions captures some but not all elements of scientific practice and each have been individually questioned and found incomplete (Fuller, 1997; Nowotny & Scott, 2001; Taylor, 1996). Siegel, for instance, in response to some of the common criticisms has attempted to argue that a central commitment of science is to evidence as the basis of belief (Siegel, 1989). Whilst that is generally unquestioned, it is also the basis of belief, at least to some extent in the social sciences and history. Donnelly, for instance, takes a different tack arguing that it is not the epistemic but the ontic nature of science which is its distinguishing feature (Donnelly, 2005). The best that the philosophy of science can offer for an account of scientific reasoning is the distinction between the three forms of argument that are commonly used in science—abductive, deductive, and inductive. Whilst school science arguably overemphasizes the inductive and deductive form of argument, philosophical analysis of this form has little substantive to offer science education in helping to identify the forms of detailed practice that would help students to develop their skill and aptitude with such forms of reasoning. Rather, it offers a meta-language for describing the broadest features of the argument and a rationale for the importance of certain activities such as modeling (Nercessian, 2008). But whereas the teacher of science needs a detailed picture of the scientific landscape and how it is mapped, the philosophy of science offers a picture sketched only in the broadest of brush strokes.

One philosopher who has been influential in this field is Toulmin. His attempt to capture the nature of informal argument as used in everyday life, as opposed to the strict requirements of logic, has helped the field to recognize that argumentation is a form of reasoning which is central to all forms of human activity (Toulmin, 1958). His field-independent notion that the essential elements of argument consist of a claim, albeit qualified, supported by data and a warrant where the warrant justifies the relevance of the data to the claim has led to an enhanced emphasis for this form of reasoning in science education (Driver, Newton, & Osborne, 2000; Duschl & Osborne, 2002; Kuhn, 1993). Its importance has been lent additional significance by work conducted in the field of science studies which has portrayed science as a practice where scientists marshal resources gathered from "inscription devices" that transform data to commonly recognized forms. This evidence is then used as a resource in developing arguments to persuade other scientists of the validity of a range of differing ontological entities and causal mechanisms (Latour & Woolgar, 1986; Traweek, 1988). Such an analysis of the practice of science has offered education a rationale for the significance of argument as a form of reasoning and its study. In addition, it provides a meta-language for describing its essential features. In that sense, the analysis of the detail of discursive practice has been useful in foregrounding the essential elements that are necessary to any account of scientific reasoning. Conjoined with the analysis offered by psychology of particular forms of argument/reasoning used within science such as the control of variables, it might be said to offer a good account of the major elements of scientific reasoning.

However, we would contend that there is still an essential element missing in all of these descriptions of scientific reasoning. This is that such accounts fail to account for the importance of criticism in the practice of science and why it is so central to scientific reasoning. Essentially, constructing an argument for the validity of a scientific claim depends as much on knowing why the wrong answers are wrong as much as it does knowing why the right answer is right. Such a position, we will show, has clear implications both for our conception of the nature of scientific reasoning and for pedagogy within science education. The substance of our argument is drawn from Bayesian accounts that see reasoning as a process not of constructing an infallible argument but rather one of drawing inferences based on the assessment of relative probabilities.

A Bayesian Perspective on Scientific Reasoning

The distinguishing feature of Bayesian inference is that it is a system of describing the certainty of knowledge. The degree of this certainty is reflected in probabilities assigned to a given hypothesis or event. As new evidence emerges, these probabilities are updated. Sometimes, the new evidence strongly favors the target hypothesis over rival hypothesis and sometimes it does not. Bayes' theorem describes mathematically how this balance of evidence changes the assigned probabilities. In other words, Bayes' theorem describes how the certainty of knowledge is updated given the new data. In this regard, Bayesian inference shares many aspects with scientific reasoning and argumentation. Both involve evaluating uncertain hypotheses and both involve weighing new evidence against target and alternative theories. In certain ways, the very process of science can be viewed as the repeated application of Bayes' theorem as data and evidence gradually change the probabilities in the minds of scientists, "convincing" them of the truth or falsity of a given hypothesis.

Bayesian inference offers a means of characterizing an individual's assessment of a hypothesis. Its tenets are derived from Bayesian *probability*, which is typically used to describe random, well-defined systems. Examples of such systems include gambling outcomes, gene assortment, and many quantum phenomena. However, whereas Bayesian *inference* is still developing as a model for scientific reasoning (Howson & Urbach, 2006), Bayesian *probability* is widely accepted as an interpretation for probabilistic systems.

Origins and an Intuitive Explanation

Bayesian probability was named after Thomas Bayes (1702–1761), an English clergyman and mathematician. Pierre-Simon Laplace (1749–1827) subsequently elaborated and popularized the field into what is known today as Bayesian probability theory (Stigler, 1986). The logic of Bayesian probabilities can be justified directly from certain requirements of rationality and internal consistency (see Cox's theorem in Cox, 1961).

An Intuitive Explainer

One of the problems confronting the wider adoption of Bayesian reasoning as a model for scientific reasoning is its expression in a mathematical formalism which is somewhat opaque. In its original mathematical form, Bayes' theorem appears as follows:

$$\mathrm{P}\,(h|e) = \frac{\mathrm{P}\,(e|h)\,\mathrm{P}\,(h)}{\mathrm{P}\,(e)}$$

In this formula, P $(h|e)$ is the probability of a hypothesis h given that some evidence e is true. This is referred to as the *posterior probability* as it is the new, updated probability assessment given the evidence e. P $(e|h)$ is the probability of the evidence e occurring given that hypothesis h is true. This is referred to as the *likelihood* of h on e because it reflects how determinate h is to explaining e. P(h) is the probability of hypothesis h being true by itself. This is called the *prior probability* since it reflects the probability of h independent of the new evidence e. Finally, P(e) is the probability of evidence e being true by itself.

This abstract formulation is the typical presentation for Bayes' theorem and while it has the advantage of being mathematically concise, the heavy use of logical symbolism does not facilitate an intuitive grasp of the meaning of the theorem. Without this, it is difficult to judge its implications or utility as a conceptual framework. It is this lack of transparency that has hindered the acceptance of Bayesian inference as a framework for science educators. To address this, we turn instead to a more intuitive example to explain how Bayesian probabilities work: the likelihood of breast cancer and mammogram tests. Both events have some associated randomness. Importantly though, the two systems are inter-related: when a woman receives a positive mammogram, her likelihood of breast cancer increases. Bayes' theorem describes how much that likelihood changes. Put differently, it explains how knowledge of the probabilities in one system changes the probabilities of a system which is related, yet distinct.

The updated probability of breast cancer (called the *posterior probability*) can be determined from three pieces of information. The first is that 1% of women, say in their 50s, have breast cancer (the *prior probability,* labeled "Info A" in Fig. 4.1). The second is that for women who definitively have breast cancer, mammograms are positive 80% of the time (the *true positives*, labeled "Info B"). The third is that for women *without* cancer, mammograms are still positive 10% of the time (the *false positives*, labeled "Info C"). Before continuing, we recommend the reader to estimate an answer: given a positive mammogram, what is the likelihood of cancer?

When phrased in this way, an alarming six out of seven doctors arrive at the *wrong* answer (Casscells, Schoenberger, & Graboys, 1978; Eddy, 1982). Most vastly overestimate the likelihood. The most typical error is to assume that a positive test implies that the individual has an 80% chance of cancer. However, this is mistaken because it neglects the large number of false positives that happen for normal individuals without cancer who are routinely tested. The correct calculation begins with the prior probability: since 1% of women have breast cancer, when testing a 1000 people, 990 will not have cancer. Of those 990, 99 will have a positive result and do not actually have cancer. Of this sample of 1000, only 10 individuals actually do have cancer and only 8 of them will be detected by the test. Therefore, given a positive test, the actual chance of having cancer is only 8 out of the 107 (99 + 8) positive results, that is, 7.5%. Whilst the answer might seem surprising, it is a

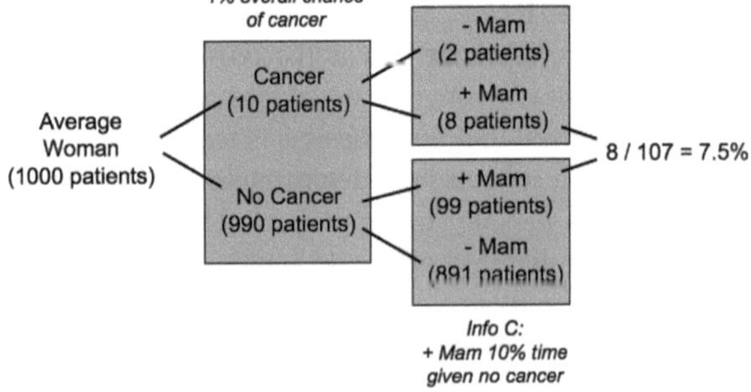

Fig. 4.1 Graphical representation of probabilityupdate calculation with Bayes' Theorem

Fig. 4.2 Simplified Bayesian probability update model

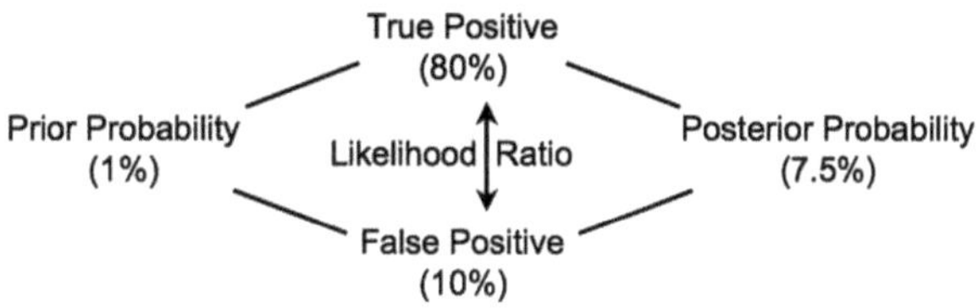

common error of logic that neglects the fact that most cases of positive mammograms actually occur when there is no cancer, that is, the false positives. A graphical demonstration of how this probability is calculated is offered in Fig. 4.1.

Bayes' theorem can be further simplified into its key conceptual components. Figure 4.2 captures the essence of what Bayes' theorem postulates: *new evidence is used to update prior probabilities to what are now posterior probabilities, a change in the degree of certainty that depends on the likelihood ratio* (*how strongly the evidence pertains to true versus false positives*). In Bayesian epistemology, this is referred to as the Simple Principle of Conditionalization (Adams, 1965).

What these examples mean is that all decisions have to be weighed not solely in terms of what information or evidence there is that they are correct but also in terms *of what the likelihood is that they might be wrong*. To do otherwise is to engage in faulty reasoning and logic and to misinterpret the inferences that can be drawn from the evidence. A patient might still opt for aggressive cancer treatment given a positive mammogram, but this is because their likelihood of a cancer has increased a little over sevenfold, and not to the commonly mistaken 80%. In reality, the unreliability of the test requires a more reliable test—a biopsy. However, it is because of the large number of false positives in women under 50 and the associated emotional turmoil that the United States Preventive Task Force recommended in its new guidelines that most women start regular screening at age 50 and not age 40 which has been the practice until now (US Dept of Health and Human Services, 2009).

Applications to the Reasoning Process

This Bayesian probability model is a widely accepted interpretation for external, objectively probabilistic systems. What is less established is using this model to describe the assessments of hypotheses by individuals. This is the key leap that characterizes the debate about the value of Bayesian inference as a model of scientific reasoning. In other words, can a probabilistic model that characterizes external, random systems be used to describe the cognitive process of belief assessment?

This application of Bayesian notions to personal degrees of belief is sometimes called the subjectivist view (De Finetti, 1974) and has been developed by certain authors such as Howson & Urbach (2006). The subjectivist use of Bayesian ideas shares the same fundamental concepts and calculus with the example above of breast cancers and mammograms. However, instead of the likelihood of cancer, Bayesian inference replaces this with an individual's likelihood assessment that a given hypothesis is correct. Such a hypothesis could be scientific such as the likelihood that the theory of dark matter is correct, or something more mundane such as the likelihood that some car mechanic is trustworthy. The subjectivist view therefore acknowledges the subjective beliefs of the individual while also claiming that updating those beliefs should follow certain elements of logic and reason.

As a model of informal reasoning, Bayesian inference provides a useful analog. When we are considering a theory, we tend to have some preconceived notions (i.e., prior probabilities). Using the car mechanic example above, we may feel that a car mechanic is trustworthy for any number of preconceived reasons. When new evidence arises, such as a friend recommending the mechanic, we are apt to update our assessment (i.e., posterior probability). That new probability, however, depends on both true and false positive considerations. If our friend is reliable and is mechanically knowledgeable, then that increases the strength of our certainty. However, if our friend is shifty and owns a stake in the mechanic's shop, it has the opposite effect enhancing the evidence of false positives. In Bayesian inference, the degree that the new data supports our target hypothesis versus alternative hypothesis is the likelihood ratio.

Bayesian inference does not explain all aspects of human thinking. Instead, it is meant as a model for *rational* thinking, namely an attempt at one's best objective assessment in contrast to a stubbornly prejudiced or capricious one. Degrees of belief are clearly individual and subjective. Nonetheless, Bayesian inference suggests that these beliefs must be updated according to the axioms of probability in order to be optimal (Maher, 1993). Support for this claim comes from the Dutch Book theorem, developed in the 1920s and 1930s by Frank Ramsey and Bruno de Finetti. They showed that violating the axioms of probability resulted in belief probabilities that were incoherent, meaning the beliefs are demonstrably irrational (De Finetti, 1937). A simple example of this theorem is a belief held that there is 70% expectation of rain which is also held conjointly with a 40% expectation of no-rain. These beliefs are incoherent when taken together because the probabilities add to 110%. If a bookie took both bets together, the combined odds would guarantee a loss of 10%. This situation, where a set of odds guarantees a loss regardless of the outcome, is known as a "Dutch Book." To prevent getting swindled by Dutch Books, wiser bookmakers are trained to build and update their odds using the rules of Bayesian calculus. These examples simply highlight the damage that irrational beliefs can have. Put another way, judging whether you will be right without judging the probability of whether you will be wrong will lead to poor assessments which are incoherent and *pragmatically self-defeating*: that is actions that, based on logical inconsistency alone, are guaranteed to make things worse than they otherwise would have been (Talbott, 2008)

The appeal then of Bayesian inference is that, in two different ways, it juxtaposes a mathematical model with intuitive experience. In one sense, it combines subjective likelihood assessments (i.e., prior probabilities) with an objective set of procedures and formula for updating those assessments (Bayes' formula). In another sense, it offers a kind of "independent opinion" about scientific reasoning since its notions are derived logically from the axioms of probability mathematics. If corroborated by empirical evidence then, Bayesian inference offers a take on scientific reasoning that arises from an independent, non-empirical source. We turn now to that empirical evidence.

Empirical Findings

In this section, we examine empirical findings about scientific reasoning from the areas of science education and psychology. Our objective is to see if the key conceptual components of Bayesian inference fit with the findings of these fields. We consider three groups of empirical results: (1) misconceptions research on students' alternative explanations; (2) findings on argumentation in classrooms; and (3) studies on coordination of theory and evidence.

Misconception Research on Students' Alternative Explanations

Numerous findings in science education have shown that providing students with correct explanations alone is inferior to also explaining why misconceptions are incorrect. For instance, Hynd & Alvermann (1986) found that physics texts that contained "refutation text" addressing common misconceptions resulted in significantly better conceptual gains. Likewise, Ames & Murray (1982) found greater learning gains among discussion groups with differing preconceptions versus those with more similar ones, even if those differences were based on incorrect premises. In short, providing information about both negative and positive cases significantly improves conceptual learning in the sciences.

These findings are consistent with Bayesian conceptions of probability updates, namely that it is not possible to develop a posterior probability without a consideration of competing alternative hypotheses. According to this view, correct explanations only provide half of the picture. They explain why the target hypothesis is right, increasing the likelihood of the true positives. However, they provide no information about why other alternative hypotheses are incorrect. This is critical because in the Bayesian model, the strength of the true positive information does not stand alone; it is always relative to strength of the false positive alternatives (Royall, 1997). As such, students need both target and competing explanations to construct assessments of the presented material. Good teachers of science recognize this need intuitively, attempting to contrast the scientific explanation with the common intuitive notions addressing why they are wrong as much as why the scientific idea is correct (Ogborn, Kress, Martins, & McGillicuddy, 1996). Likewise, the French philosopher Bachelard understood this concept when he argued that "two people must first contradict each other if they really wish to understand each other. Truth is the child of argument, not of fond affinity" (Bachelard, 1968). What both are pointing to is that it is difference which enables conceptual understanding because, as we would argue, from a Bayesian perspective it provides the individual with evidence both for the proposition and the false positives.

Argumentation in Classrooms

Whereas explanations presume truth, arguments establish it by a process of claims, counterarguments, and rebuttals (Toulmin, 1958). When utilized in the classroom, this process has been shown to result in greater conceptual learning gains (Asterhan & Schwarz, 2007; Zohar & Nemet 2001). However, the use of argument in classrooms is still not a common pedagogical practice in science education (Newton, Driver, & Osborne, 1999).

The benefits observed from argumentation for learning are also consistent with Bayesian notions of scientific reasoning. With Bayesian inference, evaluating the likelihood ratio lies at the heart of assessing a posterior probability. Therefore, evaluating a hypothesis rests critically on weighing true positive and false positive perspectives that are both consistent with the new evidence. Yet, studies have shown that individual scientists have difficulty generating alternative inductions (i.e., false positives) from data; in comparison, groups of scientists engaged in collaborative discussion are more able to do so (Dunbar, 1997). Group discussion may, therefore, enhance scientific reasoning by facilitating the otherwise difficult process of generating and evaluating false positives individually. Similar evidence comes from the work of Johnson on the history of the development of one specific engineering product—ABS braking (Johnson, 2009). In her historical account of the development of this technology, Johnson shows how knowledge sharing was essential to the process. Those who did not contribute any knowledge to the community, predominantly American engineers (regardless of whether it was right or wrong), simply did not have the information necessary to make a good judgment about the Bayesian likelihood ratio, which resulted in a loss to their European counterparts. Similar arguments can be made about Crick and Watson's development of their model for DNA. The critical pieces of information were as much the evidence why certain of their proposed structures were wrong, as it was the evidence from Rosalind Franklin's X-ray crystallography suggesting that the structure was a helix.

Coordination of Theory and Evidence

Several studies have evaluated the capability of individuals to coordinate theory and evidence (Kanari & Millar, 2004; Koslowski, 1996; Kuhn, 1991, 1993). A particularly interesting finding in this field was a study by Koslowski (Koslowski, 1996; Koslowski, Marasia, Chelenza & Dublin 2008). Koslowski and her colleagues found that information was more likely to be considered as evidence when a causal explanation was provided. In this study, subjects were provided two plausible explanations for some phenomenon. Data were presented that supported one explanation over the other. The authors observed that subjects were more likely to consider the data as evidence when given a causal framework that permitted its incorporation. Without this explanatory framework, subjects were more likely to disregard the data and did not change their evaluation of which hypothesis was better.

The results of this study can be interpreted with a Bayesian notion of likelihood ratios. By pointing out explicitly a possible explanatory framework, the likelihood of the data supporting the target hypothesis over the rival hypothesis increases. From this perspective, the data that subjects considered irrelevant may have had an evidentiary basis. However, without an explanatory framework which identifies why the data are salient to the hypothesis, the evidence is not so much discounted as simply not counted. Thus, it is not just data that matters for updating probabilities. Providing an explanatory framework which helps the individual see why the data supports the positive hypothesis enables the individual to reassess the likelihood ratio from one where the probabilities may be evenly balanced toward the target hypothesis. Such an interpretation would predict a greater change to the posterior probabilities in the subjects who were provided explanations versus those that were not, an effect that was indeed observed in the study.

Framework Comparison

In addition to empirical congruence, Bayesian inference can also be used to address problems with existing models of scientific reasoning. In this section, we compare Bayesian inference to Popper's model of falsification as well as the model of hypothesis testing known as the Frequentist probabilities.

Popperian Falsification

Falsification is a well-known concept in science and scientific reasoning developed by Karl Popper (1959). The theory of falsification states that theories can never be confirmed. Instead, confirming data merely allows a given theory to survive disconfirmation. In contrast, disconfirming data negates the theory and new theories must subsequently be developed that encompass the disconfirming case. In this way, science progressively accumulates theories of greater explanatory power. However, even theories that have survived multiple disconfirmations are never decisively proven as true.

Several aspects of Popper's model are in conflict with actual experience. The first is that falsification classifies all current theories as only having survived disconfirmation. However, scientists clearly have different certainties about different theories. No reasonable scientist would consider the theory of dark matter to be as certain as the atomic theory of matter. Popper attempts to address this by introducing the notion of "corroborated" theories. However, this effectively adds gradations in certainty, an interpretation that begins to look more like one associated with Bayesian probabilities. In fact, the very notion of degrees of corroboration is what Bayesian inference formalizes as belief probabilities (Sokal & Bricmont, 1998).

The falsification model has the additional problem of making a fundamental distinction between confirming and disconfirming evidence. With falsification, confirming evidence is not utilized in any meaningful way while disconfirming evidence has the effect of negating the theory. In actual experience, however, confirming evidence does increase the strength of a theory and multiple disconfirmations are typically needed before discarding a theory, particularly if the theory was well established (Collins & Pinch, 1993). The Bayesian model reflects both of these realities more accurately. Confirming evidence raises posterior probabilities and disconfirming evidence decreases it, reflecting the changes in certainty produced by new evidence. In addition, with Bayesian inference, no single disconfirmation is ever likely to reduce a posterior probability to zero. Instead, multiple disconfirmations are typically needed, a pattern that is more consistent with actual scientific practice.

Finally, the Bayesian model reflects a key observation of Popperian falsification, namely that disconfirmation has a more profound effect than confirmation. However, it does so under a broader explanatory framework that does not resort to fundamental distinctions between the two. In Bayesian calculus, the strength of evidence is reflected in the likelihood ratio. The numerator of this ratio is the probability that the evidence would be observed if the target theory was correct (i.e., true positives). The denominator is the probability that the evidence would be observed if some alternative theory was correct (i.e., false positives). However, in the sciences, there is almost always some alternative theory consistent with the evidence. For instance, even though Newton's theory of gravity had been confirmed by vast amounts of evidence, this evidence was also consistent with an alternative theory: general relativity, which ultimately subsumed it. As a result, the denominator for any given likelihood ratio in the sciences will always be sizeable. This limits the effect of confirming evidence: the target theory may have predicted the observed evidence, but so would various alternative theories. As a result, scientists often must address competing hypotheses when making their case.

Disconfirming evidence, however, has the opposite outcome. If a theory predicts some evidence, but that evidence is *not* observed, this results in a very small numerator. The sizeable denominator then results in a tiny likelihood ratio, amplifying the effect of disconfirming evidence. In this way, the Bayesian model reflects the Popperian observation that disconfirmation is stronger than confirmation. However, it does so by treating both of them probabilistically in contrast to the Popperian model, which treats each of them fundamentally in a different way (Yudkowsky, 2010).

Frequentist Inference

For probability mathematics, the Frequentist perspective is the other major competing notion to Bayesian probabilities. Mathematicians and statisticians consider both methods as having strong merits. However, the Frequentist approach has become the dominant approach to inferring results from data containing variability (Hacking, 1965). The Frequentist perspective presumes that multiple sampling of some phenomenon results in a distribution of possible values. The spread of these values can be estimated and compared to a null hypothesis. If the distribution of values within some confidence interval (typically 95%) does not contain the null value, the null hypothesis is said to be rejected at a certain significance level.

Proponents of Bayesian inference—as a model for reasoning—have sometimes tried to support their positions by attacking the Frequentist perspective (Howson & Urbach, 1991). This turns out to be unnecessary. The Frequentist approach to probabilities is generally used to characterize well-defined random experiments only (Hacking, 1965). It is not typically used to characterize assessment of hypothesis by individuals. The distinction lies in the Bayesian interpretation of probabilities as "a measure of a state of knowledge" (Jaynes & Bretthorst, 2003). This allows probabilities to be assigned to any statement, even one that does not involve a random process. Frequentists, on the other hand, make no such claims. The statement "I trust this car mechanic" can therefore be assigned a Bayesian probability. However, since it involves no random sampling, it cannot be assigned a Frequentist probability. An active debate may exist between Frequentist and Bayesians over probabilities for external random systems, but not over applications to individual assessments of hypothesis.

Discussion

Bayesian inference, we believe, offers a promising putative framework for scientific reasoning. It provides an alternative lens for explaining many of the empirical findings in science education and educational psychology. Yet, it arises independently from mathematical derivations that are neither empirical nor normative. Bayesian inference also addresses the shortcomings of alternative frameworks for scientific reasoning such as Popperian falsification.

Given these findings, what implications does Bayesian inference have for the practice of science education and instruction? From a curricular perspective, one immediate implication is that if individuals are to behave rationally, they need to see judgments about data and evidence being an assessment not only of the probability of the hypothesis being correct but also of it being wrong. Such evidence is essential to making an assessment of the Bayesian likelihood ratio. Within the field of argumentation, Nussbaum (2010) has proposed that Bayesian inference could be used to provide a mathematical structure to Toulmin's model for argument. For instance, he suggests that when evaluating a social issue—such as hunger—students could conduct on-line research to complete actual probability trees such as those provided in Fig. 4.1. This sort of instruction is likely to be particularly useful for students entering scientific research and practice. As mentioned earlier, most doctors are unable to make the correct assessment of risks in the breast cancer example.

More generally, Bayesian inference can also be taught as a model for the reasoning process of science. Highlighting the importance of false negatives, for instance, can improve awareness of common pitfalls to rational reasoning. In this way, Bayesian inference can help bring increased use of statistical reasoning into real-world applications. For instance, Goldacre showed the fallacy of engaging in data mining as a means of identifying terrorists simply because of the large number of false negatives identified (Goldacre, 2009).

Bayesian inference also has several potential implications for classroom pedagogy. First, it adds further emphasis to the significance of findings that alternative misconceptions must be addressed if students are to gain secure understandings of scientific concepts. Teachers need to be aware that lowering the likelihood of false positives (i.e., alternative "wrong" ideas) is as instructionally powerful as raising the likelihoods of true positives (the "correct" idea). Second, if learning does indeed occur though a Bayes-like process of data weighing and integration, this reinforces constructivist notions of knowledge acquisition. From this perspective, simply providing the correct answer is not sufficient. Students must be given evidence and allowed to grapple with assessing likelihoods in order to properly update their belief assessments (i.e., posterior probabilities). Specifically, acceptance of new concepts is a function not only of how well the teacher presents the case for a new idea (i.e., strength of the likelihood ratios), but also the extent to which they address the strength of the student's misconceptions (i.e., strength of individual prior probabilities). For students with strongly held prior misconceptions, it may take multiple exposures to evidence to change these beliefs. The Bayesian model suggests this is normal, even when the learner is evaluating the evidence rationally. Therefore, even if a student does not initially accept a new concept, instruction can still be considered a success as long as the learner is more open to the idea than they were before.

Perhaps most fundamentally, this account of scientific reasoning from a Bayesian perspective offers a rationale for why argument and critique are central and core to scientific activity. If, as we have suggested, beliefs are transformed not solely by confirming evidence but by negating alternative hypotheses, it suggests a central role for critique to the construction of knowledge both for the scientist and the learner of science. It also suggests why the few merchants of doubt who wish to cast aspersions on the scientific evidence for climate change have been so successful. In their absence, the likelihood ratio is virtually unitary. In their presence, particularly when they have scientific credibility, the existence of an alternative hypothesis which seems plausible substantially diminishes the likelihood ratio and therefore the certainty of individuals in the main hypothesis. A Bayesian perspective would suggest that the case for climate change would be made much more successfully not by asserting the validity of the scientific evidence but rather by undermining the validity of the naysayer's case. Or to put it another way, knowing why the wrong answer is wrong matters as much as knowing why the right answer is right.

References

Adams, E. (1965). The logic of conditionals. *Inquiry, 8*(1), 166–197.

Ames, G., & Murray, F. (1982). When two wrongs make a right: Promoting cognitive change by social conflict. *Developmental Psychology, 18*(6), 894–897.

Asterhan, C., & Schwarz, B. (2007). The effects of monological and dialogical argumentation on concept learning in evolutionary theory. *Journal of Educational Psychology, 99*(3), 626–639.

Bachelard, G. (1968). *The Philosophy of No* (G. C. Waterston, Trans.). New York: Orion Press.

Casscells, W., Schoenberger, A., & Graboys, T. (1978). Interpretation by physicians of clinical laboratory results. *The New England Journal of Medicine, 299*(18), 999–1001.

Chalmers, A. F. (1999). *What is this thing called science?* (3rd ed.). Milton Keynes: Open University Press.

Collins, H., & Pinch, T. (1993). *The Golem: what everyone should know about science*. Cambridge: Cambridge University Press.

Cox, R. (1961). *The algebra of probable inference*: The Johns Hopkins Press.

De Finetti, B. (1937). La pre vision: ses lois logiques ses sources subjectives.*Ann. Inst. H. Poincare* 7. English translation in H. E. Kyburg, H. E. Smokler (eds), *Studies in Subjective Probability* (pp. 1–68). New York: Wiley, 1964.

De Finetti, B. (1974). *Theory of probability*. Chichester: Wiley.

Donnelly, J. (2005). Reforming science in the school curriculum: A critical analysis. *Oxford Review of Education, 31*(2), 293–309.

Driver, R., Newton, P., & Osborne, J. F. (2000). Establishing the norms of scientific argumentation in classrooms. *Science Education, 84*(3), 287–312.

Dunbar, K. (1997). How scientists think: On-line creativity and conceptual change in science. In T. B. Ward, S. M. Smith, & J. Vaid (Eds.), *Conceptual structures and processes: Emergence, discovery, and change* (pp. 461–493). Washington, DC: American Psychological Association Press.

Duschl, R., & Osborne, J. F. (2002). Supporting and promoting argumentation discourse in science education. *Studies in Science Education, 38*, 39–72.

Eddy, D. M. (1982). Probabilistic reasoning in clinical medicine: Problems and opportunities. In D. Kahneman, P. Slovic, & A. Tversky (Eds.), *Judgment under uncertainty: Heuristics and biases* (pp. 249–267). New York: Cambridge University Press.

Ford, M. (2008). Disciplinary authority and accountability in scientific practice and learning. *Science Education, 92*(3), 404–423.

Fuller, S. (1997). *Science*. Buckingham: Open University Press.

Goldacre, B. (2009). Datamining for Terrorists would be lovely if it worked. *The Guardian*. Retrieved December 17, 2010, from http://www.badscience.net/2009/02/datamining-would-be-lovely-if-it-worked/

Gott, R., & Duggan, S. (1996). Investigative work in school science. *International Journal of Science Education, 18*(7), 791–806.

Hacking, I. (1965). *Logic of statistical inference*. Cambridge, MA: Cambridge University Press.

Howson, C., & Urbach, P. (1991). Bayesian reasoning in science. *Nature, 350*(6317), 371–374.

Howson, C., & Urbach, P. (2006). *Scientific reasoning: A Bayesian approach* (3rd ed.). Chicago: Open Court.

Hynd, C., & Alvermann, D. (1986). The role of refutation text in overcoming difficulty with science concepts. *Journal of Reading, 29*(5), 440–446.

Inhelder, B., & Piaget, J. (1958). *The growth of logical thinking from childhood to adoloscence*. London: Routledge and Kegan Paul.

Jaynes, E., & Bretthorst, G. (2003). *Probability theory: The logic of science*. Cambridge, MA: Cambridge University Press.

Johnson, A. (2009). *Hitting the brakes: Engineering design and the production of knowledge*. Durham: Duke University Press.

Kanari, Z., & Millar, R. (2004). Reasoning from data: How students collect and interpret data in science investigations. *Journal of Research in Science Teaching, 41*(7), 748–769.

Koslowski, B. (1996). *Theory and evidence: the development of scientific reasoning*. Cambridge, MA: MIT Press.

Koslowski, B., Marasia, J., Chelenza, M., & Dublin, R. (2008). Information becomes evidence when an explanation can incorporate it into a causal framework. *Cognitive Development, 23*(4), 472–487.

Kuhn, D. (1991). *The skills of argument*. Cambridge: Cambridge University Press.

Kuhn, D. (1993). Science as argument: Implications for teaching and learning scientific thinking. *Science Education, 77*(3), 319–337.

Latour, B., & Woolgar, S. (1986). *Laboratory life: the construction of scientific facts* (2nd ed.). Princeton, NJ: Princeton University Press.

Longino, H. (1990). *Science as social knowledge*. Princetown, NJ: Princetown University Press.

Maher, P. (1993). *Betting on theories*. Cambridge : Cambridge University Press.

Metz, K. E. (1995). Reassessment of developmental constraints on children's science instruction. *Review of Educational Research, 65*, 93–127.

Nercessian, N. (2008). Model-Based Reasoning in Scientific Practice. In R. A. Duschl & R. E. Grandy (Eds.), *Teaching scientific inquiry: Recommendations for research and implementation* (pp. 57–79). Rotterdam, The Netherlands: Sense.

Newton, P., Driver, R., & Osborne, J. (1999). The place of argumentation in the pedagogy of school science. *International Journal of Science Education, 21*(5), 553–576.

Norris, S., & Phillips, L. (2003). How literacy in its fundamental sense is central to scientific literacy. *Science Education, 87*, 224–240.

Nowotny, H., & Scott, P. (2001). *Re-thinking science. Knowledge in the public age of uncertainty*. Cambridge: Polity Press.

Nussbaum, E. (2010). *Argumentation, dialogue theory, and probability modeling: Alternative frameworks for argumentation research in education*. Las Vegas, NV: University of Nevada.

Ogborn, J., Kress, G., Martins, I., & McGillicuddy, K. (1996). *Explaining science in the classroom*. Buckingham: Open University Press.

Piaget, J. (1929). *The child's conception of the world*. New York: Harcourt Brace.

Piaget, J. (1953). *The origins of intelligence in children*. London: Routledge and Kegan Paul.

Popper, K. (1959). *The logic of scientific discovery*. London: Hutchinson.

Royall, R.M. (1997) *Statistical evidence: A likelihood paradigm*. London: Chapman & Hall.

Shayer, M. (1999). Cognitive acceleration through science education II: Its effects and scope. *International Journal of Science Education, 21*(8), 883–902.

Shayer, M., & Adey, P. (1992). Accelerating the development of formal thinking in middle and high school students III: Testing the permanency of effects. *Journal of Research in Science Teaching, 29*(10), 1101–1115.

Siegel, H. (1989). The rationality of science, critical thinking and science education. *Synthese, 80*(1), 9–42.

Sokal, A., & Bricmont, J. (1998). *Fashionable nonsense*. New York: Picador.

Stigler, S. (1986). *The history of statistics: The measurement of uncertainty before 1900*. Cambridge, Massachusetts: Belknap Press.

Talbott, W. (2008). *Bayesian Epistemology (Stanford Encyclopedia of Philosophy)*. Retrieved December 10, 2010, from http://plato.stanford.edu/entries/epistemology-bayesian/index.html#return-2

Taylor, C. (1996). *Defining science: a rhetoric of demarcation*. Madison: WI: University of Wisconsin Press.

Toulmin, S. (1958). *The uses of argument*. Cambridge: Cambridge University Press.

Traweek, S. (1988). *Beamtimes and lifetimes: the world of high energy physicists*. Cambridge, MA: Harvard University Press.

US Department of Health and Human Services. (2009). *Screening for breast cancer*. Retrieved from http://www.ahrq.gov/clinic/uspstf/uspsbrca.htm

Yudkowsky, E. S. (2010). *An intuitive explanation of Bayes' theorem*. Retrieved December 22, 2010, from http://yudkowsky.net/rational/bayes

Zimmerman, C. (1999). The development of scientific reasoning skills. *Developmental Review, 20*(1), 99–148.

Zimmerman, C. (2007). The development of scientific thinking skills in elementary and middle school. *Developmental Review, 27*(2), 172–223.

Zohar, A., & Nemet, F. (2001). Fostering students' knowledge and argumentation skills through dilemmas in human genetics. *Journal of Research in Science Teaching, 39*(1), 35–62.

Chapter 5
Students' Framings and Their Participation in Scientific Argumentation

Leema Kuhn Berland and David Hammer

Introduction

There is broad agreement in the research community, including among the authors of this volume, in the central importance of argumentation in science education, with argumentation generally understood to involve the articulation, comparison, and refinement of multiple theoretical perspectives and empirical findings (Berland & Reiser, 2009; Duschl, Schweingruber, & Shouse, 2007; Ford & Forman, 2006; D. Kuhn, 1991; Leitao, 2000). There is also broad agreement that argumentation rarely occurs and can be difficult to introduce in science classrooms, where students and their teachers are more likely to treat scientific knowledge as factual information for students to construct, observe, or receive from authoritative sources (Driver, Newton, & Osborne, 2000; Lemke, 1990; Weiss, Pasley, Smith, Banilower, & Heck, 2003).

At the same time, there have been multiple perspectives among researchers regarding why argumentation is rare and apparently difficult—perspectives that themselves need articulation, comparison, and refinement. Early work on the "skills of argumentation" (D. Kuhn, 1991) described developmental limitations in cognitive abilities. While this view is still guiding curriculum design, researchers have increasingly focused on the social and epistemic context of the argumentative activity. In particular, there is increasing emphasis, both in experimental studies and in the design of curriculum, on whether the context creates, in students, a need for argumentation (Chin & Osborne, 2010; Clark & Sampson, 2007; Kenyon, L. Kuhn, & Reiser, 2006; Zohar & Nemet, 2002).

This shifting emphasis brings research on argumentation into contact with research on *framing* (Goffman, 1974; Tannen, 1993), where a "frame" is an individual's sense of "what is it that's going on here" (Goffman, 1974, p. 8). More specifically, we connect this work with the construct of *epistemological framing* (as introduced in Redish, 2004), focusing on what is taking place with respect to

L.K. Berland (✉)
University of Texas, Austin, TX, USA
e-mail: leema.berland@mail.utexas.edu

M.S. Khine (ed.), *Perspectives on Scientific Argumentation*,
DOI 10.1007/978-94-007-2470-9_5, © Springer Science+Business Media B.V. 2012

knowledge. Framing presents a dynamic account of how students form an understanding of the activity at hand, and, we argue, explains why attention to the social and epistemic context is key to fostering and understanding student participation in scientific argumentation. We conclude the chapter by discussing the instructional implications of an epistemological framing account of students' participation in scientific argumentation.

Limitations in Scientific Argumentation

Argumentative discourse, in general, requires that individuals identify claims and supports for those claims. In addition, it requires that individuals evaluate and critique whether and how the supports connect to the claims. In scientific argumentation these supports are expected to include empirical evidence: "A central premise underlying science is that scientific theories stand in relation to actual or potential bodies of evidence against which they can be evaluated" (D. Kuhn, 1989, p. 674). However, a variety of studies show students having difficulty with using evidence to construct and evaluate claims—with coordinating their claims and evidence.

The first need in coordinating claims and evidence is to differentiate them, such that claims are evaluated against the evidence or supports for those claims. However, in her seminal work on children's and non-scientists' argumentation, D. Kuhn and colleagues (D. Kuhn, 1989, 1991) found that children and novice adults do not do this. For example, she presented subjects with evidence relating types of food one ate to their likelihood of catching colds. Subjects (sixth graders in this study) would defend their claims with theories (i.e., cake causes colds because it has a lot of sugar) rather than evidence, even when evidence was available and requested. In addition, subjects would alter the evidence or theory so that they matched:

> When theory and evidence are compatible, there is a melding of the two into a single representation of 'the way things are.'. . . When theory and evidence are discrepant, subjects use a variety of devices to bring them into alignment: either adjusting the theory—typically prior to acknowledging the evidence—or 'adjusting' the evidence by ignoring it or by attending to it in a selective, distorting manner. (D. Kuhn, 1989, p. 679)

In other words, they were treating the evidence as though it were a claim—as though it could be ignored or altered. This led D. Kuhn to conclude that students did not see evidence and claims as different sorts of information and that they were therefore not using the evidence to evaluate the claims.

In more recent work, Larson and Britt (2009) examined whether individuals could distinguish between arguments that were well supported and those that were not. These authors gave college students a series of simple arguments about everyday topics (i.e., "Banks shouldn't charge ATM fees because the fees make many customers unhappy," p. 348) and asked them to determine whether they were "sound" arguments. The students evaluated arguments that were structurally sound, such as the example provided, arguments that were only claims (i.e., "Banks shouldn't charge ATM fees"), and arguments that included justifications that were seemingly unrelated to the claim (i.e., Banks are financial institutions). Larson and

Britt found that college students correctly evaluated the arguments in 66% of the arguments while experts did so 80% of the time. They concluded that, given the simplicity of these arguments, students have difficulty determining whether supports are connected to their claims.

These studies reveal participants failing to evaluate whether and how well claims are supported. D. Kuhn's work discusses individuals ignoring or altering evidentiary support in order to maintain their original claims, while Larson's shows individuals failing to notice whether the supports are connected to their claims. This reveals a key challenge in argumentation: coordinating claims and evidence.

Sophisticated argumentation also involves presenting and responding to counter-arguments and rebuttals. Unfortunately, this too is seen as challenging to science students. It might be that this is challenging for precisely the reasons discussed above: if students do not evaluate whether a justification supports a claim, they will similarly not challenge that connection. However, rebuttals and counter-arguments also introduce new challenges. In particular, beyond evaluating the fit between claims and evidence, constructing a rebuttal or a counter-argument requires that one be aware of and make sense of alternatives.

Research has shown that students seldom attend to alternative arguments in such substantive ways. For example, in a study comparing adolescent and adult argumentation strategies, Felton and D. Kuhn (2001) found that adolescents focused on defending their own arguments (in this case about capital punishment) without regularly responding to their peers' arguments. Thus, they rebutted one another and directly challenged the content of one another's arguments infrequently. Similarly, in a study of junior high students, Osborne, Erduran, and Simon (2004) found that only 26% of the students' arguments about scientific topics at the beginning of the school year included substantive responses to the alternative arguments presented by their peers.

The infrequency with which students responded to one another's arguments in Felton and D. Kuhn's and Osborne and colleagues' respective studies demonstrates the second challenge seen in the literature on fostering scientific argumentation in schools: students are rarely seen attending to alternative arguments. Instead, they focus on bolstering their own argument. This limits the students' engagement in the social aspects of argumentation: they are rarely seen to interact with their peers' ideas.

Are these Limitations of Ability?

These challenges are historically viewed in terms of students lacking skills—or limitations in their abilities. For example, D. Kuhn and colleagues (1989) concluded that students are unable to use evidence to evaluate claims because two sets of skills are missing or immature:

> One is the set of skills. . .[that] pertain to the differentiation and coordination of theory and evidence. The other is the set of skills involved in understanding the meaning of evidence once it is sufficiently differentiated from theory. (p. 681)

Similarly, Felton and D. Kuhn (2001) suggested that students need to "develop more sophisticated discourse skills" (p. 151) in order to substantively respond to counter-arguments.

These claims of limited abilities have been challenged by data from classroom-based studies, which provide evidence of children using these abilities—at least in nascent form. Indeed, in a more recent study that we discuss below, D. Kuhn has discussed young children's beginning abilities for argumentation (D. Kuhn, 2010; D. Kuhn & Udell, 2007). In addition, the literature is rife with examples of argumentation among students who have had little or no argumentation instruction (Berland & Reiser, 2011; Bricker & Bell, 2007; Engle & Conant, 2002; Hammer & van Zee, 2006; Louca, Hammer, & Bell, 2002; Naylor, Keogh, & Downing, 2007; Radinsky, 2008; Warren, Ballenger, Ogonowski, Rosebery, & Hudicourt-Barnes, 2001). For example, Warren and colleagues (2001) recount a fifth-grade students' careful construction of an experiment that could provide evidence to distinguish between two competing hypotheses for whether ants favor darkness, in itself, or whether they only seem to favor darkness because it is a feature of being in dirt. Engle and Conant's (2002) well-known account of fifth graders debating whether killer whales are dolphins or whales shows students repeatedly drawing from relevant texts and discussions with experts to construct counter-arguments and rebuttals. Berland and Reiser (2011) document sixth-grade students using evidence, in the form of graphs generated by a NetLogo simulation (Wilensky, 1999) to support and challenge competing claims, with little instruction.

There is further evidence in data on student inquiry collected in other projects not specifically concerned with argumentation. For example, in a collection of case studies by Hammer and van Zee (2006), every case includes student argumentation with no explicit instruction. In one, first grader Julio thought that a crumpled sheet of paper falls to the ground faster than a flat sheet of paper because it is "kind of heavy." Brianna argued that a crumpled sheet of paper is "still the same size," un-crumpling the sheet to demonstrate, and that "it still feels light." Allison countered that "light" can mean something different from how the paper feels. In another case, second grader Taylor explained that a magnet stuck to someone's hand because people have metal in their bodies, a claim she supported with the evidence of an X-ray she had seen in the hospital. Ben disagreed, accounting for what she saw with the alternative explanation that the person "was probably wearing something."

Thus there has been split in evidence: some studies highlight students' difficulties while others reveal their abilities. Contending with the discrepant evidence has led research to focus on the learning environment and the students' interpretations of the task at hand. For example, L. Kuhn, Kenyon, and Reiser (2006) provide an example of seventh-grade students with competing claims about the reason for a decrease in seeds, coordinating claims and evidence, as well as attending to and critically evaluating each other's reasoning. These authors argued that the students' attention to one another's ideas occurred because the curriculum "created a need" for them to try to persuade each other.

In a more recent work, Berland and Reiser (2011) found variation in whether and how students respond to their classmates' arguments: Students in one class rarely challenged one another with counter-arguments or rebuttals while students in another class frequently used the available evidence to do so. The authors explained this variation in terms of differences in how the students' interpreted the activity. In the first class, it was an opportunity to learn about one another's ideas, while, in the second class, it was a kind of competition. Thus the students' apparent interpretation of the task influenced how they engaged in the argumentative activity.

In a current investigation, Berland and Lee (2010) are studying the influence of first-hand data on the students' argumentation. Preliminary results suggest that middle-school students are more likely to engage with and incorporate challenging evidence when it is evidence they can see rather than being something that is reported to them. This study underscores the affect that the learning environment has on the students' argumentation.

D. Kuhn (2010) similarly found that students would increasingly turn to evidence in their small group discussions if it were made available. Moreover, extending Felton and D. Kuhn's (2001) earlier work, D. Kuhn and Udell (2007) found that "Young adolescents, we saw, are able to attend to the other's position [in arguments regarding nonscientific topics], and even to generate an argument against it, when explicitly asked to do so" (p. 101). The authors concluded, "The challenge [for students] in this case, then, is less one of executing the skill (of addressing the opposing position) than it is one of recognizing the need to do so" (p. 102).

Engle and Conant (2002) identified that four features of the classroom environment—content was problematized; students were given authority; students were held accountable to one another and disciplinary norms; and students had the necessary resources to do the work—support students' productive engagement in disciplinary sense-making practices, such as argumentation. We suggest that these features of the classroom work precisely because they enable the students to interpret the situation as one in which engaging with data and attending to and challenging one another's ideas is appropriate.

Each of these studies suggests the importance of context on the students' work with evidence and attention to conflicting lines of reasoning: When students experience a need to persuade others who hold opposing views, they draw on their abilities for argumentation, abilities researchers now believe they have, at least in nascent forms. Thus, examining recent work in classroom argumentation, we see a move toward attending to the social and epistemic context and the purpose of the students' activity: When the situation demands it, students both coordinate claims and evidence and critically attend to alternative ideas.

This interpretation has begun to inform design research focused on fostering student participation in scientific argumentation, in which designers are moving beyond teaching argumentation skills by combining explicit instruction in the skills of argumentation with the engineering of situations that motivate argumentation. For example, Osborne, Erduran, and Simon (2004) provided students with explicit guidance, including a writing framework of sentence stems to help them identify the types of information necessary in their arguments and discussion of the criteria they should use to evaluate arguments. They also asked students questions with multiple plausible answers and placed students in small groups to argue about the questions at hand. The first half of these strategies are designed with the explicit assumption that students need to be taught how to argue: "argument is a form of discourse that needs to be appropriated by children and explicitly taught through suitable instruction, task structuring, and modeling" (pp. 996–997). The second half of the strategies, in contrast, are implemented because "there is the need to establish a social context that fosters dialogic discourse" (p. 998).

This dual focus on providing explicit instruction in the structure and components of an argument while simultaneously attending to the context is seen throughout other design work, as well (Cavagnetto, Hand, & Norton-Meier, 2010; Chin & Osborne, 2010; Clark & Sampson, 2007; Kenyon et al., 2006; Venville & Dawson, 2010; Zohar & Nemet, 2002). In fact, in a recent review of research regarding ways to foster student participation in scientific argumentation, Cavagnetto (2010) found that authors of 25 of the 54 reviewed articles revealed some combination of "scaffolds such as prompts [i.e., explicit instruction], strategic selection for group collaboration, and use of student misconceptions [i.e., engineering situations to motivate argumentation]" (p. 347). This combination is key because "One must see the point of argument if one is to invest significant effort in it and in developing the skills it entails" (D. Kuhn & Udell, 2007, p. 101). In other words, engineering classroom activities to provide a purpose for the argumentative interactions helps students understand that they should be engaging in those types of interactions—it helps cue them to engage their argumentative skills.

The need to design these situations effectively motivates further consideration and study of how students may come to "see the point of the argument." Pursuing this question connects research on argumentation to research on students' intuitive (or personal) epistemologies (Hofer & Pintrich, 2002) and, with that, to the theoretical construct of "framing" (Bateson, 1972; Goffman, 1974; Tannen, 1993). In the following section, we discuss epistemologies, framing, and "epistemological framing" (Redish, 2004; Sandoval, 2005).

Student Epistemologies

Research on learning in science has a history of attending to the students' meta-level understandings of science. This literature typically discusses these understandings under the headings of students' "epistemologies" (Carey, Evans, Honda, Jay, & Unger, 1989; Sandoval, 2005; Smith, Maclin, Houghton, & Hennessey, 2000) or understandings of the "Nature of Science" (e.g., Bell & Lederman, 2003; Duschl, 2000; Hogan, 2000).

Prominent early work approached the matter in developmental terms. For example, Carey and colleagues (1989) identified three epistemological levels. Students first view scientific knowledge as a "faithful copy of the world," such that claims and evidence are undifferentiated. At the second level, students understand that scientists construct knowledge from observations of nature; at the third they see this as a cyclical process that must account for all of the available evidence.

This view implies a developmental constraint to fostering argumentation: students will be able to "see the point of the argument" only when they have achieved an appropriately sophisticated understanding of science and science knowledge. Hammer and Elby (2002) challenged developmental accounts of epistemologies, along with accounts of coherent epistemological "theories" (Schommer, 1990; Songer & Linn, 1991), arguing that intuitive (or "personal") epistemologies are sensitive to context. This aligns with evidence of students coordinating claims and evidence and critically attending to alternative ideas in some contexts but not in others. In other words, we suggest that the evidence cited above calls a stage-based developmental view of epistemological understandings into question.

Rather than see students as constrained to a particular epistemological stage or view, we work from a view of students as having rich and varied stores of "epistemological resources" that activate differently in different situations (Hammer & Elby, 2003). For example, Lising and Elby (2005) showed how a college student "Jan" treated questions about physical situations very differently in different settings. They first observed her working in a physics course, where she used the formalism and resisted connections to her everyday experience. They then arranged interviews with her in a classroom, in the education building, posing corresponding questions in everyday language, and in that setting, she reasoned from her everyday experience.

In another study, Rosenberg, Hammer, and Phelan (2006) showed a group of eighth-grade students transitioning between two different ways of explaining the "rock cycle." At first, they approached the task as a matter of collating information from worksheets that covered prerequisite material. Working in this way, they produced an account that incorporated scientific terminology but made little sense (i.e., "the Teutonic plates move and create rock"). Their teacher, hearing how they were working, advised them "to start from what you know, not what the paper says," and this prompted them to work in a different way: instead of collating information they composed a straightforward, causal story.

These examples illustrate instances in which students' epistemologies could not be characterized by a single general theory or stage of development; nor could they be characterized as "incoherent and fleeting" (Rosenberg et al., 2006, p. 288). Rather, they showed distinct, contrasting stabilities, or multiple local coherences, in how they treated knowledge and learning. Redish (2004) observed that these phenomena of multiple coherences could be analyzed in terms of how students "framed" what they were doing with respect to knowledge. He therefore proposed the notion of "epistemological framing" to connect this work on intuitive epistemologies to a perspective with an extensive history in anthropology (Bateson, 1972), sociology (Goffman, 1974), linguistics (Tannen, 1993), and cognitive science (Minsky, 1975). We review that perspective now.

Framing

Bateson (1972), an anthropologist, first formed the idea of a "frame" as he watched monkeys play fighting and considered how they knew and let each other know that it was only play. He wrote that the monkeys do "not quite mean what they are saying" (p. 319), in their displays of aggression, and that they signaled that "this is only play" in subtle "metacommunicative messages," or implicit cues for how an explicit communicative act should be interpreted. Bateson extended this insight into an account of how people also experience situations and communicate with each other at a meta-level.

Goffman (1974), a sociologist, is widely known for his work on framing, which he described as being concerned with

> what one individual can be alive to at a particular moment, this often involving a few other particular individuals.... I assume that when individuals attend to any current situation, they face the question: "What is it that's going on here?" Whether asked explicitly, as in times of confusion and doubt, or tacitly, during occasions of usual certitude, the question is put and the answer to it is presumed by the way the individuals then proceed to get on with the affairs at hand. (p. 8)

For example, as we discuss below, framing a situation as a school assignment, a student might be less "alive to" differences of viewpoints by other students, with attention more directed on what the teacher has instructed the class to do.

Tannen (1993) discussed frames in terms of *schemas* (Bartlett, 1932), or "structures of expectation based on past experience." She gave examples from a study in which she and her colleagues asked subjects from different countries to watch a short movie and then to describe it to someone who had not seen it. She identified evidence of how subjects' expectations affected their interpretation of the film as well as of the task at hand. For example, some subjects showed awareness of the movie as a film, commenting on the sound track, the colors, and what the "point" might be; others spoke only of the story within the movie, focusing on the depicted events and the characters' feelings and motivations. Tannen described how there were some differences by country; for example, Americans were divided between seeing the task as film criticism or as recounting a story, while all of the Greek subjects treated the task as the latter. Tannen concluded that the different experiences of the American and Greek participants influenced how they interpreted—or framed—the task of describing a movie.

Some frames (or aspects of frames) are epistemological in substance. Thus Redish (2004) suggested, the multiple local coherences in Jan's or the eighth-grade students' approaches to reasoning could be understood to reflect different "epistemological frames." In this case, the students' sense of "what is it that's going on here" with respect to the knowledge changed across contexts, for Jan as she moved from the physics course to the interviews, and for the eighth graders, after the teacher's intervention.

How individuals frame what is taking place influences their forming objectives, within that framing. For example, students framing an activity as an exchange of ideas might coordinate objectives of hearing and being heard. In contrast, if they are framing an activity as a "lesson," they might be aware of the teacher's objective of evaluating their contributions and so they may have the objective of earning points. Students' framing also influences how they interpret the actions of others; for example, a student framing a class discussion in one way would hear a teacher's question as a genuine request for clarification and information, but framing it in another way would suggest that the same question was a correction or a "test." That is, how students frame what is taking place is directly related to their understanding "the point."

Framing as a Dynamic Process

Throughout this literature, framing is implicitly treated as both psychological constructs that can be distinguished and as a dynamic process. That is, framings are schemas or structures that organize one's past experiences, and they are a process by which one interprets their current experience. For example, Bartlett (1932) described a schema as an "active organization ... of past experiences" (p. 201). That schema are active is essential to the account: No new situation will precisely match previous experience, so, to be functional, any structure that organizes past experience must be flexible and responsive. Thus, the literature presents frames as dynamic structures, active moment-to-moment, which is why researchers (as we will do here) often speak of "framings" rather than "frames" (MacLachlan & Reid, 1994).

As well, the dynamics of framing often involve shifts among recognizably distinct frames. Tannen and Wallat (1993) demonstrate this dynamic process in their account of multiple, interacting frames for a doctor conducting a medical examination. This account reveals the meta-communicative moves the doctor makes—such as shifts in her vocal register or terminology—to signal the shifts of framing to others in attendance.

We cited a simple example of the dynamics of epistemological framing above, in the group of eighth graders shifting in their framings after the teacher's intervention (Rosenberg et al., 2006). Frank (2009) and Scherr and Hammer (2009) add to this literature, demonstrating shifts in framing within collaborative group work among college students. These shifts are often in response to subtle features of the immediate context, including the arrangement of materials (Frank, 2009) or an overheard remark. Thus, these authors argue for

> a dynamic systems account of framing; [in which] coherences emerge from the activations and interactions of many cognitive elements. They may involve resources within an individual's mind or across multiple individuals or a group. (Scherr & Hammer, 2009, p. 151)

We (Berland & Hammer, under review) present a similar analysis of how a middle-school science classroom community converged (or did not) around a shared framing of their class discussions. There, the classroom community maintained their shared framing through both explicit and implicit rejections of behaviors that conflicted with those framings. In this emerging account, the stability of an epistemological framing may be an individual phenomenon or it may be distributed, in Hutchins's (1995) sense, across a group and materials (Conlin, Gupta, & Hammer, 2010; Frank, 2009).

Hutchison and Hammer (2010) illustrate college students' shifting frames as a result of much more subtle and unintended cues. In the moment they consider, the instructor shifted from a stance of listener to one of presenter, moving to the blackboard and presenting an idea for the students to consider. The students seemed to respond with a shift in their framing, from "making sense of phenomena" to "the classroom game" (Lemke, 1990) of trying to produce formally correct answers. This shift is akin to what Jiménez-Aleixandre, Rodríguez, and Duschl (2000) call "doing the lesson" as opposed to "doing science." Thus the shift in the instructor's behavior seemed to cue or resonate with one of the "basic frameworks" (Goffman, 1974) the students had formed from previous experiences in science classes—that of being receiving and reiterating facts. Hutchinson and Hammer (2010) conclude that "Individuals are accomplished at attending to what is going on around them for signals that indicate the type of activity and altering the framing when it appears appropriate" (p. 518). The signals, in this case, were that the lesson had become more like something to which they were accustomed in science class.

Framing and Argumentation

While there has been extensive attention and discussion of framing in anthropology, linguistics, sociology, and cognitive science for years, the notion is only recently becoming prominent in the learning sciences. Engle (2006), for example, applied framing to reconsider the concept of transfer, discussing how framing can connect previously distinct contexts.

A great deal of existing work, however, analyzes classroom interactions in ways consistent with the framing literature (e.g., Leander & Brown, 1999; Lemke, 1990; Lidar, Lundqvist, & Ostman, 2006; Mehan, 1979). For example, Enyedy and Goldberg (2004) wrote

> People use their understandings of what type of activity they are doing and what their role is to limit what they do and say—what people do and say has direct implications for the meaning that they take away from the experience. (p. 910)

In a sense they were discussing framing, the "understandings of what type of activity they are doing." We see a similar trend in the research on argumentation: As it moves to focus on how participants understand what they are doing and what is taking place, it has entered the domain of framing.

With that perspective, it is easy to see a variety of ways in which typical class discussions could cue student framings that are at odds with argumentation. A teacher may signal, in various ways, that she knows and will be the judge of the correct answer to a question, and therefore tip students into framing the activity as a performance for the teacher, or "doing the lesson." This is suggested by Hutchison and Hammer's (2010) account of the instructor's behavior influencing the students' framing of the class discussion. In fact, Tabak and Baumgartner (2004) found that cues as subtle as the teacher's use of pronouns ("I, you, and we") contributed to the students interpretation of whether the teacher was a partner or mentor, with the former more conducive for argumentation.

In this way, too, framing may account for the conflicting patterns of evidence in research on argumentation. That is, rather than progress through developmental stages of epistemological understandings, students come to the classroom able to frame what is taking place in a variety of ways, with a variety of understandings of what the knowledge is, where it comes from, and the students' relationship to that knowledge. However, even if the teacher or researcher asks students or subjects in clinical studies to argue, they may not always frame what they are doing as argumentation. Instead, subtle, unintended, "meta-communicative messages" may tip them into framing the activity in another way.

Implications for Education

To summarize, we have discussed the trend in research on argumentation to increasingly emphasize exploring whether students "see the point of the argument" when they are engaged in argumentative activities. This work demonstrates that, when students see their discussions as argumentative, they begin to engage in argumentative behaviors that are typically seen as challenging, such as coordinating claims and evidence and attending to alternative lines of reasoning. Thus, for curriculum designers, we see a decreasing emphasis on teaching students argumentative skills and an increasing focus on creating situations in which students see the need to engage in these argumentative behaviors. This shift brings argumentation research into contact with the notion of framing, and recent work on epistemological framing. In the following section, we use the research on framing to highlight the difficulties of this new approach as well as the resources there might be to meet it.

Difficulties of Designing for Argumentation

As discussed above, students do not typically frame their classroom activities as instances in which they should coordinate evidence and claims or attend to alternative perspectives. The challenges arise because framing is a subtle, dynamic process, and students, like everyone, pick up on the "meta-communicative messages," as well as the explicit messages. That is, students are aware of both the explicit instructions from their teacher and also the meta-instructions that are communicated implicitly, and these meta-instructions influence students' interpretations of the task at hand.

Consider the following illustrative example taken from observations of a sixth-grade science class in 2006 (Berland, 2008). The teacher designed an activity for her students that she hoped would prompt them to use evidence and to substantively engage with one another's ideas. Each student was assigned an atomic element (e.g., helium) to investigate for the purpose of presenting evidence that would convince his or her classmates to "buy" that element. On the day of the presentations, to encourage critical interaction, the teacher required her students to critique one another. After each presentation she called on a few students and asked them to evaluate it.

One might suppose the teacher in this example was creating a need for students to attend to others' ideas by requiring that they respond to one another's presentations. In addition, one might suppose she was helping students learn to be better critics by modeling it herself and by evaluating whether the critiquing student had responded to appropriate elements of the presentation. Of course, it did not work. First, few of the presentations addressed the goal of "selling" an element; instead students generally gave presentations that focused on facts, such as the number of neutrons, protons, and electrons in their respective elements, and the everyday objects that include their elements. Second, when called on to critique presentations, students often had nothing to say. When they did comment, they responded to the teacher directly by identifying surface features of the presentation that they did or did not like (i.e., font color). In this way, the students seemed to see the critique as an obligation to demonstrate that they were listening, rather than as an opportunity to communicate something meaningful to the presenter.

Here then is an example of the influence of meta-instructions on students' framing: There were, clearly, explicit instructions that the students should critique and discuss each other's ideas, but there were meta-instructions as well. For example, the teacher evaluated students' feedback to one another. This evaluation sent the message that the critiques she required were performances for her. We suggest that, partially as a result of these meta-instructions, the students framed what was taking place as "doing the lesson" (Jiménez-Aleixandre, Rodriguez, & Duschl, 2000), one of the basic frameworks individuals have and apply to understand what is it that is going on within schools. We could tell a corresponding story regarding the storyline for the activity of "selling" the different elements: in the context and with the meta-instructions signaling the existence of "correct" answers, students framed the task as a recital of the standard information they had learned about elements, with the advertising storyline only a veneer.

We could tell a similar story about how the teacher framed the task—how the teacher formed a sense of what was taking place. This would also be a dynamic involving her explicit goals for the activity, her implicit framework of expectations built from past experiences, and the students' reactions to the activity. In this case, the teacher had hoped for student-to-student interaction, yet her role during the activity was that of the traditional evaluator, a role the students expected. In this way, the students and teacher reinforced each other's framing the task as a traditional presentation of facts.

Resources for Meeting the Challenge

As illustrated by the above vignette, part of the challenge in designing for argumentation is that it contrasts with the familiar expectations of what happens in science class and these expectations influence students' framings of their argumentative activities in science class. However, students are able to frame discussions in ways that align with argumentative interactions. That is, the same students who perform for the teacher in the classroom, rather than engage with each other, can walk into the hallway and argue about which movie to see later, whom to support in a school election, which team has the best chances of winning the division, and so forth. In fact, Bricker and Bell (2007) reveal students constructing complete arguments, that include evidence, in a nonschool context. Moreover, recent work by Berland and Hammer (under review) demonstrates the same students shifting between argumentative and nonargumentative framings of their class discussions, based on the social context, including their teacher's meta-messages. In addition, the studies we cited above revealed students arguing about science in situations when they experienced differences of belief that they cared to resolve, and felt they could resolve. Thus, students have versions of argumentation among their basic frameworks, and these are resources for argumentation teachers and designers can access and upon which they can build.

One implication for educators interested in promoting argumentation is that we must be aware of when it is occurring, when it is beginning, and whether the argumentation arises organically or by design. Indeed, students may independently initiate what educators hope to design. Thus, much of the challenge in fostering argumentation is helping teachers recognize when it occurs and giving teachers the autonomy to take advantage of such opportunities. Here we focus more on the challenge of designing materials and strategies to promote argumentation—to create such opportunities.

The field has come to recognize that promoting argumentation requires designing situations that make "the point of argument" clear. In earlier work, the first author and colleagues (L. Kuhn et al., 2006) addressed this challenge by building on Edelson's (2001) Learning-For-Use approach to design. In particular, we discussed "creating a need" for students to engage in scientific argumentation. Tacit in this approach was the expectation that it must be an authentic need to persuade, not a "need" to perform for the instructor. Thus, fostering argumentation requires designing situations in which the questions that are asked, the process through which students answer the questions, the criteria students use to evaluate answers, and the way the teacher evaluates them combine to make argumentation a sensible and purposeful thing to do. The perspective of framing adds to this approach an understanding that educators cannot attend only the particular need they intend to create: educators must attend to how students might understand the activity as a whole, taking into account the basic frameworks they bring, from within class and without. Truly, designers cannot create needs; they can only design materials that might induce those needs in students, and students do not experience the needs for coordinating claim and evidence or for attending to others' views in the absence of a larger framing of what is taking place. It is with this in mind that we discuss four design strategies for developing learning environments to promote argumentation.

Arrange Situations in Which Multiple Perspectives Are Salient

Students come to science class with basic frameworks from their experiences of argumentation in everyday life. These may depend on their particular cultural backgrounds (e.g., Moje, Ciechanowski, Kramer, Ellis, Carrillo, & Callazo, 2004), but the evidence of students engaging in argumentation with little instruction discussed above demonstrates that the resources for argumentation are there. One aspect of situations students frame as instances in which argumentation would be appropriate is the salience of multiple, apparently conflicting, perspectives that are interesting. For designers, this means creating contexts that are rich enough to enable multiple perspectives (de Vries, Lund, & Michael, 2002).

A variety of design work for scientific inquiry creates rich contexts that have salient multiple perspectives. For example, Blumenfeld, Soloway, Marx, Krajcik, Guzdial, and Palincsar, (1991) describe students designing artifacts that require them to integrate multiple pieces of information. In addition, in the work by both Hatano and Inagaki (1991), and Osborne, Erduran, and Simon (2004), we see the researchers presenting students with differing claims in which each claim is plausible, depending on your interpretation of the evidence. D. Kuhn (2010) asked students questions with multiple plausible answers and then made the evidence available for their investigation. In each of these examples, the multiple plausible answers created a need for the students to coordinate their claims and evidence: the students needed to use the evidence to resolve disputes.

In the vignette above, the students were presenting facts (or data) about elements, but the data was not in support of a claim. That is, their task did not require that the facts be applied to solve a larger problem. Moreover their task (i.e., "selling" their elements) could have been accomplished without evidence (i.e., they could have used advertising strategies that appealed to their audiences' emotions). Beyond affecting the presentations themselves, this question focus also affected the audience members and their feedback: the students were evaluating surface features of the presentation instead of the evidence partly because evidence was not necessary for them to engage with the question of whether they would "buy" the element. Perhaps a rewording would have helped students frame the activity in a more meaningful way. For example, if the presenters were asked to describe what would happen if their element were no longer available students might have seen information about the element—its presence and role in food, human physiology, and in everyday material objects— as meaningful to their case for its importance.

Let Students Be Responsible for Resolving Disagreements

Another aspect of situations that encourage framings that align with argumentation is enabling students to see themselves as having the "epistemic agency" (Scardamalia, 2002) to resolve the disagreements, reasoning with knowledge they have available or could obtain. To do this, students must be accountable to more than just the teacher; they must be accountable to each other as well (Brown & Campione, 1996; Engle & Conant, 2002; Sohmer & Michaels, 2005).

We suggest that one way to accomplish this is to create situations in which students are likely to experience a need to engage with each another's ideas. This design strategy goes beyond the typical classroom activity of asking students to critique or discuss one another's ideas, as the teacher in the above vignette did when she required students to critique one another's presentations. In that activity, the students experienced a need to attend to alternative ideas, but the need was within the traditional framing of "doing the lesson" (Jiménez-Aleixandre et al., 2000): They needed to attend because the teacher required it of them. The challenge for designers is to create situations in which students are likely to experience the need as meaningfully inherent to their activity. This occurs when understanding and responding to alternative perspectives accomplishes some larger goal, such as determining the best choice or persuading others of an idea.

One way to do this is to engage the students in what the first author and colleagues (Berland & Reiser, 2011; L. Kuhn et al., 2006) call an "argument jigsaw." In this two-step activity structure, pairs of students construct an explanation and then two pairs combine to form a group of four. The group of four is asked to converge onto a single explanation. This activity moves beyond telling students to evaluate one another's explanations by requiring them to reach consensus. As such, the activity *makes them responsible for resolving disagreements.* Moreover, by asking the students to construct preliminary arguments before joining the larger group we are giving students an opportunity to develop ownership over their ideas, thereby *making multiple perspectives salient* and giving them a reason to defend their own perspectives, in the larger group. Other authors similarly use consensus building to promote argumentation. For example, Clark and Sampson (2007) and de Vries, Lund, and Baker (2002) both created software environments in which students engaged in scientific argumentation in order to reach consensus, and Chin and Osborne (2010) grouped students to maximize disagreement.

Beware of Cuing "The Classroom Game" and Pseudoargumentation

There is a consensus in the literature that sophisticated practices of argumentation involve explicit awareness of objectives, components, and criteria. At some point, students should not only coordinate claims and evidence, generate and evaluate hypotheses, consider theoretical coherence and so on but also be able to say what they are doing. That is, they should know the terminology of scientific argumentation. Developing this terminology enables students to discuss how and why they are evaluating competing arguments and to ensure that they are using common criteria to do so.

Much of the design work fostering student participation in scientific argumentation has introduced that vocabulary from the outset, providing explicit instruction in the "skills" or "criteria" of argumentation by telling students the components of a good argument. For example, as discussed above, Osborne and colleagues (2004) supported students by giving them sentence stems that identified the components of a complete argument. In more recent work, Chin and Osborne (2010) asked teachers to introduce an argumentative activity by defining the components of an argument and emphasizing the importance of justifications before grouping disagreeing students together. Design work on which the first author has worked uses a similar strategy of giving students an instructional framework that highlights the components of an argument (Finn, L. Kuhn, Whitcomb, Bruozas, & Reiser, 2006; McNeill, Harris, Heitzman, Lizotte, Sutherland, & Krajcik, 2004; McNeill, Lizotte, Krajcik, & Marx, 2006).

Consideration of students' framings raises concerns about these strategies that provide students with information before the students are likely to see a meaningful need for it. The risk is in the meta-message—in what the instruction signals to the students. In particular, it is likely to signal to students that they should be following instructions, which could promote their framing of the activity as a "classroom game" (Lemke, 1990). As well, by calling students' attention to the expectations of the structure of their arguments, these approaches may distract from the substance, which is what would naturally motivate framing an activity as argumentative. In both of these ways, introducing argumentation through explicit instruction in how to argue might undermine framings that are more consistent with scientific argumentation and therefore inhibit student engagement in this practice. For example, Berland (2008) shows a class of students interpreting an argumentative discussion as an opportunity to present their arguments to their teacher–audience rather than an opportunity to work collaboratively with their peers to make sense of the evidence and alternative ideas.

We refer to the problem as "pseudoargumentation." In this we follow research on composition (Paretti, 2009; Petraglia, 1995; Spinuzzi, 1996) that distinguishes *transactional* and *pseudotransactional* writing. Pseudotransactional writing occurs when the purpose of the genre differs from the purpose of the assignment. For example, the purpose of a letter to the editor is to share an opinion with the public, but the purpose of an assignment to write a letter to the editor is typically to demonstrate an ability to do so (Petraglia, 1995). In other words, pseudotransactional writing is writing that occurs when students have explicit instruction in the expectations for the assignment but those expectations are not fulfilling an authentic goal. Similarly, pseudoargmentation is argumentation that can occur in schools when the students' attention is on doing what they expect the teacher will value rather than on the substance of the ideas at hand.

Thus, we suggest designing learning environments and activities that help students engage in scientific argumentation requires that we, as educators and researchers, attend carefully to when and how explicit instruction is used: in order to avoid pseudoargumentation, we must engage in explicit instruction if and only if it provides information that solves a problem students are experiencing rather than providing external criteria for them to meet. For example, we suggest creating a need for argumentation (through the first two strategies) and engaging in an explicit discussion of how to evaluate and compare alternative ideas only after students have struggled with doing so themselves.

Design Curricula to Be Responsive

We mention one final implication before we close, and that is the need for humility in the extent to which we can control what will happen in students' minds. The dynamics of framing are subtle and sensitive to context. Thus, apparently minor aspects of curriculum and activity design could, in some classrooms, have a strong effect on framing. Moreover, regardless of the design of the materials, if students are granted epistemic agency, much of the action will be in the particular ways inquiry unfolds in particular classrooms.

This, in part, motivates extensive pilot testing of materials in a variety of settings with careful attention to variations in the classroom culture, such as work done by Berland (in press). It also motivates designing materials in a way that supports teachers to attend and respond to what arises. The second author and colleagues are exploring ways to design "responsive curricula," with embedded "teacher guides" as support or professional development.[1] These would include examples of what occurred in pilot tests of the curriculum, along with commentary about the student thinking and ideas for how a teacher might respond.

Closing Remarks

Research on student argumentation in science probably owes its existence, and certainly a great deal of its progress, to the seminal work of the cognitive psychologist Deanna Kuhn. Cognitive psychology has a rich, extensive intellectual history of conceptualizing development in stages of successively more sophisticated structures, and Kuhn's work began there, conceptualizing abilities in argumentation that come into being over time. In her original formulation, learning to argue required "strong restructuring" (D. Kuhn, 1989) of the cognitive machinery students have available.

As discussed in this chapter, research in classroom-based scientific argumentation, including Kuhn's work, has progressed from focusing on students' argumentation skills to whether they recognize the "point of argument" (D. Kuhn & Udell, 2007). We argue that this shift brings research on argumentation into contact with another set of ideas that have rich, extensive intellectual history: framing. Unlike cognitive psychology, framing expects a substantial variability in individuals' reasoning and behavior and explains that variability in terms of the situation and how individuals understand it.

It is important to note, in closing, that the perspectives framing and cognitive psychology share common origins in schema theory (Bartlett, 1932). Within the tradition of cognitive psychology, perhaps driven by Piaget's biological account of structure, schemas are discussed in terms of structures within individual's heads; within the tradition of framing, schemas have remained more process-like, dynamic, and evolving. Clearly the synthesis and possible reconciliation of these accounts go beyond the scope of this chapter,[2] but they are not necessarily in opposition: What begins as a local framing may become structurally stable over time.

---

[1] Goldberg, F. M., Hammer, D. M., Bendall, S., & Coffey, J. (2008–2011). Learning Progression for Scientific Inquiry: A Model Implementation in the Context of Energy. San Diego State University and University of Maryland: Project funded by the National Science Foundation, Grant DRL 0732233.

[2] We note the dynamic systems approach of Thelen and Smith (1994), which considers on local and developmental time scales. See also Conlin et al. (2010) work for early thoughts on theoretical continuity.

References

Bartlett, F. C. (1932). *Remembering: An experimental and social study*. Cambridge: Cambridge University Press.

Bateson, G. (1972). A theory of play and fantasy (Originally published in 1954). In G. Bateson (Ed.), *Steps to an ecology of mind: Collected essays in anthropology, psychiatry, evolution, and epistemology* (pp. 177–193). San Francisco: Chandler.

Bell, R. L., & Lederman, N. G. (2003). Understandings of the nature of science and decision making on science and technology based issues. *Science Education, 87*(3), 352–377.

Berland, L. K. (2008). *Understanding the composite practice that forms when classrooms take up the practice of scientific argumentation.* Evanston, IL: Northwestern.

Berland, L. K. (in press). Explaining variation in how classroom communities adapt the practice of scientific argumentation. *Journal of Learning Sciences.*

Berland, L. K., & Lee, V. R. (2010). *Anomalous graph data and claim revision during argumentation.* Paper presented at the International Conference of the Learning Sciences, Chicago, IL.

Berland, L. K., & Reiser, B. J. (2009). Making sense of argumentation and explanation. *Science Education, 93*(1), 26–55.

Berland, L. K., & Hammer, D. (under review). Framing for scientific argumentation. *Journal of Research in Science Teaching.*

Berland, L. K., & Reiser, B. J. (2011). Classroom communities' adaptations of the practice of scientific argumentation. *Science Education, 95*(2), 191–216.

Blumenfeld, P., Soloway, E., Marx, R. W., Krajcik, J. S., Guzdial, M., & Palincsar, A. (1991). Motivating project-based learning: Sustaining the doing, supporting the learning. *Educational Psychologist, 26*, 369–398.

Bricker, L., & Bell, P. (2007, April). *"Um...since I argue for fun, I don't remember what I argue about": Using children's argumentation across social contexts to inform science instruction.* Paper presented at the National Association of Research in Science Teaching, New Orleans, LA.

Brown, A. L., & Campione, J. C. (1996). Psychological theory and the design of innovative learning environments: On procedures, principles, and systems. In L. Schauble & R. Glaser (Eds.), *Innovations in learning: New environments for education* (pp. 289–325). Mahwah, NJ: Lawrence Erlbaum Associates.

Carey, S., Evans, R., Honda, M., Jay, E., & Unger, C. (1989). "An experiment is when you try it and see if it works": A study of grade 7 students' understanding of scientific knowledge. *International Journal of Science Education, 11*, 514–529.

Cavagnetto, A. R. (2010). Argument to foster scientific literacy. *Review of Educational Research, 80*(3), 336–371.

Cavagnetto, A., Hand, B. M., & Norton-Meier, L. (2010). The nature of elementary science discourse in the context of the science writing heuristic approach. *International Journal of Science Education, 32*(4), 427–449.

Chin, C., & Osborne, J. (2010). Supporting argumentation through students' questions: Case studies in science classrooms. *The Journal of the Learning Sciences, 19*, 230–284.

Clark, D. B., & Sampson, V. (2007). Personally-seeded discussions to scaffold online argumentation. *International Journal of Science Education, 29*(3), 253–277.

Conlin, L., Gupta, A., & Hammer, D. (2010). *Where to find the mind: Identifying the scale of cognitive dynamics.* Paper presented at the International Conference of the Learning Sciences, Chicago, IL.

de Vries, E., Lund, K., & Michael, B. (2002). Computer-mediated epistemic dialogue: Explanation and argumentation as vehicles for understanding scientific notions. *Journal of the Learning Sciences, 11*(1), 63–103.

Driver, R., Newton, P., & Osborne, J. (2000). Establishing the norms of scientific argumentation in classrooms. *Science Education, 84*, 287–312.

Duschl, R. A. (2000). Making the nature of science explicit. In R. Millar, J. Leach, & J. Osborne (Eds.), *Improving science education: The contribution of research* (pp. 187–206). Buckingham: Open University Press.

Duschl, R. A., Schweingruber, H. A., & Shouse, A. E. (Eds.). (2007). *Taking science to school: Learning and teaching science in grades K-8*. Washington, DC: National Academies Press.

Edelson, D. C. (2001). Learning-for-Use: A framework for the design of technology-supported inquiry activities. *Journal of Research in Science Teaching, 38*(3), 355–385.

Engle, R. A. (2006). Framing interactions to foster generative learning: A situative explanation of transfer in a community of learners classroom. *Journal of the Learning Sciences, 15*(4), 451–498.

Engle, R. A., & Conant, F. R. (2002). Guiding principles for fostering productive disciplinary engagement: Explaining an emergent argument in a community of learners classroom. *Cognition and Instruction, 20*(4), 399–483.

Enyedy, N., & Goldberg, J. (2004). Inquiry in interaction: How local adaptations of curricula shape classroom communities. *Journal of Research on Science Teaching, 41*(9), 905–935.

Felton, M., & Kuhn, D. (2001). The development of argumentative discourse skill. *Discourse Processes, 32*(2&3), 135–153.

Finn, L. E., Kuhn, L., Whitcomb, J. L., Bruozas, M., & Reiser, B. J. (2006). Where have all the creatures gone? In J. Krajcik & B. J. Reiser (Eds.), *IQWST: Investigating and questioning our world through science and technology*. Evanston, IL: Northwestern University.

Ford, M. J., & Forman, E. A. (2006). Redefining disciplinary learning in classroom contexts. In J. Green & A. Luke (Eds.), *Review of educational research* (Vol. 30, pp. 1–32). Washington, DC: American Education Research Association.

Frank, B. W. (2009). *The dynamics of variability in physics students' thinking: Examples from kinematics*. Unpublished doctoral thesis, University of Maryland, College Park.

Goffman, E. (1974). *Frame analysis: An essay on the organization of experience*. Cambridge, MA: Harvard University Press.

Hammer, D., & Elby, A. (2002). On the form of a personal epistemology. In B. K. Hofer & P. R. Pintrich (Eds.), *Personal epistemology: The psychology of beliefs about knowledge and knowing* (pp. 169–190). Mahwah, NJ: Lawrence Erlbaum Associates.

Hammer, D., & Elby, A. (2003). Tapping epistemological resources for learning physics. *Journal of the Learning Sciences, 12*(1), 53–90.

Hammer, D., & van Zee, E. H. (2006). *Seeing the science in children's thinking: Case studies of student inquiry in physical science*. Portsmouth, NH: Heinemann.

Hatano, G., & Inagaki, K. (1991). Sharing cognition through collective comprehension activity. In L. B. Resnick, J. M. Levine & S. D. Teasley (Eds.), *Perspectives on socially shared cognition* (pp. 331–348). Washington, DC: American Psychological Association.

Hofer, B. K., & Pintrich, P. R. (2002). *Personal epistemology: The psychology of beliefs about knowledge and knowing*. Mahwah, NJ: Lawrence Erlbaum.

Hogan, K. (2000). Exploring a process view of students' knowledge about the nature of science. *Science Education, 84*(1), 51–70.

Hutchins, E. (1995). *Cognition in the wild*. Cambridge, MA: MIT Press.

Hutchison, P., & Hammer, D. (2010). Attending to student epistemological framing in a science classroom. *Science Education, 94*(3), 506–524.

Jiménez-Aleixandre, M. P., Rodriguez, A. B., & Duschl, R. A. (2000). "Doing the lesson" or "doing science": Argument in high school genetics. *Science Education, 84*(3), 287–312.

Kenyon, L., O., Kuhn, L., & Reiser, B. J. (2006). Student epistemologies of science to guide the practice of argumentation. In S. A. Barab, K. E. Hay, & D. T. Hickey (Eds.), *7th annual international conference of the learning sciences* (pp. 321–327). Mahwah, NJ: Lawrence Erlbaum Associates.

Kuhn, D. (1989). Children and adults as intuitive scientists. *Psychological Review, 96*(4), 674–689.

Kuhn, D. (1991). *The skills of argument*. Cambridge: Cambridge University Press.

Kuhn, D. (2010). Teaching and learning science as argument. *Science Education, 94*(5), 810–824.

Kuhn, D., & Udell, W. (2007). Coordinating own and other perspectives in argument. *Thinking and Reasoning, 13*(2), 90–104.

Kuhn, L., Kenyon, L. O., & Reiser, B. J. (2006). Fostering scientific argumentation by creating a need for students to attend to each other's claims and evidence. In S. A. Barab, K. E. Hay, & D. T. Hickey (Eds.), *7th annual international conference of the learning sciences* (pp. 370–375). Mahwah, NJ: Lawrence Erlbaum Associates.

Larson, A., & Britt, A. (2009). Improving students' evaluation of informal arguments. *The Journal of Experimental Education, 77*(4), 339–365.

Leander, K. M., & Brown, D. E. (1999). "You understand, but you don't believe it": Tracing the stabilities and instabilities of interaction in a physics classroom through a multidimensional framework. *Cognition and Instruction, 17*(1), 93–135.

Leitao, S. (2000). The potential of argument in knowledge building. *Human Development, 43*, 332–360.

Lemke, J. L. (1990). *Talking science: Language, learning, and values*. Norwood, NJ: Ablex.

Lidar, M., Lundqvist, E., & Ostman, L. (2006). Teaching and learning in the science classroom: The interplay between teachers' epistemological moves and students' practical epistemology. *Science Education, 90*(1), 148–163.

Lising, L., & Elby, A. (2005). The impact of epistemology on learning: A case study from introductory physics. *American Journal of Physics, 73*(4), 372–382.

Louca, L., Hammer, D., & Bell, M. (2002). Developmental versus context-dependant accounts of abilities for scientific inquiry: A case study of 5–6th grade student inquiry from a discussion about a dropped pendulum. In P. Bell, R. Stevens, & T. Satwicz (Eds.), *Keeping learning complex: Fifth international conference of the learning sciences* (pp. 261–267). Mahwah, NJ: Erlbaum.

MacLachlan, G. L., & Reid, I. (1994). *Framing and Interpretation*. Carlton, VIC: Melbourne University Press.

McNeill, K. L., Harris, C. J., Heitzman, M., Lizotte, D. J., Sutherland, L. M., & Krajcik, J. (2004). How can I make new stuff from old stuff? In J. Krajcik & B. J. Reiser (Eds.), *IQWST: Investigating and questioning our world through science and technology*. Ann Arbor, MI: University of Michigan.

McNeill, K. L., Lizotte, D. J., Krajcik, J., & Marx, R. W. (2006). Supporting students' construction of scientific explanations by fading scaffolds in instructional materials. *Journal of the Learning Sciences, 15*(2), 153–191.

Mehan, H. (1979). *Learning lessons: Social organization in the classroom*. Cambridge, MA: Harvard University Press.

Minsky, M. (1975). A framework for representing knowledge. In P. Winston (Ed.), *The psychology of computer vision* (pp. 211–277). New York: McGraw-Hill.

Moje, E. B., Ciechanowski, K. M., Kramer, K., Ellis, L., Carrillo, R., & Callazo, T. (2004). Working toward third space in content area literacy: An examination of everyday funds of knowledge and discourse. *Reading Research Quarterly, 39*(1), 38–70.

Naylor, S., Keogh, B., & Downing, B. (2007). Argumentation in primary science. *Research in Science Education, 37*, 17–39.

Osborne, J., Erduran, S., & Simon, S. (2004). Enhancing the quality of argumentation in school science. *Journal of Research in Science Teaching, 41*(10), 994–1020.

Paretti, M. (2009). When the teacher is the audience: Assignment design and assessment in the absence of 'real' readers. In M. Weiser, B. Fehler, & A. Gonzalez (Eds.), *Engaging audience: Writing in an age of new literacies* (pp. 165–185). Urbana, IL: National Council of Teachers of English.

Petraglia, J. (1995). Spinning like a kite: A closer look at the pseudotransactional function of writing. *Journal of Advanced Composition, 15*, 19–33.

Radinsky, J. (2008). *Students' sense-making with visual data in small-group argumentation*. Paper presented at the International Conference of the Learning Sciences, Utrecht, The Netherlands.

Redish, E. F. (2004). A theoretical framework for physics education research: Modeling student thinking. In E. F. Redish & Vicentini (Eds.), *Proceedings of the Enrico Fermi Summer School Course, CLVI* (pp. 1–63). Bologna, Italy: Italian Physical Society.

Rosenberg, S., Hammer, D., & Phelan, J. (2006). Multiple epistemological coherences in an eighth-grade discussion of the rock cycle. *The Journal of Learning Sciences, 15*(2), 261–292.

Sandoval, W. A. (2005). Understanding students' practical epistemologies and their influence on learning through inquiry. *Science Education, 89*(4), 634–656.

Scardamalia, M. (2002). Collective cognitive responsibility for the advancement of knowledge. In B. Smith (Ed.), *Liberal education in a knowledge society* (pp. 67–98). Chicago: Open Court.

Scherr, R., & Hammer, D. (2009). Student behavior and epistemological framing: Examples from collaborative active-learning activities in physics. *Cognition and Instruction, 27*(2), 147–174.

Schommer, M. (1990). Effects of beliefs about the nature of knowledge on comprehension. *Journal of Educational Psychology, 82*, 498–503.

Smith, C., Maclin, D., Houghton, C., & Hennessey, M. G. (2000). Sixth-grade students' epistemologies of science: The impact of school science experiences on epistemological development. *Cognition and Instruction, 18*(3), 349–422.

Sohmer, R., & Michaels, S. (2005). The "two puppies" story: The role of narrative in teaching and learning science. In U. Quasthoff & T. Becker (Eds.), *Narrative interaction* (pp. 57–91). Philadelphia, PA: John Benjamins.

Songer, N. B., & Linn, M. C. (1991). How do students' views of science influence knowledge integration? *Journal of Research in Science Teaching, 28*(9), 761–784.

Spinuzzi, C. (1996). Pseudotransactionality, activity theory, and professional writing instruction. *Technical Communication Quarterly, 5*(3), 295–308.

Tabak, I., & Baumgartner, E. (2004). The teacher as partner: Exploring participant structures, symmetry, and identity work in scaffolding. *Cognition and Instruction, 22*(4), 393–429.

Tannen, D. (1993). *Framing in discourse*. New York: Oxford University Press.

Tannen, D., & Wallat, C. (1993). Interactive frames and knowledge schemas in interaction: Examples from a medical examination/interview. In D. Tannen (Ed.), *Framing in discourse* (pp. 57–76). New York: Oxford University Press.

Thelen, E., & Smith, L. B. (1994). *A dynamic systems approach to the development of cognition and action*. Cambridge, MA: MIT Press.

Venville, G. J., & Dawson, V. M. (2010). The impact of a classroom intervention on grade 10 students' argumentation skills, informal reasoning, and conceptual understanding of science. *Journal of Research in Science Teaching, 47*(8), 952–977.

Warren, B., Ballenger, C., Ogonowski, M., Rosebery, A. S., & Hudicourt-Barnes, J. (2001). Rethinking diversity in learning science: The logic of everyday sense-making. *Journal of Research in Science Teaching, 38*(5), 529–552.

Weiss, I. R., Pasley, J. D., Smith, P. S., Banilower, E. R., & Heck, D. J. (2003). *Looking inside the classroom: A study of K–12 mathematics and science education in the United States*. Horizon Research.

Wilensky, U. (1999). *NetLogo [computer program]: Center for connected learning and computer-based modeling*. Northwestern University, Evanston, IL.

Zohar, A., & Nemet, F. (2002). Fostering students' knowledge and argumentation skills through dilemmas in human genetics. *Journal of Research in Science Teaching, 39*(1), 35–62.

Part II
Practice Perspectives in Argumentation

Chapter 6
The Design and Enactment of Argumentation Activities

Shirley Simon, Katherine Richardson, and Ruth Amos

Introduction

In recent years new approaches to the teaching of school science have focused on extending the goals of science education beyond the learning of a body of established knowledge to encompass cognitive, epistemic and social aims. School science has become more concerned with foregrounding the epistemic basis of science, providing opportunities for the development of scientific reasoning through the co-ordination of theory with evidence (Kuhn, 1991), and of epistemological understanding through the evaluation of scientific knowledge claims (Sandoval & Reiser, 2004). By engaging collaboratively in argumentation activities that make reasoning public, students can gain experience of constructing arguments, justifying arguments with evidence, evaluating alternative arguments and reflecting on the outcomes of argumentation. Experience of argumentation in different contexts can equip students with the skills to make decisions about controversial issues in science, to understand how evidence is used to construct explanations and to understand the criteria that are used in science to evaluate evidence. Though the role of argumentation has become more highly valued in science education, research shows that only if it is specifically addressed in the curriculum and explicitly taught through task structuring and modelling will students gain the skills needed to explore its use in science and socio-scientific contexts (Erduran & Jiménez-Aleixandre, 2008; Osborne, Erduran & Simon, 2004a).

The development of argumentation activities in science education, and strategies for implementing such activities in science classrooms, has emerged alongside and been influenced by a global research programme into students' argumentation and teachers' professional development in argumentation. The work presented in this chapter has its origins in three UK-based research and development projects that grew from recognition of the importance of argument in science education (Driver, Newton, & Osborne, 2000) and a lack of discursive practice in school science in

---

S. Simon (✉)
Institute of Education, University of London, London, UK
e-mail: s.simon@ioe.ac.uk

M.S. Khine (ed.), *Perspectives on Scientific Argumentation*,
DOI 10.1007/978-94-007-2470-9_6,

the UK (Newton, Driver, & Osborne, 1999). The first of these projects focused on enhancing the quality of argumentation in school science (EQUASS) and was concerned with the development of argumentation activities by researchers in partnership with teachers. As teachers implemented the activities over a nine-month period, a quasi-experimental study showed that the quality of student argumentation improved, but that the change was not significant (Osborne et al., 2004a). The authors concluded that developing the skill and ability to argue effectively is a long-term process requiring many opportunities to engage in argumentation throughout the curriculum. A focus on the teachers' pedagogic skills in using argumentation activities also showed that though some teachers became more skilled at scaffolding argumentation, many found the enactment of argumentation in school science challenging and did not change their practice (Simon, Erduran, & Osborne, 2006). The second project was concerned with developing a pack of materials, called IDEAS, that consolidated the teachers' activity designs within a set of resources, each with learning objectives and a set of procedures (Osborne, Erduran, & Simon, 2004b). These resources were accompanied by training sessions for teachers to develop their strategies in using the activities.

Recent observations of teachers using IDEAS and an analysis of the interpretations required for effective use of the resources (Simon & Richardson, 2009) now suggest that the design and enactment of argumentation activity requires more attention and could be a critical factor in developing effective pedagogy of argumentation. The third project, called Talking to Learn in science (TTL), is currently studying the enactment of argumentation activities by groups of teachers working together in school science departments. These teachers are drawing on IDEAS resources and other sources of activities, and embedding these within the science curriculum for all teachers to use across the age range of students within the school. Observations of these teachers' practice confirm the complexity of the interpretive steps teachers need to take for effective enactment of argumentation activities (Simon & Richardson, 2009). Further work on activity design has taken place with some teachers from the TTL project; in this chapter we report on one study of argumentation design, interpretation and enactment to highlight the issues involved in transferring design between teachers.

Background

To promote the use of argumentation requires activities that are designed to achieve the cognitive and epistemic aims referred to above and an understanding of how these aims can be realised through student engagement. Argumentation activities set in social contexts can be the vehicle for developing students' epistemological understanding as with appropriate scaffolding by teachers and peers students can be encouraged to value the role of evidence in knowledge claims. Although we now have an established body of work on the value of argumentation and small group discussion in science education, few studies have attempted to unpack the nuances of how argumentation activities are designed (Howe and Mercer, 2007), as research has tended to focus on evaluating argumentation outcomes. Activities need to engage students' interest, stimulate discussion and provide resources that can be used as the basis of evidence in constructing and evaluating arguments, provide alternative choices or positions, involve a solution that is not obvious and be manageable for teachers.

Argumentation activities were developed in the EQUASS project by working together with teachers, according to the curriculum needs they identified and their interpretation of the meaning of argumentation. Frameworks were used for developing argumentation activities that involved the generation of differences, for example, presenting competing theories for students to examine and evaluate. In addition to providing stimulus material, the activities included small group discussion so that students would co-construct arguments justifying their case for alternative positions. As students require data to construct arguments, the activities also included information that could be used as evidence to support different theories. In this research nine generic frameworks were developed from literature sources (see Osborne et al., 2004a for the sources) that included concept mapping, designing an experiment, constructing an argument and different designs involving competing theories.

Building on this original research, the IDEAS pack was developed (Osborne et al., 2004b) to include a resource of 15 lessons that incorporate a variety of frameworks, including examples based on the frameworks introduced in the original research. One of the lessons is set in the context of a socio-scientific issue (SSI), requiring students to make a decision about the funding of a new zoo, and was intended to be accessible to most students aged 11–14 years. The framing of this lesson involves weighing up evidence for and against the new zoo and reaching a decision with reasons. The activity thus requires students to take a position and justify their claims with data, warrants and backing. The arguments for and against the zoo can be equally weighted, that is students can be asked to construct both sides of the argument thus forcing consideration of counter-claims, so the zoo activity can provide a good stimulus for argumentation involving rebuttals and counter-argument. The EQUASS research used this zoo lesson as a means of collecting data on teachers' developing pedagogy, so it was observed many times, at intervals with the same teacher and across 12 different teachers (Simon et al., 2006). Each teacher interpreted the lesson guidance and resource differently, so even though they worked from the same basic design, enactment showed differences in how they organised and managed group discussion, for example paired work, large group brainstorm or role-play. Teachers drew on and encouraged the use of different sources of evidence, from website information on breeding and habitats to students' own experiences and emotive responses. Teachers also interacted differently with students to scaffold argumentation processes such as justification and counter-argument. The findings showed that there was more variation in interpretation and enactment of the activity across teachers than for an individual teacher across time, even though there was evidence of developing pedagogic skills.

In our analysis of the IDEAS lessons (Simon & Richardson, 2009), we focused on a selection of lessons to examine the design framework, the science context used, lesson planning notes and the teacher's role. This analysis revealed the importance of purposeful design throughout the initial activity construction, to align the design framework with a suitable science context and create resources to help students reason argumentatively in that context. It also identified purposeful design by the teacher as a key process for ensuring effective enactment, both in interpreting the layers of the activity design to plan a specific teaching episode and in their role during teaching. The complexity of the interpretation required suggests a need for closer study of the design, enactment and transfer of argumentation activities.

Designing argumentation activities is an important aspect of supporting teachers in teaching argumentation (Osborne et al., 2004a). Teachers lack time to fully design their own argumentation activities, which often require more preparation than teacher-led lessons. When teachers draw on existing activities or share their own activity designs with other teachers, this time burden is reduced. Further, argumentation activities often serve an educative purpose, supporting teacher learning by outlining teaching strategies which are considerably different from standard practice (Davis & Krajcik, 2005). Since most argumentation activities are designed for other people to use, it is crucial to support the successful enactment and transfer of argumentation activities within the design itself.

Activity Design, Enactment and Transfer

We frame *activity design* as the creation of a tool which can be used by teachers to achieve their goals. Activities embed the author's purpose in representations which are used by teachers or students. These may include teaching notes, lesson plans and sample student responses as well as resources for direct use by students.

The initial design of argumentation activities involves several layers of purpose (Simon & Richardson, 2009). In some cases, the design begins by specifying learning aims for argumentation and science, considering the nature of the scientific knowledge involved. These aims are embedded in a student activity which supports the intended learning. Alternatively, the design may begin with a student activity, which is then analysed to identify potential learning outcomes in argumentation and science. In either design sequence, teaching notes are often layered onto the design, to provide procedural guidance intended to help teachers align their actions with the activity goals. To help teachers to interpret the activity, it is useful to share the rationale and purpose behind the activity design as well as the activity itself. This involves a shift from 'teacher-proof' activity design, which seeks to minimise teacher agency by providing highly detailed procedure-centric resources, to 'teacher-transparent' materials which support teachers in designing appropriate instruction by providing resources which can be used flexibly, and which are annotated to help teachers identify their affordances and constraints (Davis & Krajcik, 2005).

The *enactment* of an activity describes how teachers use an activity to design and implement instruction. The term 'enactment' acknowledges that practice involves the translation of beliefs, knowledge and experience into action (Clarke &

Hollingsworth, 2002). In enactment, the teacher draws on the argumentation activity, their existing practice and their intentions, knowledge and beliefs. Teacher use of argumentation activities therefore involves a collaborative design process where the teacher influences the use of the activity, and the activity design influences the teacher (Brown, 2009). Teachers' use of activities can be characterised according to the distribution of agency between teacher design and activity design. The teacher may *offload* design agency to the activity, relying on the design embedded in the activity to achieve their goals. As the teacher increases their design agency, they may *adapt* the activity, or *improvise*, creating their own design with the activity as a starting point (Brown, 2009). Since no activity design can completely specify classroom practice, the teacher always retains some agency in interpreting and reconstructing the activity (Ben-Peretz, 1990).

The successful transfer of activities between curriculum designers and teachers, or between different teachers, has been studied primarily as *fidelity to* or *adaptation from* the pre-existing design. However, the framework of distributed agency allows us to see both these options as principled design decisions by a teacher, and so focuses attention on how to help teachers creating principled teaching designs using the activities available to them. Teachers' agency in interpreting and adapting argumentation activities is supported by their pedagogical design capacity, their 'ability to perceive and mobilize existing resources in order to craft instructional episodes' (Brown, 2009, p. 29). For an individual teacher, this is likely to vary depending on their familiarity with 'reading' different types of curriculum material. For example, teachers who are new to argumentation may begin with minimal pedagogical design capacity for argumentation, and so may benefit from more detailed support in understanding the activity materials. This support can be provided through discussion with more experienced teachers or curriculum developers, included in the activity design via annotations or more extensive teaching notes, or provided via generic supports for analysing curriculum materials (Beyer & Davis, 2009; Brown, 2009). Since some teachers exhibit resistance to overt analysis of curriculum materials as they believe that good use of materials is often based on implicit and intuitive notions of teaching (Schwarz, Gunckel, Smith, Covitt, Bae, Enfeild, et al., 2008), the more subtle supports of discussion and teaching notes may be more fruitful in aiding the transfer process.

The Olympic Activity

Background to the Design

One of the authors, Ruth Amos, was asked to design an argumentation activity for an out-of-classroom learning initiative being led by the UK's Field Studies Council (FSC) in a small venue overlooking the construction site of the London 2012 Olympic Park in Stratford, East London. The FSC aims to bring groups of 11–14-year-old students to the site and provide them with opportunities to explore how it is being developed, including impacts on the local community and environment. The construction site was originally a busy industrial area, supporting small businesses, some residential housing and amenities such as allotments (local authority allocation of space for community horticulture) and local shops. The initiative runs from 2010 until 2013, therefore activity designs needed to encompass as many aspects of continuous site development as possible. The argumentation activity was intended to be part of a suite of science activities, alongside others such as those involving the physics of the structures in the Olympic stadium.

At the initial project meeting, with engineers and planners from the Olympic Delivery Authority (ODA), discussion centred on the ODA's strategy for making London 2012 the most sustainably constructed Olympic site yet. Their sustainability strategy has won a national award (London, 2012), and an impressive catalogue of sustainable construction practices has been presented to the public. However, there are some voices of discontent in the background, and the clean, green corporate image being portrayed by the ODA has been questioned to see if the claims being made are accurate and truthful (Slavin, 2006). The existence of these opposing sources of 'evidence' for the sustainable construction of the London 2012 Olympic Park offers the opportunity for the creation of a potentially engaging socio-scientific argumentation activity which would enable students to critically examine the case being made by the ODA for sustainable development. Such an SSI affords argumentation activity design involving multiple stances and opposition, similar to that of the zoo lesson.

Approach to Activity Design

An FSC tutor, Heather, with a geography background, was leading all the FSC activity sessions. Though she was an experienced fieldwork tutor, she had not led argumentation activities before and had received no training in supporting successful argumentation in a science setting where she would be teaching a whole class of students. She was apprehensive about the prospect and it became evident that the activity needed to be designed by Ruth to incorporate clear guidance on purpose and enactment, and on the nature of students' argumentation. The guidance in the IDEAS pack (Osborne et al., 2004b) provided a useful foundation for the creation of the Olympic activity.

The design of the activity was influenced by the findings of earlier research on the challenges faced by teachers using argumentation (Simon et al., 2006), the problems of interpretation of IDEAS (Simon & Richardson, 2009), and by Ruth's experience of training pre-service teachers in the use of argumentation activities. To build Heather's confidence a detailed approach was provided, including a list of resources to support initial engagement with the activity, evidence statements, tutor guidance and student guidance. Specific suggestions for activity stages, group design and outcomes were made (see Appendix 1). Thus procedural guidance was used to support low pedagogical design capacity for argumentation associated with lack of experience.

When designing a context for a socio-scientific argumentation, the inclination of students towards engaging with contexts and their ability to access evidence, either from their own experience or from a resource, needs to be taken into account. Issues around the sustainable building of an Olympic site may not be universally appealing, and students would need to have sufficient resources to stimulate interest and be able to take part successfully. Ruth therefore took a resource-centric approach by providing a range of supports which could be selected by Heather and other teachers according to their aims and student needs. She drew on aspects of activity design from IDEAS resources, including evidence statements to support different claims, role-play, and a decision-making brief. Strategies such as the use of argument prompts and writing frames were also included to support students in organising their thinking (see Appendix 2). As the designer, Ruth took the view that all the available resources would need to be used effectively during the session, but there was potential for them to be used in a number of ways (e.g. by giving students written or oral guidance). Ruth then organised a pilot session with a local teacher, Alan, and a class of 14-year-old students, all of whom were already experienced in the enactment of argumentation in school, in order to allow Heather to focus fully on developing her embryonic skills for leading such an activity.

Planning the Pilot Session

A planning meeting for the pilot session was held between Ruth, Heather and Alan, in which Ruth briefly outlined the initial purposes of the activity and the activity design was discussed. Subsequent design choices reflected the purposes of all three agents involved and how these were aligned. For example, Alan was already envisaging creating groups of three which blended students of differing confidence together, thus bringing his experience of group work into the design. He also anticipated students' perspectives with respect to the activity and considered drafting an introductory activity to 'inspire them beforehand' with recourse to the various website sources, thus building on his experience of needing to provide a stimulus for engagement. Heather gave Alan information about alternative useful websites (London 2012, Get Set), in doing so bringing her agency to the design based on her desire that students should look at both sides of the sustainability argument. Heather was conscious that all the official Olympic websites portrayed only the positive side of the argument.

Heather asked Alan what he felt were the important outcomes of the activity. This was an interesting question as it revealed a limited sense of purpose on Heather's part and an offloading response by Alan who asked 'why are you looking at me?'. This interaction probably arose from the use of a third-party design rather than something self-initiated by either Heather or Alan. Ruth as the original designer suggested one purpose might be for students to explore their understanding of the term 'sustainability' in the light of its very frequent, contemporary use by the media. She also hoped that students might be able to look 'behind' the evidence statements, particularly the pro-case, to contemplate what had been deliberately excluded. Both

Alan and Heather agreed to this suggestion, thus accepting and reinforcing Ruth's agency in the design. Another feature of the discussion was how the design included a distinction between types of evidence, as Ruth's earlier work with SSI had shown that students focus more readily on economic and social evidence rather than scientific evidence, and she had thus included a 'type of evidence' writing frame to explore this dimension. This purpose of the design would not have been clear to the teachers without the opportunity to discuss the intention behind it.

The design involved encouraging students in their groups of three to identify stronger pieces of evidence from statements provided, discarding others, and then to compare and argue for their position with another group of three using the evidence they kept. It was agreed that students would pick the 10 strongest pieces of evidence; however, Alan was concerned that the choice of evidence would be too consensual and students would have minimal differences once they came together from threes to sixes. He proposed that groups of three be assigned a position, that is either a pro-case or an against- case, and asked to identify the strongest pieces of evidence for their position, so that when threes came together as a group of six there would be the need for persuasion and counter-argument. In Alan's experience, students all tend to arrive at the same position quickly if left to their own devices, he also felt that students would not be challenged if following their natural inclination towards a particular position. Heather listened carefully to this experience and was keen to learn from it. She and Alan continued to discuss further aspects of the activity design, Alan focusing on small group procedures and outcomes, Heather on organizational features such as timings and phases of the activity. Eventually they agreed on a design based on his pedagogical experience and her desire to create an activity that would be effective for her purpose. Alan made the point that any argumentation activity tends to develop itself depending on what the students bring up, thus demonstrating the improvised aspect of design based on responses to the situation.

As a result of her planning discussions with Alan, Heather adapted the student guidance sheet for the activity to facilitate the pre-assignment of a position either 'for' or 'against' for the small groups of three, as well as to direct them to reach agreement upon the 10 'strongest' pieces of evidence to support this position. She also prepared an introductory session of PowerPoint slides for students based on the question 'A Sustainable Olympic Park?' drawing on geographical, historical and social issues relating to the site. Heather invested much of her pre-session preparation time in finalising the resources and in the preparation of her introductory scene-setting role. The focus of her preparations accords with one of her main purposes in the design of the activity, to provide interest in the topic and to have the procedures well-structured for small-group discussion. In this way she would rely on the resources to guide her through the lesson, thus offloading design agency to the activity during enactment. Her reliance on the resources became more apparent through her preparation of 'student packs' for the session, which included 30 evidence cards, labelled 'for' or 'against', so that students were told how the evidence supported each position, and colour-coded for altering group composition according to positional role-play. This aspect of the design added to the structure Heather thought necessary for controlling small group discussion and role-play.

Alan set up a group of 27 students from several different teaching groups (13–14-year-olds) who were experienced in argumentation. He had arranged for the group to meet three days before the session and had asked them to select their own groups of three for the activity. This meeting gave the students a chance to look at some of the Olympic websites that showed different sources of evidence to support, and to challenge, the issues around sustainable construction of the site. Alan gave the students a short worksheet of information about the session, which included an instruction to prepare their own understanding of 'sustainability' in relation to construction. Alan's preparatory focus thus adopted the resource design created by Ruth in terms of exploring their understanding of sustainability and using different sources of evidence. It was also aligned to his own purpose of focusing on the nature of evidence the students might use in their argumentation.

The Pilot Session

The argumentation session was video-recorded by Ruth, who observed the session and also made field notes. Heather introduced the session, reviewing its purpose, the issue of sustainability and the role-play context. She built a structure to the session that would enable her to rely on the activity itself to facilitate smooth running of the session once argumentation began, which initially focused on providing interest to the students, an aspect of the design where she exercised high agency. She took the students out on to a balcony overlooking the site, then showed them photos and video-clips of people talking about the site, for example from people who would lose their allotments and local businesses who were being relocated. Students were interested and attentive throughout the introduction.

After the scene setting was complete, Heather gave out the student packs, which the students opened immediately to discover whether they were 'for' or 'against' the sustainability issues, and she explained that they had to develop their argument using 10 pieces of evidence. From the start, all students were actively discussing evidence with one another. Some of the groups of three also started listening to, and counter-arguing with, another group of three who were focusing on an opposite position. Thus students of their own accord began the next stage of the activity, a common feature we have observed in highly structured argumentation lessons. Heather's resource-reliant approach meant that she subsequently introduced the use of a worksheet or writing frame to explore the evidence further. This was a rather mechanistic and unnecessary embellishment of what the students had just accomplished through discussion. The students completed it very quickly and two groups began to lose concentration. Heather became concerned that the students were working too fast and she moved on to the opposing arguments stage. Heather's design was to use the colour-coded cards to set-up oppositional groups, an aspect of the design that did in fact afford an almost seamless, uncomplaining movement of students, something she had been concerned about before the session. Once everyone was re-positioned, students immediately started to relate their positions to one another. The debates became quite heated at times and students started to use personal evidence and opinion, that is, not presented on the evidence cards. Heather tried to focus students' attention on the provided evidence, as well as asking them to think about the accuracy of their own sources of personal evidence, thus showing some doubts about offloading agency to students.

After a period of fairly heated debates within groups of six, Heather gave each group two minutes to present their consensus position via a selected spokesperson, praising their 'fantastic debating skills', and invited them to think of questions as they listened to each group. Whilst arguments were presented by each group, Heather:

- answered any questions raised;
- encouraged applause and praise 'now this group lasted longest in their attempt to convince their opposition, well done, but . . .';
- agreed with evidence cited 'yes, that's a very interesting one about moving animals . . .';
- embellished evidence 'but the newts and frogs will just make their way back to where they were before, so. . .';
- asked questions in response to evidence cited 'well, how many trees were cut down to make way . . .?' to model countering a group citing the positive effects of the proposed new tree-planting on site;
- encouraged some counter-argument 'anyone else, points 'against'?'

Very few students asked questions or raised counter-arguments during this stage. Heather invited anyone who was originally on an 'against team' to make a strong point against the 'pro-sustainability' argument. No one responded, as students had inclined mostly to the 'for' position, finding the evidence for the 'against' position to be weak and based on peoples' opinion. Heather then summarised some of the shorter- and longer-term goals for sustainability from ODA sources.

Discussion

Through the design, planning and implementation of this Olympic argumentation activity, we can see how the framework of distributed agency can be used to interpret actions of the designer, (Ruth), new tutor (Heather) and experienced teacher of argumentation (Alan), and how those actions have implications for the way in which an activity is enacted with students. One key issue is how purpose, or intended learning outcomes, guide the actions of each person during design, planning and implementation stages. Ruth had a range of purposes informing her resource-centric approach to designing the argumentation activity. Her concern was to create resources and guidance that could be effectively used by a tutor inexperienced in facilitating argumentation. She also wished to include a role play scenario for the activity, with appeal for students coming from anywhere in the UK, and to provide prompts and writing frames for students based upon IDEAS to enhance argumentation and debate. Both Heather and Alan offload agency to the designer and the activity with respect to identifying learning outcomes, but accept more agency with respect to design choices for scaffolding argumentation, such as assigning stances and providing ways for groups to learn from others' discussions. The resistance to clearly identifying learning outcomes suggests that Heather and Alan assume these are either embedded in the activity design, or they expect them to be provided by the designer.

The purposes of both Heather and Alan became more apparent during the enactment of the argumentation activity. During small group discussion Alan became more focused on encouraging students in their approaches to argumentation. Whilst circulating amongst the small groups of students, he continuously:

- embellished the provided evidence;
- explained the scientific and technological processes being used in the construction processes;
- rephrased students' developing arguments 'so what you are saying is...'
- prompted students to justify their arguments, using phrases such as 'what if...'.

Thus Alan's contribution to the design was focused on the nature of the evidence and its use in providing justification for the students' position. In contrast, Heather continuously:

- kept an eye on the time, and communicated time targets;
- encouraged by nodding, smiling, praising and injecting a little humour;
- focused students on their assigned positions and on one of the goals of Task 2 'remember you are trying to decide on the 'strongest' 10 pieces of evidence for if you are 'for' or 'against'';
- shared progress 'ok, we have a couple of groups who are down to their final 10...';
- answered task-related questions.

Heather's contribution was focused on the structure of the session and the control of the discourse, rather than the nature of argumentation. Her primary purpose was managing the process of group work, while Alan's was improving the quality of argumentation. The contrast between the novice and the experienced teacher of argumentation is striking and shows the importance of building experience over time in the teaching of argumentation, in order to purposefully include scaffolding of argumentation processes. Alan's design choices are much richer because of his knowledge of argumentation and the students – he anticipates students' responses. During the episode where students had formed into larger groups with opposing positions, Heather continued to focus on procedural aspects and also basic scaffolding processes such as:

- reminding students to listen well to one another;
- defining the process: 'you argue 'for', you argue 'against'';
- reminding them to persuade their opponents;

- focusing them on the evidence;
- embellishing upon evidence;
- keeping an eye on the time, but less so than in the first stage.

Alan, on the other hand, concerned himself with modelling how both sides could go back and forward across the debate, countering each others' arguments by suggesting 'come on, offer a piece of evidence against that'. He continued to adopt the role he had taken in the first argument phase, in addition focusing on:

- giving more encouragement to considering counter-arguments to their own position;
- encouraging students to consider the validity of their own arguments 'so what you are saying is that the soil was already contaminated, but now the contamination is concentrated into a smaller area, so that's better?';
- reminding students about the rules for effective small group discussion 'excuse me, listen to . . . .'.

To explore how both Heather and Alan reflected on the activity afterwards, Ruth conducted telephone interviews with each teacher. Heather expressed a positive reaction overall and was pleased with how the students engaged. When Ruth pointed out the limited time spent in whole-class debate, Heather expressed less confidence about encouraging prolonged counter-argument with the whole class, which is why she had not adopted suggestions made by Alan for students recording their final argument and presentations on posters for viewing and discussion. Heather suggested adapting the design by introducing 'blank' cards to allow students to input their own personal evidence. Heather could see the potential, through her experience of the argumentation activity, for marketing sessions in a more cross-curricular way involving geography, history and citizenship. Teaching the activity had helped her to articulate a better way forward for her project design, thus creating new ways in which she would exercise her design capacity. Alan thought that all the students definitely needed time to 'get into the evidence' though he felt that the language was appropriate and ideas accessible. Even though he had brought a group experienced in argumentation, his view was 'learning to argue is only part of the jigsaw of their development in science and all the other things they need to do'. He had perceived his role as encouraging students to think about the argument they were building by directing the students back to the original evidence cards at all stages and anticipating counter-arguments and rebuttals. He had witnessed even the four/five 'good debaters' were slow to do this. Though students were definitely engaged with the activity throughout, they were still facing challenges in terms of demonstrating higher-level argumentation skills. Alan was a little disappointed that Heather did not take up his suggestion about poster presentations as in his view better opportunities for counter-arguing would have ensued. Though students had needed a lot of time to get their own arguments sorted out, Alan knew from experience of argumentation in school that plenary strategies, with a reasonable time allocation, are needed to explore thinking. Thus we can see a contrast between the novice teacher,

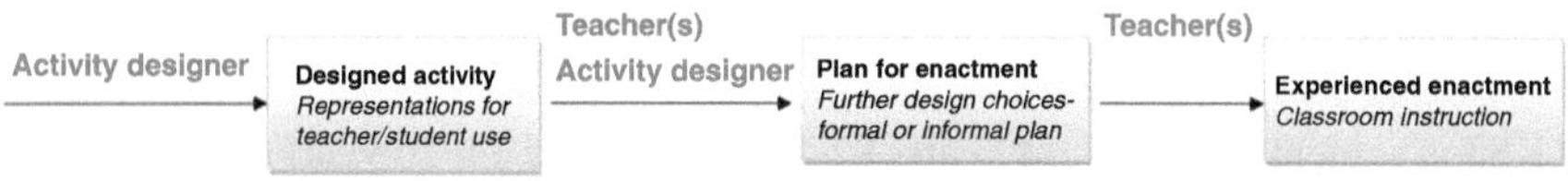

Fig. 6.1 Purpose into practice: Design, enactment and transfer diagram

Heather, and the more experienced teacher, Alan, in their reflection of salient outcomes (Clarke & Hollingsworth, 2002), as Heather's was focused on engagement, whereas Alan's was on higher-level argumentation.

From our interpretations of the Olympic activity we have summarised the process of design, enactment and transfer that took place in this study. In Fig. 6.1 design outputs are represented by boxes, and the processes between them are represented by arrows showing who was involved at each stage of the study.

The first stage draws on the designer's knowledge of the teaching context, learning aims and pedagogical content knowledge. The output is a designed activity, which usually includes representations for students to work with, such as evidence cards or writing frames, and representations for teachers to use directly, such as teaching notes or lesson plans. In the second stage, teachers use their pedagogical design capacity to transform the activity design into a plan for an instructional episode. The teacher may significantly modify the activity, and if the activity serves an educative purpose for the teacher, the activity design may significantly influence the teacher to modify their normal practice. Even if the activity is not substantially modified, this stage usually involves further design choices to help teachers craft the lesson, such as deciding how to group pupils, when and how to present evidence, and how different groups might learn from each other. In the case of our study, two teachers co-planned the teaching, with the designer in attendance at the planning meeting. The output is a plan for enactment, which may or may not be formally represented. The final stage is the classroom enactment of the argumentation activity. Design choices at this stage may be triggered by student responses to the activity or unanticipated constraints such as a shortened timeframe, or may arise spontaneously. The design output is the instructional episode itself, as experienced by the teacher(s) and students.

The message from this small-scale study for designers of argumentation activities is that teachers need careful guidance, including clarity of authorial intentions. Unless these are flagged – for example the purpose of sorting evidence into different types, such as economic, social, environmental – teachers may not enact the activity as intended. Whatever guidance is provided, a certain level of pedagogical design capacity is required for teachers to realise the potential of their own design choices even in 'ready-to-go' activities. In our case, Alan problematised student grouping and how to ensure multiple stances, in doing so adding further layers of design to the lesson structure. Heather is less experienced and does not initially pick these out as issues. Thus annotated guidance will need careful construction for inexperienced teachers of argumentation.

Appendix 1

(*KS3 is students aged 11–14 years, KS4 students aged 14–16 years)

The Legacy of the Olympic/Paralympic Games London 2012: The Story of the Sustainability of the Olympic/Paralympic Stadium

Activity Teaching Notes – Second Draft Version

In order to engage students in an argumentation activity looking at environmental impacts and potential sustainability issues, questions need to be raised which encourage them to defend/argue about positions for and against such an endeavour. The outline below is intended for middle–higher attaining students in KS3*. It can be modified for lower attainers (simplifiying statements, reducing the number of statements, using writing frames, using images) and particularly for KS4* (by bringing in 'late evidence', by playing devil's advocate etc.).

Setting the Scene

The London 2012 Olympic and Paralympic Games will take place from 27 July–12 August and 29 August– 9 September respectively, a total of 4 weeks.

Twenty-six sports in the Olympics and 20 in the Paralympics will take place.

The estimates for the final costs of the Olympic Park are, at the time of writing (January 2010), running at between £3 and 9 billion.

The Olympic Delivery Authority (ODA), the organisation responsible for building the facilities and so on, say that London 2012 is aiming to set 'new standards' of sustainability and to 'create positive, lasting changes for the environment and communities'.

Possible Scenario Questions

Does holding the Olympic/Paralympic Games make environmental sense?

Is the Olympic Park being built in a sustainable way?

Learning Objectives/Goals

The learning objectives for the students are to:

- explore the nature of materials being used to construct the Olympic Park;
- recognise the environmental impacts that a project such as this may have;
- distinguish between scientific, environmental, social and economic evidence when constructing an argument;
- construct arguments to justify their position with respect to the progress of the building of the Park from an environmental impact perspective.

(Note: The potential outcomes are in line with current science NC PoS at KS3 and KS4 for How Science Works and Assessing Pupils' Progress (APP) – AF2 Implications and Applications particularly.)

Teaching Sequence (2 Hour Session)

Students begin the activity by looking out over the Olympic site and undertaking the 'what do you notice most/most interesting/spot the . . . .' activity.

- Introduce the argumentation activity – Story of the Sustainability of the Olympic Park (whilst looking out over the site). What impressions do they have of potential environmental impacts and so on?
- Divide students into scientific adviser teams of three.
- Students watch the PP/video sequences to set the scene (when available), possibly with a prompt sheet for noting interesting aspects.
- Distribute and go through the activity handouts, telling the students that their task is to decide whether or not the Olympic Park is being built in a sustainable way (some discussion of what this means will be needed).
- Explain to the students that they should provide reasons for choosing their most important statements, supporting or challenging the sustainability claims being made by the ODA. The group should discuss the reasoning behind their choices and put together a coherent argument. One person in the group acts as scribe; one could put forward positive arguments, the other negative arguments and all to try to anticipate counter-arguments. Alternatively, all the evidence cards could be 'dealt' out so that all three in a group have some and they then go about discussing the various pros and cons and so on (Various prompts/scaffolds can be used if groups need them). Allow 20 minutes for the discussion.
- At the end of 20 minutes, ask the groups to try to decide which pieces of evidence are scientific, environmental, social or economic (some definitions need to be agreed here) and get them to categorise the evidence they are focusing on most strongly.
- The groups of three join another group of three (threes to sixes) and put forward their arguments. Where do they agree, disagree? Which 'strongest' evidence have they chosen? The larger group needs to formulate an agreed argument to present to the rest of the class; they will be able to speak for 2 minutes. Allow 15 minutes for the discussion/decision.
- Ask students to choose a representative who will present their case to the class.
- Run the presentation of arguments (high-attaining students can try to respond to the previous groups as it progresses). Allow 15 minutes for the debate.
- Finish by conducting a plenary discussion on the outcome(s) of the debate. Did groups agree; where/why did they differ? Do they recognise the kinds of evidence that they were drawing upon (scientific, environmental, social, economic)? Try to encourage a class consensus as to whether the ODA's sustainable development claims are justified.

A possible activity sheet for the students is outlined below.

London 2012 – Sustainable Development or Not?

The activity – student version

You are a team of three scientific experts advising on the building of the Olympic Park. The construction is now well underway and you are looking out on to the Olympic Site with your colleagues, trying to decide whether the Olympic Development Agency (ODA) are committed to sustainable development. Just how environmentally friendly is all the building? Is it really justifiable to spend all that money for an event that will only last about 4 weeks in 2012?

Your task

You have to compile a short scientific report for the Government (Department for Culture, Media and Sport) showing whether or not the Park is being built in a sustainable way, with the intention of making the impacts on the environment, and local people, as positive as possible.

You will work as a team of three initially; listen carefully to the instructions, some of which are summarised below.

- Look out over the Olympic Park and/or watch the video/PP presentation to gain a sense of all the building work that is going on at the moment across the Olympic Park.
- Examine the evidence that you have gathered, on the evidence cards, about the use of materials across the Park, the claims of the construction companies, the thoughts of local people and so on.
- Divide up the evidence cards between you. Decide in your team how to build your argument as you decide whether the Olympic Park is being built in a sustainable way. What does 'sustainability' mean here?
- Decide who is doing what – perhaps one of you select all the evidence 'for' sustainability and someone else select the evidence 'against' sustainability; the third person could be the writer, recording your decisions and the building of your argument. Or you could divide all the evidence up randomly first between you and then everyone decide on the pros and cons of some of the evidence.
- Record the start of your argument on Evidence Sheet A.
- Think about the following questions:
  - Where do you agree, disagree?
  - Which evidence is the 'strongest' you have chosen 'for' and 'against' sustainable development? Which evidence is the most convincing to you?
- Using Evidence Sheet B, sort your evidence into 'scientific', 'environmental', 'social' or 'economic' and see which seems to be the strongest – perhaps use different colours to show the different types of evidence.

When instructed you will join another team of three and try to persuade them that your arguments 'for' and 'against' the sustainable development of the Park are good arguments.

Select a spokesperson to present your case.

- In your team of six, try to reach agreement about the strongest arguments 'for' and 'against' that you have, and prepare a short presentation, choosing a spokesperson, to give your overall decision to the DCMS at the end of this session. In other words, you now all have to decide whether the Park is being built in a sustainable way, or not, based on the evidence you have discussed and what you have seen today.

Either write out your own final argument, or use Evidence Sheet C to help you.

Appendix 2

Argumentation Prompts (Adapted from the IDEAS Project, Kings College)

- What makes you think that?
- What is your reason for that?
- Can you come up with another argument for your point of view?
- Can you think of an argument against your point of view?
- How do you know that?
- What is your evidence for ...?
- Why do you feel that ... is the most important evidence?

Evidence Sheet C – Your Final Argument London 2012: Sustainable Development or Not?

Our Argument ... Team ....

Our argument / position is that ....

Our reasons are that ....

Arguments against our idea might be that ....

We would convince somebody who does not believe us by ....

The evidence we would use to convince them is that ....

London 2012 Olympic Park Sustainability Evidence Cards

Rain water will be collected from the roof of the Velodrome to use to flush toilets inside and to water plants and trees	No materials leave the site if they can be re-used or recycled here
10,000 tonnes of steel will be used in the main stadium compared to 40,000 tonnes used in Beijing 2008	Two camps of travellers (gypsies) had to move when the Olympic Park site was chosen
The dust raised during the building works is a constant potential hazard for local residents	The 80,000 seat Olympic stadium will be reduced to 25,000 afterwards, and the steel will be re-used
60% of all the original materials on the site have been used again or recycled (bricks, cobbles, man-holes)	All the soil on site is being washed at the on-site 'soil hospital' – oil/contaminants are shaken free
Washed soil from the site is being used to landscape the area; 4,000 new trees will be planted	Some old businesses did not want to move from the site in 2005 and 2006
The site was contaminated with metals like mercury (from an old battery factory)	The bridges across the site will be made smaller after the Games and materials re-used
Radioactive material was found on site; it probably came from an old watch and clock factory	The radioactive material on site was sealed and buried under one of the bridges, rather than taking it away
Health and safety is very high on the agenda on the Olympic site	The energy centre on the site will use biomass boilers to generate some of the site's electricity
A new 120 metre high wind turbine will generate some of the site's electricity	The ODA hope 20% of all the energy used in the Olympic Park will be generated from renewable resources
To reduce carbon emissions, 50% of all the building materials being delivered to the site are coming by train	Some waste materials are being removed from the site by boat to reduce carbon emissions
In November 2009, just 2 of 9 sites being monitored for increased dust levels around the Park showed greater than normal levels on 2 days or more	New cycle routes and paths are being built to encourage people to walk or cycle to the Games
All the timber (wood) being used on site is coming from sustainably managed forests	Originally, there were about 500 objections raised to the building of the Olympic site by local people, businesses and environmental groups
Existing wildlife habitats are at risk from being disturbed by the building work for at least 5 years	Contractors are moving thousands of newts and hundreds of frogs to new habitats on site during building work
Part of Hackney Marshes (currently football pitches) will become a huge car park for the Games	People in Clay Lane, Marshgate Lane and Dace Lane (3 local roads) had to be re-housed to make way for the site
Each year 50–75 people die in the UK in accidents on building sites. So far, no one has died on the Olympic site	Allotments originally on the site were moved to another space, but the soil wasn't good enough to grow vegetables there. Some people lost their livelihoods as a result

References

Ben-Peretz, M. (1990). *The teacher-curriculum encounter: Freeing teachers from the tyranny of texts*. Albany, NY: State University of New York.

Beyer, C., & Davis, E. A. (2009). Supporting preservice elementary teachers' critique and adaptation of science lesson plans using educative curriculum materials. *Journal of Science Teacher Education, 20*(6), 517–536.

Brown, M. (2009). The teacher-tool relationship: Theorizing the design and use of curriculum materials. In J. Remillard, G. Lloyd, & B. Herbel-Eisenmann (Eds.), *Mathematics teachers at work: Connecting curriculum materials and classroom instruction*. New York: Routledge.

Clarke, D., & Hollingsworth, H. (2002). Elaborating a model of teacher professional growth. *Teaching and Teacher Education, 18*, 947–967.

Davis, E. A., & Krajcik, J. S. (2005) Designing educative curriculum materials to promote teacher learning. *Educational Researcher, 34*(3), 3–14.

Driver, R., Newton, P., & Osborne, J. (2000). Establishing the norms of argumentation in classrooms. *Science Education, 84*(3), 287–312.

Erduran, S., & Jiménez-Aleixandre, M. (Eds.). (2008). *Argumentation in science education*. Heidelberg: Springer.

Howe, C., & Mercer, N. (2007). *Children's social development, peer interaction and classroom learning* (Primary Review Research Survey 2/1b). Cambridge: University of Cambridge.

Kuhn, D. (1991). *The skills of argument*. Cambridge: Cambridge University Press.

London 2012 Olympic Games website http://www.london2012.com/news/2009/11/london-2012s-green-build-on-track.php. http://getset.london2012.com/en/home

Newton, P., Driver, R., & Osborne, J. (1999). The place of argumentation in the pedagogy of school science. *International Journal of Science Education, 21*(5), 553–576.

Osborne, J., Erduran, S., & Simon, S. (2004a). Enhancing the quality of argument in school science. *Journal of Research in Science Teaching, 41*(10), 994–1020.

Osborne, J., Erduran, S., & Simon, S. (2004b) *Ideas, evidence and argument in science*. In-service Training Pack, Resource Pack and Video. London: Nuffield Foundation.

Sandoval, W., & Reiser, B. (2004). Explanation driven inquiry: Integrating conceptual and epistemic scaffolds for scientific inquiry. *Science Education, 88*(3), 345–372.

Schwarz, C., Gunckel, K., Smith, E., Covitt, B., Bae, M., Enfield, M., et al. (2008). Helping elementary preservice teachers learn to use curriculum materials for effective science teaching. *Science Education, 92*(2), 345–377.

Simon, S., Erduran, S., & Osborne, J. (2006). Learning to teach argumentation; Research and development in the science classroom. *International Journal of Science Education, 28*(2–3), 235–260.

Simon, S., & Richardson, K. (2009). Argumentation in school science: Breaking the tradition of authoritative exposition through a pedagogy that promotes discussion and reasoning. *Argumentation, 23*(4), 469–493.

Slavin, M. (2006). *Olympic Games Monitor website*. Retrieved February 15, 2010, from http://www.gamesmonitor.org.uk/

Chapter 7
Argumentation and Reasoning in Life and in School: Implications for the Design of School Science Learning Environments

Leah A. Bricker and Philip Bell

Many science education scholars, predominantly using Toulmin's argumentation framework (1958/2003) as a design template, have created learning environments to engage youth with what it means to argue scientifically (Bell, 2004; Clark & Sampson, 2008; Kelly & Bazerman, 2003; Kuhn, 1992, 1993; Kuhn, Kenyon, & Reiser, 2006; Newton, Driver, & Osborne, 1999; Osborne, Erduran, & Simon, 2004; Sandoval, 2003). We argue, using Tilly's (2006) framework that categorizes people's reasons, that there is also promise in utilizing the everyday argumentative competencies of youth as a design template in argument-focused school science curricula. In this chapter, we synthesize our work on argumentation. As part of his dissertation research, Bell created an argument-mapping environment called SenseMaker and then studied middle school science students' use of SenseMaker as part of conceptual change instruction over six classroom design experiments. As part of her dissertation research, Bricker used data from a long-term team ethnography of youth science and technology learning across settings and timescales to examine youth's everyday argumentative practices. After explicating our research, we then describe design-based research that is needed to investigate young people's appropriation of scientific argumentation using their everyday argumentative competencies as a springboard.

> Because you always, always, always have to explain your yourself.

The reason that opens this chapter was uttered by a middle-school youth who was a participant in the long-term, team ethnography that Bell, Bricker, and colleagues conducted to explore young people's science and technology learning across settings and over time (see Bell, et al., 2006, in press; Reeve & Bell, 2009). Because Toulmin has played such an important role in shaping the learning environments in school science that researchers have developed to engage young people with what it means to argue scientifically (see Bell, 2004; Clark & Sampson, 2008; Kelly & Bazerman, 2003; Kuhn, 1992, 1993; Kuhn et al., 2006; Newton et al., 1999; Osborne

L.A. Bricker (✉)
University of Washington, Seattle, WA, USA
e-mail: lbricker@u.washington.edu

M.S. Khine (ed.), *Perspectives on Scientific Argumentation*,
DOI 10.1007/978-94-007-2470-9_7,

et al., 2004; Sandoval, 2003), we begin with Toulmin and colleagues' ideas about reasoning. Toulmin, Rieke, and Janik (1984) ask: "What does it mean to ask if someone's statement or argument or advice is sensible or well reasoned, sound or logical? What do these demands for 'good reasons' and 'sound arguments' amount to? And how are we to judge this kind of goodness and soundness? (p. 4)." Toulmin and colleagues define reasoning as a central activity in claim-making. They take care to acknowledge that reasoning differs from situation to situation and what counts as a good reason is dependent on situation and context.

In his book *Why?: What happens when people give reasons...and why* (2006), Charles Tilly outlines a framework for categorizing people's reasons and notes the necessity of paying attention to context and power relations when analyzing reason giving. He explicates four overlapping broad categories of reasons: (1) conventions; (2) stories; (3) codes; and (4) technical accounts. *Conventions* are commonplace reasons that are almost always accepted but do not necessarily involve adequate causal accounts and do not claim to provide them. Examples include, "I am tired" and "I am not good at this." As Tilly notes, "Conventions claim, confirm, repair, or deny social relations...[and] therefore differ greatly depending on the social relations currently in play" (p. 16). *Stories* are narratives that are explanatory reasons, usually of extraordinary events or phenomena that are unexpected. Tilly gives examples that include the events of September 11, 2001 and the phenomenon of running into a high school classmate in Egypt at the Great Pyramid of Giza. Stories work to diminish cause and effect so that events are simplified, provide subtle justifications, blame, and so on, and minimize any possible impact of factors such as environment and error. However, the claims put forth in stories are a bit stronger than those implied by conventions.

Codes are reasons that embody actions that govern society, such as legal documents, religious doctrine, and government policy. Codes do not have to have explanatory qualities because they typically convey that rules are being followed. As Tilly describes, "...for those who play the game, codes have an air of inevitability, even of sanctity" (p. 18). Lastly, *technical accounts* are reasons that vary greatly depending on institution, field, and so on (and then with respect to their structure and content) but they claim reliable and durable ties to cause and effect. Tilly explains that entire professions and their associated professional knowledge serve as warrants for technical reasons. Tilly diagrams the reason space he envisions as follows (p. 19):

	Popular	**Specialized**
Formulas	Conventions	Codes
Cause-Effect Accounts	Stories	Technical Accounts

He explains his diagram with the following text:

> From left to right, the diagram represents the extent to which ordered, disciplined, internally coherent schemes dominate reason giving, with "popular" reasons being widely accessible, and "specialized" reasons relying on extensive training in the discourse. Top to bottom, the diagram runs from X-to-Y matching, in which criteria of appropriateness rather than causality prevail (formulas), to tracing of causal processes from X to Y (cause-effect accounts). Obviously, the scheme orders claims made by givers and/or accepted by receivers rather than any judgment of their adequacy by third parties, including you and me. (p. 19)

Tilly notes that when people give reasons, relational work is ultimately accomplished whether that relational work falls into confirmation of, establishment of, negotiation of, and/or repair of relationships between people. The coding of reasons as appropriate or acceptable varies widely with respect to relationship and context. Tilly explains: "When someone offers you codes or technical accounts in unfamiliar idioms, you rapidly choose between two interpretations [with respect to your relationship with that person]: either this person has misunderstood the relationship between you, or she is claiming superiority and demanding deference by virtue of esoteric knowledge (p. 180)." With respect to this idea, one can draw parallels to school-based studies examining the various categories of social and intellectual disconnect between teacher and student, as well as what school-based discourse conveys to some students (Brice Heath, 1983; Nespor, 1997; Sarangapani, 2003; Willis, 1997/2000). In other words, one can examine the types of reasons used in schools to study power hierarchies, hidden meanings, and social disparities. We return to this issue with respect to argumentation in our concluding remarks. Next however, we briefly explicate our conceptions of argumentation and where we believe reasoning fits into the process.

Argumentation and Reasoning

Keith and Rehg (2008) explicate understandings for the concepts of "argument" and "argumentation." An argument is usually understood as a product and reasons, evidence, and justifications are the material that substantiates that product. In the sciences, that product might be a journal article, technical report, or a conference paper. As Keith and Rehg state, "We can...imagine argument as circulating – as a set of texts and utterances that circulate through society, in different forms and modalities, modifying and being modified as they go" (p. 212). This is an image that they attribute to Warner (2002) and one that they note is also found frequently in Latour's text (Latour, 1987). Argumentation then is the process of constructing arguments. Far from being a decontextualized process, however, argumentation is situated and as Toulmin (1958/2003) notes, it is field-dependent. Toulmin called for anthropologic and historical studies of argumentation in order to explicate a contextualized conceptualization of argumentation.

Keith and Rehg explain that some theorists (Tindale, 1999; Wenzel, 1990) further dissect the concept of argumentation to include not only process (i.e., the process of communicating an argument as it unfolds temporally) but also procedure.

They describe argumentation procedure as "...a discursive structure that normatively guides a process, determining (in part) the order in which participants speak or communicate, the allowable or relevant content at each stage, role divisions, and the like (e.g., trial procedures...)" (p. 213). In this sense, argumentation is akin to Tilly's concepts of reasons as codes and technical accounts; governed by in-field norms and hierarchical relationships.

Because Toulmin's micro-structural account of argumentation is so heavily leveraged in the science education community, we turn now to his ideas. Even though Toulmin developed his ideas about the structures of arguments in the context of legal argumentation, his framework has been widely generalized (Bricker & Bell, 2008). Toulmin's (1958/2003) conception of an argument's structure is as follows: One has data (D) so one is able to make a qualified (Q) claim (C), unless one's qualified claim is rebutted (R). One is able to connect D and Q,C because of a warrant (W) that has been backed (B). Toulmin gives the following example of this argument structure: "Harry was born in Bermuda [D] so presumably [Q], Harry is a British subject [C] unless both his parents were aliens/he has become a naturalised American/...[R]" (p. 97). In this example, one is able to connect the data that Harry was born in Bermuda with the qualified claim that Harry is, presumably, a British subject because of a warrant that explicates the fact that "a man born in Bermuda will generally be a British subject" [W] (p. 97) and furthermore, one can back that warrant by noting that the warrant is reasonable because of legal provisions, statues, and so on.

Toulmin's micro-structural account of argumentation is an example of the procedural accounting of argumentation discussed above and thus affiliated with reasons that are codes and technical accounts in Tilly's framework. This is important because when Toulmin's framework is applied to everyday argumentation, the everyday argumentation tends to look impoverished because some of Toulmin's structures are missing. However, Simosi (2003) for example, claims that this is to be expected because everyday argumentation is of a qualitatively different kind than the technical argumentation Toulmin built his model to explore. Simosi notes, "These elements [Toulmin's structures] may be missing because the arguer considers them to be well-known – or assumed – by his interlocutor, and, thus, he does not regard it necessary to refer to them explicitly in his attempt to persuade the other" (p. 188). Returning to Tilly, everyday argumentation relies much more heavily on reasons situated in the realms of conventions and stories and as we have discussed above, conventions and stories are of a different kind than codes or technical accounts. One could argue that it is nonsensical to analyze and evaluate them utilizing a framework developed to highlight reasoning situated in codes and technical accounts.

To highlight some of the ideas above, we now turn to a discussion of our dissertation research. We both wrote our dissertations about youth argumentation. Bell's design-based research was situated in middle school science classrooms and involved the design of a computer program called *SenseMaker* to scaffold students' abilities to coordinate evidence with theory. Bricker's ethnographic research was situated in the everyday lives of youth and involved documenting their everyday argumentative practices and their perceptions of those practices. Bell's research is an exemplar of engaging youth with reasoning in Tilly's codes and technical accounts categories; specifically, helping them argue like scientists using *SenseMaker* as a tool to guide them in their learning about how to build Toulmin style structural arguments. Bricker's research is an exemplar of how reasoning in Tilly's conventions and stories categories looks and sounds in everyday life. After summarizing our dissertation research, we discuss design principles that could inform design-based research needed to investigate young people's appropriation of scientific argumentation using their everyday argumentative competencies as a springboard.

SenseMaker: Scaffolding Students' Scientific Argumentation and Reasoning

At the time of Bell's dissertation research, various researchers were constructing computer-learning environments to support discipline-specific collaboration and learning, making significant use of the emerging networked information infrastructure (Pea and Gomez, 1992). The Knowledge Integration Environment effort was funded to explore educational possibilities of the Internet in science education. Those involved in this work developed and studied project-based experiences where students engaged in various epistemic practices (debate, critique, design) and used online information resources that were treated as scientific evidence (see Linn, Davis, & Bell, 2004 for a summary of the research). Within the context of this effort, Bell focused on the scaffolding of argumentation and debate in a middle school physical science context. While recognizing the need for and possibilities associated with constructing a knowledge representation tool to support students in argumentation—and drawing inspiration from early work on the Belvedere system—Bell created an argument mapping environment called SenseMaker (Bell, 1997). He then studied it over six classroom design experiments (Bell, 2004).

Over the course of these six iterations, Bell designed a curricular and instructional intervention around SenseMaker to scaffold middle school students' construction of argument forms stemming from Toulmin's model. Students then used their argument structures to debate the question: Does light go on forever until absorbed or does light die out as one moves farther away from a light source? The purpose for engaging students with this activity was twofold: to engage them with the science of light and to help them build understandings about the scientific enterprise and its practices. Students in the classrooms Bell studied participated in a five-week unit about light, learning information about light sources, reflection, and absorption, for example. Students then participated in the aforementioned debate activity, which lasted eight to ten days. Bell's writings about this work detail the design specifications of several iterations of this debate activity, as well as document implementation data collected during the enactments. Subsequent iterations were revisions of previous models, with implementation data informing the revisions.

The first iteration consisted of students selecting one of the light theories (i.e., light goes on forever until absorbed or light dies out) and then creating an argument by selecting pieces of provided evidence that both grounded and refuted students' stated positions. The evidence available for students to use consisted of brief segments of text, usually describing a phenomenon (e.g., a person looking at the sky at night and not seeing any stars but then looking through a telescope and seeing stars). At the end of their argument construction work, each pair of students presented their argument to the class and then responded to questions from their classmates. The second iteration consisted of introducing evidence in a multi-media format (instead of on paper). The rest of the work then proceeded as it did in iteration one.

In the third iteration, Bell introduced SenseMaker, which allowed students to sort pieces of evidence into theory "boxes" (e.g., light goes on forever until absorbed, light dies out, irrelevant to either theory). One purpose of the SenseMaker argument maps was to make students' thinking visible. Other purposes of the tool included helping students understand the relationship between claim and evidence, helping them coordinate claims and evidence, and helping them attend to the entire evidentiary corpus instead of paying attention only to a subset within the corpus. The remaining three iterations focused on how to offer more support to students as they utilized SenseMaker to construct arguments and then present those arguments in a debate with their classmates.

Bell saw improvement for many students on several fronts: (a) students utilized the entire evidence corpus versus only the pieces that were thought to support their chosen theory, (b) students learned how to coordinate evidence with theory, (c) students made gains in their conceptual understanding of the behavior of light, and (d) students were able to defend both theoretical positions. Bell expressed his overall conclusion as such: "Students may not have been able to spontaneously engage in these performances, but when they are tuned onto the epistemic game at hand and supported in their inquiry, they can indeed engage in such intellectual activities and develop a more integrated understanding of complex science topics in the process (Bell, 2004, p. 140)." This is significant, he claimed, because traditionally it has been recognized that students have a great amount of difficulty engaging in argumentation, including coordinating evidence with theory and considering the merits of various perspectives, given evidence for those perspectives (Kuhn, 1992).

Finally, Bell claimed that the argumentation intervention he designed and implemented also impacted students' epistemological commitments. During the sixth iteration, he asked students epistemological questions, both before and after their participation in the debate activity. These questions focused on the role of argumentation and debate in science. Besides finding that students were better able to understand the role of evidence in scientific argumentation, Bell and colleagues also found that students came to appreciate the usefulness of debate as a learning practice, as well as the usefulness of social processes in general for articulating and clarifying ideas. Bell and colleagues interpreted students' insights as outcomes of the scaffolded instruction that Bell and his team designed. In subsequent years since his dissertation, Bell wonders if the student outcomes he documented were solely the result of SenseMaker or whether students also recognized an opportunity through Bell's design to utilize their existing argumentative and debate competencies. In other words, did Bell and his colleagues create a space in the science education classroom for students to feel confident and comfortable using their already well-honed argumentative practices? Bricker's dissertation explored those everyday practices so that Bell and Bricker might think about the affordances of those practices as design tools when creating curricular and instructional models that engage young people with what it means to argue scientifically.

Documenting Youths' Everyday Argumentation and Reasoning

Youth already bring a rich set of argumentative practices to science education (cf. Corsaro, 2003; Corsaro & Maynard, 1996; Goodwin & Goodwin, 1987; Hudicourt Barnes, 2003; Kyratzis, 2004; Ochs, Taylor, Rudolph, & Smith, 1992). They routinely interpret and produce arguments as they navigate the social settings and activities of their lives. Bricker sought to richly document youths' everyday argumentative practices and to understand their perspectives about those practices in part to think deeply about implications for the design of school science learning environments that engage young people with what it means to argue scientifically. Her research questions focused on what youth count as argumentation (i.e., what the word means to them), how their characterize their argumentative practices, including their use of evidence, whether, and if so how, youth believe argumentation plays a role in learning, and the relationships between argumentation and youth, family, and community cultural practices.

Theoretical Framework

To engage her research questions, Bricker utilized a framework she constructed using theoretical lenses from (a) Irving Goffman's (1974) ideas about frames on activity and the importance of orienting to sustained activity and collectivity; (b) Kenneth Burke's (1969) ideas about language use as a window through which one can view people's sense-making within activity, as well as his ideas relative to describing people's motives in activity using his theater metaphor; and (c) ideas from sociocultural historical activity theorists, starting with Lev Vygotsky (1978), about sociohistorical time, enculturation, and mediation via signs and tool systems (e.g., language). Bricker analyzed youth argumentation as practices situated in a multitude of activity systems in which youth participate over time. For example, youth might engage in different activities (e.g., school science and recess) within the same setting (e.g., school) and they might engage in the same activity (e.g., playing soccer) across settings (e.g., neighborhood and school). Regardless, youth report using argumentation within that network of activities and Bricker attempted to understand the fine-grained details of their argumentation use, including linguistic details and the work that argumentation was doing relative to the interaction within those various activity systems. The framework also allowed Bricker to capture youths' ideas about operating frames on the situated activity systems in which they participate and their perspectives on the roles of argumentation within those systems—an emic perspective (Harris, 1987; Pike, 1954), which allowed youth voices to play a prominent role in Bricker's dissertation.

Methods and Analysis

The data for Bricker's dissertation were collected as part of a three-year team ethnography (cf. Erickson & Stull, 1998). Researchers followed the same youths across the settings of their lives to study how these youths learn about science and technology (Bell et al., 2006). In the spring of 2005, researchers formed a partnership with a local elementary school (pseudonym Granite), which caters to a student body that is diverse with respect to ethnicity, nationality, languages spoken, and socioeconomic status. In the fall of 2005, researchers began recruiting families into the ethnographic study. Thirteen families agreed to participate and the sample of focal participants from each of those families was balanced for age (six youths were in fourth grade and seven were in fifth grade at the beginning of the study) and gender (seven boys and six girls). Besides the focal participants and their immediate family members, extended family members (e.g., grandmothers, cousins), teachers, and peers were consented into the study. As of December 2007, 128 people were consented into the ethnography.

A guiding methodological principle of this research was to follow the same people as they navigate various settings. The majority of the observations of the focal participants took place in school and at home. However, focal participants were also observed participating in activities and interacting with others in a multitude of additional settings, such as religious institutions, after school clubs, museums, sporting events, camping excursions/vacations, neighborhoods, and parks. Regardless of setting, data collection methods included (a) observation and participant observation; (b) interviews (both ethnographic and clinical); (c) self-documentation techniques (e.g., Glesne, 1999; Reeve & Bell, 2009), where focal participants were given digital cameras and asked to document various objects and phenomena (e.g., argumentation) and then answer questions about their photographs; and (d) document collection. Two surveys, designed to gather information about socioeconomic status and ethnic identity and participation in science respectively, were administered. Researchers also conducted analyses of public census tract data for the neighborhoods in which participant families live. Data sources include (a) *field notes* of all observations, interviews, participant self-documentation assignments, and documents collected; (b) *videotape and audiotape* of all observations and interviews (when in settings that allowed video and/or audio taping); (c) *digital photographs* taken during observations and interviews; (d) *video and/or digital photographs* taken by participants as part of their self-documentation tasks; (e) *documents* collected during family visits (e.g., magazines, school work, writing samples from clinical interviews, written survey responses); and *survey results*.

As is the case with ethnography, data analysis began while fieldwork was still in progress in an effort to be as prepared as possible for future data collection and to guide that data collection. All video and audio taped segments (that did not overlap with videotape) were content logged and both logs and field notes were initially coded using a set of conceptual tags (e.g., [arg] for argumentation events in the data corpus). Transcripts were created of any video or audio taped segments that were of analytical interest. These transcripts were then coded with codes created based on meanings generated by participants, patterns found in the data, and also based on important constructs from the relevant literature. Patterns and assertions were generated from coding exercises and triangulated using other data sources, when applicable. As part of searching for disconfirming evidence (Erickson, 1986), assertions and representations were member checked with research participants in order to solicit their thoughts about the assertions and representations generated (cf. Heyl, 2001). Assertions and representations were also checked by other researchers who were present during any given data collection moment.

Categories of Findings—A Summary

As shown across the learning cases highlighted in Bricker's dissertation, without asking youth about their argumentative practices as associated with *specific* activity in *specific* settings, youth tended to associate the word "argument" with behaviors, such as fighting and yelling, a finding explicated often in the literature (cf. Tannen, 1999). One might be tempted to conclude that youth associate argumentation with fighting in all instances. Yet, some participants also associated the word "argument" with practices such as debate, decision-making, and discussion. Because argument involves language, we can turn to Wittgenstein's (1953/2001) notion of "language games." As he explains, "...the *speaking* of language is part of an activity..." (p. 10^{e}[1], emphasis in the original). He reminds us that "...naming is [solely] a preparation for description. Naming is so far not a move in the language-game—any more than putting a piece in its place on the board is a move in chess. We may say: *nothing* has so far been done, when a thing has been named. It has not even *got* a name except in the language-game (p. 21^{e}, emphasis in the original)" Participants were able to describe what they associated with the word "argument" in the abstract but as Wittgenstein reminds us, something's name starts to become significant only when that name is coupled with how the something actually plays out in context. As participants began to describe argument use within the context of their activities, their ideas were more broad and nuanced than images of fighting and yelling.

Given that various forms of evidence play a critical role in warranting scientific claims (cf. Tufte, 2006), Bricker chose to explore participating youths' perceptions of evidence. Popular culture (e.g., television shows such as CSI) seemed quite influential relative to their understandings about evidence and its roles in argumentation. The participants' school science experiences at Granite Elementary also appeared to play a role because youth reported that they learned to provide

[1] The "e" superscript represents page numbers containing the English translations from the original German.

evidence for the conclusions they drew in science class, for example, and were asked repeatedly to link evidence with claims by utilizing such linguistic markers as "because." Findings indicated that participating youth used linguistic elements (both verbal and non verbal), such as discourse markers, evidentials, and indexicals when bringing evidence to bear on their claims (cf., Aikhenvald, 2004; Schiffrin, 1987). Furthermore, findings showed that some of these linguistic elements mark sources of evidence and are helpful in identifying when participating youth learned something in one setting and transferred it to another setting.

Findings also highlighted that some youth identified argumentation as a learning practice (Billig, 1987/1996) and they noted its similarity to critique and its role in helping to make ideas visible so that others can learn from those ideas. There are important implications for looking at the relationships between communicative practices, such as argumentation, and learning. There is significant research to show that learners learn through meaningful, relevant interactions with people, environments, and objects and materials found in those environments (cf. McDermott, 1996). They learn because of the necessity to participate in various communities, solve problems, and pursue interests. It might not be enough, however, to simply embed learner-associated ways of talking, problem solving, and acting, for example, as well as learner interests and identified problems of relevance into curriculum and instruction. The learners themselves might not recognize any of this as (a) related to their lives (because it will all be taking place in a setting—i.e., school—that is quite different from the usual settings where they see these things utilized) or (b) helpful to the learning they find meaningful. Given that caution, we turn now to design principles gleaned from our dissertation research. We believe that these can inform design-based research efforts that investigate whether leveraging youths' everyday argumentative practices, which involve giving reasons situated in Tilly's conventions and stories categories, helps scaffold youths' learning with respect to arguing scientifically, a practice that is mostly situated in Tilly's technical accounts reasoning category.

Implications for Design-Based Research: A Work in Progress

In this section, we begin to enumerate some assertions for the design of argumentative learning environments that stem from our dissertation research. Although the following design assertions are based on empirical research, they should still be interpreted as conjectural and interconnected features of a designed learning environment. For that reason, we frame them as a numbered and bulleted list of abstracted and interconnected elements in three spheres.

1. Technology rich learning environments

   1.1. Computer environments can provide unique affordances for and constraints on the learning of science. The influences on learning are about as variable as the technological forms in question.

- Computer environments are best considered to contribute structural elements of the functional systems that come to be associated with learning (or distributed cognition). As described by Hutchins: "The real power of human cognition lies in our ability to flexibly construct functional systems that accomplish our goals by bringing bits of structure into coordination. That culturally constituted settings for activity are rich in precisely the kinds of artifactual and social interactional resources that can be appropriated by such functional systems is a central truth about human cognition" (Hutchins, 1995, p. 316).

1.2 Technology-rich learning environments, when customized or designed specifically for use in science education, can be a strongly contributing element of educational packages that scaffold the construction of arguments and that use such arguments as focal objects or contexts associated with learning.

2. The social framing of argumentation and debate in the classroom

2.1. The framing of the learning enterprise is of crucial importance. A classroom idioculture (cf. Fine, 1979) can be cultivated which constitutes argumentation and debate in long-term collaborative terms—in that the ultimate, overarching goal is to develop collective understanding and a roughly-agreed-to social consensus on knowledge claims (or areas of uncertainty)—although it may involve competitive and confrontational, as well as collaborative features in the short term. We term this ideoculture as one that engages by-and-large in "collaborative debate." When engaged in collaborative debate, students are more capable of sharing, exploring, testing, refining, and integrating their scientific ideas without an overriding fear of *ad hominem* attacks.

- Argumentation can become manifest in the classroom in divergent ways. A fundamental complexity of education derives from the situated nature of learning as it comes to be accomplished by specific communities. Through substantial effort in terms of coordination and comparison, progress can be made in promoting roughly similar learning communities across classrooms.
- Pop culture accounts of argumentation frequently depict socially undesirable quarrelling, partisan bickering, or intractable articulation of differences in opinion. Collaborative debate is socially rare. Intellective identities and reputations are almost always implicated and on-the-line during everyday social dispute processes.

2.2. Cultivating and enforcing an identity-safe learning environment is a principal responsibility of learners within a collaborative debate community—teachers and students alike. Given the social position typically afforded teachers, they should seek to dampen socially adverse interactions, as interpreted in local cultural terms. Generally, they should treat the various theoretical positions associated with the scientific debate in parallel terms while modeling the use of epistemic practices and application of epistemological criteria to deliberations.

- Selecting unsettled (i.e., open) controversies in science has the unique educational advantage of circumventing the work-around pursued by many students who go on the prowl to determine and support the "right answer" within a debate whether or not they come to understand the choice or not (Bell, 2004). Such topics also highlight the dynamic nature of scientific understanding and help students develop that epistemological understanding of the enterprise (Linn, Shear, Bell, & Slotta, 1999).

3. Introducing scientific argumentation and collaborative debate

 3.1. Students can be oriented to a collaborative debate approach by coordinating educational supports and expectations with rendered accounts of the desired epistemic practices. Students need to be brought into the epistemic game that is desired of them (see discussion of leveraging their everyday argumentative practices below).

 - Collaborative debate ideocultures can be cultivated over the course of a small number of months in the classroom (Herrenkohl & Guerra, 1998; Sandoval & Reiser, 2004). As shown in Bricker's dissertation, students bring the social and cognitive competencies needed to do the intellectual work associated with collaborative debate and can refine their competencies under specific, supportive conditions.
 - Introducing argumentation and collaborative debate through the exploration of a historical debate between scientists might allow students to understand the social and intellectual purposes of scientific argumentation, the creativity and conceptual precision associated with theorizing and coordinating theory with evidence, as well as the theory-laden nature of observation and argumentation.
 - Another possible approach for introducing argumentation would map the epistemics directly onto an inquiry project to be pursued by the students (e.g., to make sense of experiments).

Code Switching: Leveraging Conventions and Stories When Learning About Codes and Technical Accounts

We have argued in this chapter that reasoning is a critical aspect of argumentation, whether scientific or otherwise. Using Tilly's (2006) categories of reasons, we have made the case that most curricular and instructional efforts aimed at helping young people argue scientifically in school science classrooms have employed Toulmin's micro-structural framework of argumentation in which reasons given are codes and/or technical accounts (the reasons almost exclusively used in the sciences). We summarized Bell's dissertation as an example of this type of effort.

Additionally, we have made the case that young people are adept at mounting and interpreting arguments in their everyday lives in a variety of settings. The reasons most frequently embedded in youths' argumentative practices are conventions and stories. We summarized Bricker's dissertation as an example of research documenting young people's naturalistic argumentation. Given this landscape—the types of reasons employed in scientific argumentation and everyday argumentation, as well as the characteristics of most research efforts aimed at engaging young people with how to argue scientifically—we have proposed principles for future design-based research. We turn now to a discussion of why we believe that a fruitful category of design-based research involves leveraging aspects of young people's argumentation, focused specifically on the types of reasoning embedded in that argumentation, when engaging them with how to argue scientifically.

Argumentation and Learning

Several scholars have proposed that argumentation is a learning tool because it helps make ideas visible and public (Billig, 1987/1996). In what follows, we discuss one learning framework and the implications for designing learning environments to engage young people with what it means to argue scientifically. Banks, Au, Ball, Bell, Gordon, & Gutiérrez et al. (2007) propose a framework for learning that takes into account three important understandings about learning: (1) that it is life-long; (2) that it is life-wide; and that it is (3) life-deep. Life-long learning describes the learning in which we engage longitudinally, from birth until death. Life-wide learning comprises the breadth of our experiences from across the settings of our lives. Life-deep learning "embraces religious, moral, ethical, and social values that guide what people believe, how they act, and how they judge themselves and others" (p. 12). In addition to explicating these three components of learning, Banks et al. outline four principles with respect to learning, one of which is particularly germane to our discussion about argumentation. This specific principle notes that learning is facilitated when learners are encouraged to use language resources from their home and community contexts.

This report was written to specifically discuss diverse learners (e.g., students from underrepresented groups, English language learners), but we argue that the report's tenants and the learning principles we cite here are highly relevant to our arguments in this chapter. As the report reminds us: "When [children] arrive at school. . .[they] have to become conscious of the fact that, in informal settings, we tend to focus more on the *communicative intent* (the message that the speaker is trying to convey) than on the *code* (the form of language the speaker is using)" (emphasis in original, p. 21). Argumentation is a discursive practice that does learning work but as Tilly (2006) highlighted, it also does relational work. By way of reminder, he claims that when a person offers another person codes and technical accounts in unfamiliar ways and structures, that speaker could be positioning the hearer in a deficit fashion and communicating as much—both to the learners and to others involved in the interaction. Banks et al. explain that, "Every society has a culture of power, and students must learn the languages or codes of the culture of power to advance to higher education, to obtain good jobs, and to experience social-class mobility (Delpit, 1988)" (2007, p. 22).

As noted in the beginning sections of this chapter, many scholars have noted the power differentials in place in formal schooling and how implicit messages of power and position are conveyed through the formal structures and languages of school. Additionally, the argument is often made that the sciences' languages—their codes and technical accounts—serve to establish the sciences' positions of power in society and to establish the superiority of scientific ways of knowing and doing (cf. Harding, 1993). In order to engage all youth with the languages of the sciences, in hopes that the sciences become more democratic and representative, we believe strongly that young people should be learning the various practices of the sciences, including the central discourse practices, of which argumentation is one (cf. National Research Council (NRC), 2007 and 2009). As a community of science educators, however, we might want to rethink the practice of bringing young people into the culture of scientific argumentation solely by teaching them the affiliated structural forms and the proper codes and technical accounts. These aspects of scientific argumentation are of course critical and important but if they are the starting point and the sole basis for curriculum and instruction, the oft-stated finding that young people have a difficult time arguing scientifically (e.g., coordinating evidence and theory) may not be surprising. It is an empirical question but might we see youth have more success if, as a design principle, we started with young people's everyday argumentative forms, as well as the conventions and stories embedded within them, as leverage points when teaching them how to argue scientifically? In other words, might we see more success if we taught them how to code switch (cf. Gumperz & Hymes, 1986)? In the process, we might better facilitate youth learning and help flatten the hierarchies found in schools by showing them that we value their everyday argumentative practices enough to leverage them in service of helping them learn how to employ scientific argumentation when in situations that warrant that type of discourse practice.

References

Aikhenvald, A. Y. (2004). *Evidentiality*. Oxford, UK: Oxford University Press.

Banks, J. A., Au, K. H., Ball, A. F., Bell, P., Gordon, E. W., Gutiérrez, K. D., et al. (2007). *Learning in and out of school in diverse environments: Life-long, life-wide, life-deep*. Seattle, WA: The LIFE (Learning in Informal and Formal Environments) Center & The Center for Multicultural Education; University of Washington.

Bell, P. (1997). Using argument representations to make thinking visible for individuals and groups. In R. Hall, N. Miyake, & N. Enyedy (Eds.), *Proceedings of CSCL '97: The second international conference on computer support for collaborative learning* (pp. 10–19). Toronto, ON: The University of Toronto.

Bell, P. (2004). Promoting students' argument construction and collaborative debate in the science classroom. In M. C. Linn, E. A. Davis, & P. Bell (Eds.), *Internet environments for science education* (pp. 115–143). Mahwah, NJ: Erlbaum.

Bell, P., Bricker, L. A., Lee, T. R., Reeve, S., & Zimmerman, H. T. (2006). Understanding the cultural foundations of children's biological knowledge: Insights from everyday cognition research. In S. A. Barab, K. E. Hay, & D.T. Hickey (Eds.), *Proceedings of the Seventh International Conference of the Learning Sciences* (pp. 1029–1035). Mahwah, NJ: Lawrence Erlbaum Associates.

Bell, P., Bricker, L. A., Reeve, S., Zimmerman, H. T., & Tzou, C. (in press). Discovering and supporting successful learning pathways of children in and out of school: How everyday expertise develops across places and pursuits. In B. Bevan, P. Bell, & R. Stevens (Eds.), *LOST Learning: Learning about out-of-school time learning opportunities.* New York, NY: Springer.

Billig, M. (1987/1996). *Arguing and thinking: A rhetorical approach to social psychology* (new ed.). Cambridge, UK: Cambridge University Press.

Brice Heath, S. (1983). *Ways with words: Language, life, and work in communities and classrooms.* New York: Cambridge University Press.

Bricker, L. A., & Bell, P. (2008). Conceptualizations of argumentation from science studies and the learning sciences and their implications for the practices of science education. *Science Education, 92*(3), 473–498.

Burke, K. (1969). *A grammar of motives.* Berkeley, CA: University of California Press.

Clark, D. B., & Sampson, V. (2008). Assessing dialogic argumentation in online environments to relate structure, grounds, and conceptual quality. *Journal of Research in Science Teaching, 45*(3), 293–321.

Corsaro, W. A. (2003). *We're friends right? Inside kids' culture.* Washington, DC: Joseph Henry Press.

Corsaro, W. A., & Maynard, D. W. (1996). Format tying in discussion and argumentation among Italian and American children. In D. I. Slobin, J. Gerhardt, A. Kyratzis, & J. Guo (Eds.), *Social interaction, social context, and language: Essays in honor of Susan Ervin-Tripp* (pp. 157–174). Mahwah, NJ: Erlbaum.

Delpit, L. (1988). The silenced dialogue: Power and pedagogy in educating other people's children. *Harvard Educational Review, 58*(3), 280–298.

Erickson, F. (1986). Qualitative methods in research on teaching. In M. Wittrock (Ed.), *Handbook of research on teaching* (pp. 119–161). New York: Macmillan.

Erickson, K., & Stull, D. (1998). *Doing team ethnography: Warnings and advice.* Thousand Oaks, CA: Sage.

Fine, G. A. (1979). The idioculture of little league baseball teams. *American Sociological Review, 44*(5), 733–745.

Glesne, C. (1999). *Becoming qualitative researchers: An introduction* (2nd ed.). New York: Longman.

Goffman, E. (1974). *Frame analysis: An essay on the organization of experience.* Boston: Northeastern University Press.

Goodwin, M. H., & Goodwin, C. (1987). Children's arguing. In S. U. Phillips, S. Steele, & C. Tanz (Eds.), *Language, gender, and sex in comparative perspective* (pp. 200–248). Cambridge, UK: Cambridge University Press.

Gumperz, J. J., & Hymes, D. (Eds.). (1986). *Directions in sociolinguistics: The ethnography of communication.* Oxford, UK: Basil Blackwell.

Harding, S. (1993). Rethinking standpoint epistemology: What is "strong objectivity"? In L. Alcoff & E. Potter (Eds.), *Feminist epistemologies* (pp. 49–82). New York: Routledge.

Harris, M. (1987). *Cultural anthropology* (2nd ed.). New York: Harper & Row.

Herrenkohl, L. R., & Guerra, M. R. (1998). Participant structures, scientific discourse, and student engagement in fourth grade. *Cognition and Instruction, 16*, 433–475.

Heyl, B. S. (2001). Ethnographic interviewing. In P. Atkinson, A. Coffey, S. Delamont, J. Lofland, & L. Lofland (Eds.), *Handbook of ethnography* (pp. 369–383). London, UK: Sage.

Hudicourt-Barnes, J. (2003). The use of argumentation in Haitian Creole science classrooms. *Harvard Educational Review, 73*(1), 73–93.

Hutchins, E. (1995). *Cognition in the wild*. Cambridge, MA: The MIT Press.

Keith, W., & Rehg, W. (2008). Argumentation in science: The cross-fertilization of argumentation theory and science studies. In E. J. Hackett, O. Amsterdamska, M. Lynch, & J. Wajcman (Eds.), *The handbook of science and technology studies* (3rd ed., pp. 211–239). Cambridge, MA: MIT Press.

Kelly, G. J., & Bazerman, C. (2003). How students argue scientific claims: A rhetorical-semantic analysis. *Applied Linguistics, 24*(1), 28–55.

Kuhn, D. (1992). Thinking as argument. *Harvard Educational Review, 62*(2), 155–178.

Kuhn, D. (1993). Connecting scientific and informal reasoning. *Merrill-Palmer Quarterly, 39*(1), 74–103.

Kuhn, L., Kenyon, L., & Reiser, B. J. (2006). Fostering scientific argumentation by creating a need for students to attend to each other's claims and evidence. In S. A. Barab, K. E. Hay, & D. Hickey (Eds.), *Proceedings of the seventh international conference of the learning sciences (ICLS)* (pp. 370–375). Mahwah, NJ: LEA.

Kyratzis, A. (2004). Talk and interaction among children and the co-construction of peer groups and peer culture. *Annual Reviews, Anthropology, 33*, 625–649.

Latour, B. (1987). *Science in action*. Cambridge, MA: Harvard University Press.

Linn, M. C., Davis, E. A., & Bell, P. (Eds.). (2004). *Internet environments for science education*. Mahwah, NJ: Lawrence Erlbaum Associates.

Linn, M. C., Shear, L., Bell, P., & Slotta, J. D. (1999). Organizing principles for science education partnerships: Case studies of students' learning about 'rats in space' and 'deformed frogs'. *Educational Technology Research and Development (ETR&D), 47*(2), 61–84.

McDermott, R. P. (1996). The acquisition of a child by a learning disability. In S. Chaiklin & J. Lave (Eds.), *Understanding practice: Perspectives on activity and context* (pp. 269–305). Cambridge, UK: Cambridge University Press.

National Research Council. (2007). *Taking science to school: Learning and teaching science in grades K-8*. Washington, DC: The National Academies Press.

National Research Council. (2009). *Learning science in informal environments: People, places, and pursuits*. Washington, DC: National Academies Press.

Nespor, J. (1997). *Tangled up in school: Politics, space, bodies, and signs in the educational process*. Mahwah, NJ: Lawrence Erlbuam and Associates.

Newton, P., Driver, R., & Osborne, J. (1999). The place of argumentation in the pedagogy of school science. *International Journal of Science Education, 21*(5), 553–576.

Ochs, E., Taylor, C., Rudolph, D., & Smith, R. (1992). Storytelling as a theory-building activity. *Discourse Processes, 15*, 37–72.

Osborne, J., Erduran, S., & Simon, S. (2004). Enhancing the quality of argumentation in school science. *Journal of Research in Science Teaching, 41*(10), 994–1020.

Pea, R. D., & Gomez, L. M. (1992). Distributed multimedia learning environments: Why and how? *Interactive Learning Environments, 2*(2), 73–109.

Pike, K.L. (1954). *Language in relation to a unified theory of the structure of human behavior*. Ann Arbor, MI: Braun-Brumfield.

Reeve, S., & Bell, P. (2009). Children's self-documentation and understanding the concepts 'health' and 'unhealthy.' *International Journal of Science Education, 31*(14), 1953–1974.

Sandoval, W. A. (2003). Conceptual and epistemic aspects of students' scientific explanations. *The Journal of the Learning Sciences, 12*(1), 5–51.

Sandoval, W. A., & Reiser, B. J. (2004). Explanation-driven inquiry: Integrating conceptual and epistemic scaffolds for scientific inquiry. *Science Education, 88*, 345–372.

Sarangapani, P. M. (2003). *Constructing school knowledge: An ethnography of learning in an Indian village*. New Delhi, India: Sage.

Schiffrin, D. (1987). *Discourse markers*. Cambridge, UK: Cambridge University Press.

Simosi, M. (2003). Using Toulmin's framework for the analysis of everyday argumentation: Some methodological considerations. *Argumentation, 17*, 185–205.
Tannen, D. (1999). *The argument culture: Stopping America's war of words.* New York: Ballantine Books.
Tilly, C. (2006). *Why?: What happens when people give reasons. . .and why.* Princeton, NJ: Princeton University Press.
Tindale, C. W. (1999). *The act of arguing*. Albany, NY: State University of New York Press.
Toulmin, S. E. (1958/2003). *The uses of argument.* Cambridge, UK: Cambridge University Press.
Toulmin, S., Rieke, R., & Janik, A. (1984). *An introduction to reasoning* (2nd ed.). New York: Macmillan.
Tufte, E. (2006). *Beautiful evidence*. Cheshire, CT: Graphics Press LLC.
Vygotsky, L. S. (1978). *Mind in society: The development of higher psychological processes*. Cambridge, MA: Harvard University Press.
Warner, M. (2002). *Publics and counterpublics*. New York: Zone Books.
Wenzel, J. A. (1990). Three perspectives on argument. In R. Trapp & J. Schuetz (Eds.), *Perspectives on argumentation: Essays in honor of Wayne Brockriede* (pp. 9–26). Prospect Heights, IL: Waveland.
Willis, P. (1997/2000). *Learning to labor: How working class kids get working class jobs.* Burlington, VT: Ashgate.
Wittgenstein, L. (1953/2001). *Philosophical investigations* (3rd ed.) (G.E.M. Anscombe, Trans.). Malden, MA: Blackwell.

Chapter 8
Argumentation and Evaluation Intervention in Science Classes: Teaching and Learning with Toulmin

Janis A. Bulgren and James D. Ellis

Introduction

A major challenge for teachers in our schools is to help students engage in scientific reasoning (National Research Council (NRC), 1996; National Research Council (NRC), 2007). One aspect of scientific reasoning is the ability to evaluate claims or statements made about scientific issues in a variety of fields. To evaluate claims, students must use reasoning skills associated with *argumentation*. The types of thinking associated with argumentation are often incorporated into state standards and assessments. As a result, students must engage in inquiry procedures as they evaluate the quality of evidence and reasoning presented in support of a claim. Furthermore, standards from the National Research Council (1996, 2007) emphasize the need for students to make connections between explanations and evidence, and to take ownership and responsibility for their decisions.

Various authors have explored components of argumentation. For example, Toulmin, Rieke, and Janik (1984) defined *argumentation* as "the whole activity of making claims, challenging them, backing them up by producing reasons, criticizing those reasons, rebutting those criticisms, and so on" (p. 14). This is an important activity in the development of scientific literacy. Wallace, Hand, and Yang (2004) contended that an essential characteristic of scientific literacy is the ability to evaluate a scientific knowledge claim. To do this, students must understand the relationships between questions, data, claims, and evidence. This is the guiding mindset of our project, resulting in instructional procedures, an associated graphic organizer, and an embedded strategic approach to evaluation of claims and arguments.

This material is based upon work supported by the National Science Foundation under Grant Number 0554414. Any opinions, findings, and conclusions, or recommendations expressed in this material are those of the authors and do not necessarily reflect the views of the National Science Foundation.

J.A. Bulgren (✉)
University of Kansas, Lawrence, KS, USA
e-mail: jbulgren@ku.edu

M.S. Khine (ed.), *Perspectives on Scientific Argumentation*,
DOI 10.1007/978-94-007-2470-9_8, © Springer Science+Business Media B.V. 2012

Purpose of the Study

By way of response to the challenge of helping students engage in higher-order thinking associated with argumentation, the purpose of this study was to develop a set of instructional tools that would support science teachers in helping middle and secondary students improve their scientific argumentation.

The project developed an Argumentation and Evaluation Intervention (AEI) and associated graphic organizer, the Argumentation and Evaluation Guide (AEG). We designed these materials to assist students in the following argumentation activities: (1) identifying a claim presented in a written science-based report or from an inquiry activity, and analyzing the claim for qualifiers; (2) identifying evidence, labeling the evidence by type, and judging the quality of the evidence; (3) identifying the reasoning that allowed the claimant to make the claim based on the evidence presented, labeling the reasoning by type, and judging the quality of the reasoning; and (4) drawing a conclusion about the quality of the claim and explaining the reasoning that supported the conclusion, including presenting rebuttals or counterarguments.

The design study included two parts: a quantitative and a qualitative study. The results of the quantitative study demonstrated the efficacy of the AEI for use by teachers in inclusive middle and secondary school science classrooms that contain students of diverse abilities, including students with learning disabilities and those who have been identified as gifted (Ellis & Bulgren, 2009). Analyses indicated that significant differences were found for mean total test scores in favor of the students who were taught with the AEI over students taught with a traditional lecture-discussion format. In addition, for three of the four subscale scores, significant differences were found for students in the AEI condition over the comparison condition. These were the second, third, and fourth subscales described earlier.

The design study is built upon the research and recommendations of Bannan-Ritland (2003) and Kelly (2004). Central to the iterative design process are activities such as analyzing what needs to be developed, implementing iterative development until solutions are developed, and analyzing the effect of the solutions on teacher and student performance data. The design and development process begins with informed exploration to understand the situation (Bannan-Ritland, 2003). Then, the development process in an iterative approach includes: (a) identifying the design principles, (b) operationalizing the target cognitive processes, and (c) balancing the theoretical model with real-world requirements. This is a cyclical process with prototyping and testing continuing until ease of use and intended functionality is achieved. This process includes iterative user-centered design, interaction analysis, and usability and feasibility analysis. The process also involves evaluating the impact, which includes feasibility analysis, fidelity of implementation, analysis of the effect of the instructional process and materials on students' and teachers' skills and understanding, and efficacy studies.

Theoretical Framework

The conceptualization for evaluating claims and arguments is based on the theories of Toulmin (1958), who defined the central components of *argumentation* as reasoning from grounds or data (evidence) to knowledge claim (conclusion) with warrants (links from the evidence to the claim with principles and underlying assumptions), possibly additional backings, and qualifiers and rebuttals. In addition, others have contributed research and commentary on argumentation. For example, Driver, Newton, and Osborne (2000) noted the need to emphasize the correctness of judgments about arguments in addition to the structure of an argument. Lawson (2003) agreed with Driver et al. (2000), but noted the need for attention to real-world issues related to argumentation as students evaluate their own claims and conclusions.

The evaluation component of the strategy was based on components of scientific thinking that Kuhn (1991) calls the "skills" of argumentation. Evaluation components include, among others, an appreciation of the role of empirical evidence (Kuhn, Amsel & O'Loughlin, 1988) and judging the credibility of evidence in terms of (1) reliability (Schauble, 1996), (2) experimental control (Koslowki, Okagaki, Lorenz & Umbach, 1989; Kuhn Garcia-Mila, Zohar & Andersen, 1995; Schauble, 1996), and (3) objectivity (Klahr, Fay & Dunbar, 1993; Kuhn et al., 1995; Penner & Klahr, 1996; Schauble, 1996). We incorporated these evaluation components into the instructional procedures for use by science teachers in inclusive general education classes.

This project also incorporated the research-to-practice interface and supports as discussed by Klahr, Chen, and Toth (2001). These involve the use of direct instruction to prepare students for evaluation of evidence, support for transfer and generalization to other experiences, the provision of strategic skills that help students acquire domain knowledge, the ability to evaluate one's own use of these skills as well as those of others, and the goal of raising new issues for future research. These components fit well with explanations about how people learn and how they construct knowledge (Bransford, Brown & Cocking, 2000), and also include intervention strategies suggested by Carver (2001).

This project can also be built on the work of others relative to designing education *curriculum materials*. In terms of designing educative curriculum materials to promote teacher learning, this proposal incorporates heuristics that help teachers engage students in asking and answering scientific questions and making explanations based on evidence. This process is supported by procedural guides and professional development activities. These guides and activities may be applied to a variety of chapters, texts, and curricula, such as promoted by Davis and Krajcik (2005). However, while others (Linn, Clark & Slotta, 2003) provide already-prepared content and context for students to analyze for scientific arguments, our approach provides guides for students and teachers to use with a variety of student-generated and teacher-identified arguments and claims. These claims may be found in a range of scientific sources as well as real-world issues.

Products of the Design Study

We produced two major products from the design study, the AEI and the AEG with its embedded reasoning strategy. These products are instructional materials that include strategic-thinking approaches to support science teachers in improving the science argumentation knowledge and skills of middle school and secondary school students. The authors and teacher-researchers collaboratively developed these products through the design study process over a period of 3 years.

Description of the Argumentation and Evaluation Intervention

Content Enhancements. The AEI builds on a collection of instructional resources called Content Enhancement Routines developed by the Center for Research on Learning at the University of Kansas (KU-CRL) (Bulgren & Lenz, 1996). The Content Enhancement instructional interventions help students to process complex information using higher-level thinking skills. Previous studies indicated that students of diverse ability levels can learn content information using Content Enhancement procedures and that teachers can learn the instructional procedures easily. Researchers have found statistically significant results in favor of students who received instruction using Content Enhancements Routines when compared to students receiving traditional lecture-discussion instruction. These results were found for a number of routines including those designed to enhance concept acquisition (Bulgren, Schumaker & Deshler, 1988), learning by analogy (Bulgren, Deshler, Schumaker & Lenz, 2000), and manipulation of content information such as making comparisons (Bulgren, Lenz, Schumaker, Deshler & Marquis, 2002). This project was designed to move Content Enhancement research to a focus on higher-order thinking associated with reasoning about and evaluation of argumentation—a logical progression in the line of research.

The graphic organizer. The graphic organizer is the AEG (see Fig. 8.1 for an example AEG). The AEG contains a flexible cognitive reasoning strategy (the Argumentation and Evaluation Strategy) that guides students as they evaluate the components of arguments made in support of claims. Specifically, the strategy consists of the following steps: (1) identify the claim and qualifiers; (2) identify the evidence presented; (3) identify the type of evidence as data, fact, opinion, or theory; (4) evaluate the quality of the evidence; (5) explore the reasoning that connects the evidence to the claim; (6) identify the type of reasoning as theory, authority, or logic; types of logic include reasoning by analogy, cause–effect, correlation, or generalization; (7) evaluate the quality of the reasoning; (8) explore rebuttals, counterarguments, or new questions; and (9) draw a conclusion accepting, rejecting, or withholding judgment about the claim and explain the reasoning for the conclusion. We included supporting questions on the Guide as prompts for the learner. During the instructional process, students write information associated with each of the nine steps of the strategy on the AEG. The teacher guides the students to reach consensus on a class rendition of the AEG.

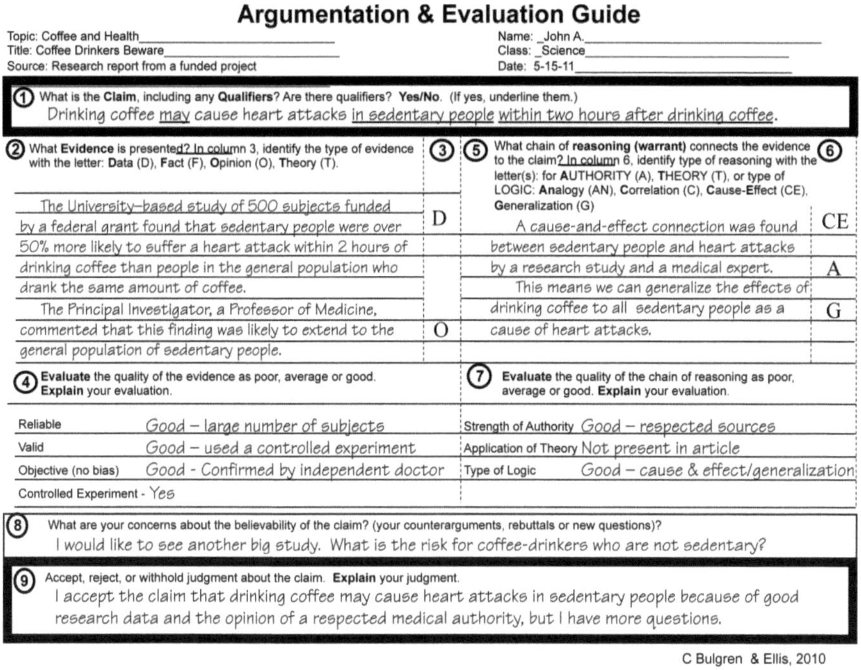

Argumentation & Evaluation Guide

Topic: Coffee and Health
Title: Coffee Drinkers Beware
Source: Research report from a funded project

Name: John A.
Class: Science
Date: 5-15-11

① What is the **Claim**, including any **Qualifiers**? Are there qualifiers? **Yes/No**. (If yes, underline them.)
Drinking coffee may cause heart attacks in sedentary people within two hours after drinking coffee.

② What **Evidence** is presented? In column 3, identify the type of evidence with the letter: **Data** (D), **Fact** (F), **Opinion** (O), **Theory** (T).	③	⑤ What chain of **reasoning (warrant)** connects the evidence to the claim? In column 6, identify type of reasoning with the letter(s): for AUTHORITY (A), THEORY (T), or type of LOGIC: **Analogy** (AN), **Correlation** (C), **Cause-Effect** (CE), **Generalization** (G)	⑥
The University-based study of 500 subjects funded by a federal grant found that sedentary people were over 50% more likely to suffer a heart attack within 2 hours of drinking coffee than people in the general population who drank the same amount of coffee.	D	A cause-and-effect connection was found between sedentary people and heart attacks by a research study and a medical expert.	CE, A
The Principal Investigator, a Professor of Medicine, commented that this finding was likely to extend to the general population of sedentary people.	O	This means we can generalize the effects of drinking coffee to all sedentary people as a cause of heart attacks.	G

④ **Evaluate** the quality of the evidence as poor, average or good. **Explain** your evaluation.		⑦ **Evaluate** the quality of the chain of reasoning as poor, average or good. **Explain** your evaluation.	
Reliable	Good – large number of subjects	Strength of Authority	Good – respected sources
Valid	Good – used a controlled experiment	Application of Theory	Not present in article
Objective (no bias)	Good - Confirmed by independent doctor	Type of Logic	Good – cause & effect/generalization
Controlled Experiment - Yes			

⑧ What are your concerns about the believability of the claim? (your counterarguments, rebuttals or new questions)?
I would like to see another big study. What is the risk for coffee-drinkers who are not sedentary?

⑨ Accept, reject, or withhold judgment about the claim. **Explain** your judgment.
I accept the claim that drinking coffee may cause heart attacks in sedentary people because of good research data and the opinion of a respected medical authority, but I have more questions.

C Bulgren & Ellis, 2010

Fig. 8.1 A sample Argumentation and Evaluation Guide

The AEG graphic organizer provides a space for each step (and associated question) of the argumentation process, starting with step one (designated by the number 1) and the question, "What is the claim, including any qualifiers?" Then, step two prompts, "What evidence is presented?" There is an adjacent space for step three, "Identify the type of evidence" with letters that represent data, fact, opinion, or theory. Then, there is step four to evaluate the evidence, prompted by, "Evaluate the quality of the evidence as poor, average, or good." Next, there is step five to identify the chain of reasoning, with the prompt, "What chain of reasoning (warrant) connects the evidence to the claim?" Following step five is step six, a column for students to "identify the type of reasoning" with letters that represent authority, theory, or type of logic (i.e., analogy, correlation, cause–effect, or generalization). Then, there is a space for step seven to evaluate the reasoning, prompted by the challenge, "Evaluate the quality of the reasoning as poor, average, or good." Then, there is a step eight with the question, "What are your counterarguments, rebuttals, or new questions related to this claim?" Finally, in step nine, students are guided to arrive at their conclusion, "Accept, reject, or withhold judgment about the claim, and explain your judgment." In addition, a Scoring Rubric was developed for use in analyzing students' evaluation of a claim (see Fig. 8.2).

The instructional procedures. The AEI project materials support teachers by explaining instructional procedures (the Argumentation and Evaluation Routine

Argumentation and Evaluation Scoring Rubric

Name: ______________________ Date: ______________________
Teacher: ______________________ Topic: ______________________

Scoring Guidelines for each Step	0 Poor	1 Needs Improvement	2 Good Progress Toward Improvement	3 Very Good Meets Standards	Score
1 Claim	Student gives no response.	The student response inaccurately identifies the claim being made or writes a response not structured as a claim.	The student partially identifies the claim being made.	The student accurately identifies the claim being made.	
1 Qualifier	Students gives no response.	The student response fails to accurately identify qualifier(s) within the claim OR fails to state there are no qualifiers present.	The student partially identifies qualifier(s) within the claim that are present.	The student accurately identifies most of the qualifier(s) within the claim OR correctly states that none are present.	
2 Evidence	Student gives no response.	The student response identifies evidence that fails to support the claim.	The student accurately identifies some evidence used to support the claim.	The student accurately identifies most of the evidence used to support the claim.	
3 Identifying Types of Evidence: Data, Fact, Theory or Opinion	Student gives no response.	The student response fails to accurately identify any types of evidence.	The student accurately identifies some types of evidence.	The student accurately identifies all of the evidence as data, fact, theory or opinion.	
4 Evaluation of Quality of Evidence	Student gives no response.	The student response fails to accurately evaluate OR discuss the quality of the evidence.	The student evaluates and discusses some of the quality of evidence OR indicates that quality was not relevant.	The student evaluates and discusses the quality of evidence (i.e., validity, reliability, objectivity/bias or controlled experiment.)	
5 Chain of Reasoning (Warrant)	Student gives no response.	The student response fails to explain the author's reasoning connecting the evidence to the claim.	The student explains some of the author's reasoning connecting the evidence to the claim.	The student explains the author's reasoning connecting the evidence to the claim (i.e., authority, theory, or types of logic such as generalization analogy, correlation, or cause and effect).	
6 Identification of Types of Reasoning	Student gives no response.	The student response fails to accurately identify types of reasoning.	The student accurately identifies some types of reasoning.	The student accurately identifies types of reasoning (i.e., authority, theory, or types of logic such as generalization, analogy, correlation, or cause and effect).	
7 Evaluation of Quality of Reasoning	Student gives no response.	The student response fails to accurately evaluate the quality of reasoning OR explain his/her evaluation.	The student evaluates some of the quality of reasoning and/or explains some of his/her evaluation.	The student evaluates the quality of reasoning AND explains his/her evaluation (i.e., authority, theory, or types of logic, such as generalization, analogy, correlation, or cause and effect).	
8 Concerns of the student	Student gives no response.	The student response raises no new relevant concerns.	The student raises some new relevant concerns.	The student clearly raises new relevant concerns AND expresses them as counterarguments, rebuttals or new questions OR states there are none.	
9 Conclusion and explanation about the Claim	Student gives no response.	The student response neither makesa conclusion to accept, reject or withhold a decision about the claim NOR provides an explanation of his or her reasoning.	The student makes a conclusion to accept, reject or withhold a decision about the claim OR provides an explanation about his or her reasoning.	The student makes a conclusion to accept, reject or withhold a decision about the claim AND provides an explanation for his or her reasoning.	

Fig. 8.2 Rubric for scoring Argumentation and Evaluation Guides

[AER]) to use in instruction and discussion, and to guide student dialogue in whole group and small group cooperative structures. These methods have a sequence of three instructional phases: "Cue," "Do," and "Review." During the "Cue" phase, the teacher (a) introduces the scientific argument either through a reading or through reviewing the results of an inquiry investigation; (b) explicitly informs students about the importance and benefits of learning scientific argumentation; (c) distributes and explains a one-page graphic organizer, called the AEG; and (d) prompts the students to take notes on the guide and to participate in the discussion.

During the "Do" phase, the major part of the routine, all students complete all of the parts of the guide by following the set of nine strategic thinking steps; these are the steps of the strategy that are cued on the AEG described previously. The teacher scaffolds the process of learning to analyze and construct scientific argumentations by first developing the AEG as a whole-class, teacher-guided activity, second by having students work collaboratively on the AEG with substantial teacher guidance, sharing and discussing group work with the class to create a class version of the AEG, and finally to work in small groups to collaboratively construct the AEG and then to present and defend their AEGs to the class.

Finally, in the "Review" phase, the teacher reviews the information covered in the "Do" phase and the process that the students have used to analyze and evaluate the claim and supporting argument. In this phase, students reflect on their understanding of the process of scientific argumentation, and the teacher identifies areas for additional attention.

The teacher might begin the process by reading an article that includes a science claim and evidence (either the teacher or the students reading it aloud or students reading the article individually). The teacher can then engage in whole group instruction. In addition, the teacher has the option of forming students into small, collaborative research teams and guiding them through the process of constructing a graphic organizer to support the exploration, development, analysis, and evaluation of a claim.

The teacher always guides the students in the development of understanding by co-construction of the ideas within the class, beginning with a blank AEG and interactively constructing the AEG elements based on students' prior knowledge, insights, discourses, and explanations. It is important to note that the teacher *never* hands out this graphic study guide to the students in a completed form. Rather, the teacher completes a guide before the class only as an instructional plan to organize and clarify his or her own understanding, and the teacher and all students start with a blank guide and interactively discuss and complete the guide as a whole class or in collaborative groups. Therefore, the final guide might differ from what the teacher planned, because the teacher incorporates students' contributions, questions, and insights into the final construction of the guide developed through dialogue and a consensus-building approach.

The AEI is meant to support rather than replace the ways that teachers teach critical science content. Therefore, although the intervention includes components of direct instruction as recommended by Klahr et al. (2001) and Carver (2001), it must be stressed that the components of the intervention are flexible and do not replace hands-on inquiry experiences. However, the intervention does help teachers and students organize, synthesize, record, discuss, co-construct, and review understandings from a variety of sources and experiences, including the results of student science investigations.

Lessons Learned from the Design Study

We have separated the qualitative findings from this study into three general areas. First, the teacher-researchers provided insights into their views of argumentation, perceptions of their own abilities to teach argumentation, and their views about students' abilities to engage in argumentation. Second, we gathered information about classroom implementation of the AEI as teacher-researchers used the intervention in classroom contexts; this included suggested revisions to the instructional procedures. Third, we gathered information about general strategies that teachers used to support the instruction, and about the "big picture" insights into the cross-curricular use of the AEI.

Teachers' Views of Argumentation

Background. An impetus for this study came from reports that stress the need to help our students engage in higher-order thinking. Data from national assessments, such as the National Assessment of Educational Progress Report Card (NAEP) (Applebee, Langer, Mullis, Latham & Gentile, 1994), from the Program for International Assessment (PISA) (IES, 2007), and from some research projects (Kuhn, 1991; Means & Voss, 1996), have indicated that most young Americans do not have a firm grasp of higher-order reasoning such as that associated with argumentation. This is an important issue for all students in education today. However, according to the National Research Council (2007) in *Taking Science to School*, the norms of scientific argument, explanation, and evaluation of evidence differ from the norms students encounter in everyday life. As a result, an important goal in science education is that teachers are able to support students as they learn appropriate norms and language for productive participation in the discourses of science. This goal, in itself, is challenging. Added to this, however, is the need to teach a wide range of students of differing abilities in science classes. Students may include those who have disabilities, are gifted, or are average achieving.

This idea of science for all students fits with the views of the National Science Education Standards (NRC, 1996), of AAAS Project 2061 (1993), and a multitude of national reports calling for science literacy as a basic requirement for all citizens. The goal is that efforts to improve scientific literacy be infused throughout the K-12 curriculum for all students, not just for the best and brightest. Our design study addressed the challenge of developing instructional resources that promote the acquisition of higher-level thinking skills by all students. Unfortunately, all too often materials for academically challenged students focus only on lower-level knowledge and skills and deprive them of the opportunity to acquire the critical higher-level thinking skills required to engage in quality scientific reasoning.

Ysseldyke (2009) raised an issue of importance for this study related to quality science for all students. Ysseldyke contended that *all* students should be challenged. This would require a shift in focus from providing remediation for struggling students to making sure that all students struggle. He contended that all students should be working at what Vygotsky and Michael (1978) called their zone of proximal development. A critical feature of addressing the zone of proximal development is involving students in social learning with their peers through collaborative discourse on argumentation and evaluation.

Teachers' views at the beginning of the study. At the outset, after the authors had modeled and explained scientific argumentation to the teacher-researchers during a 2-week summer institute, the teacher-researchers agreed that higher-order reasoning associated with scientific argumentation is one of the most important scientific abilities. Furthermore, they agreed that it would help students respond to real-world needs, become informed citizens who are not gullible, and succeed on state and other assessments requiring higher-order analysis and evaluation. They, therefore, recognized that the skills of argumentation would easily impact multiple areas of science literacy. They recognized that students need enough content knowledge to engage in higher-order thinking, but also need general processes and procedures such as those used in the AEI.

Despite their general support for this type of instruction, the teacher-researchers were unfamiliar with details of the higher-order reasoning as it is characterized in Toulmin's approach to argumentation or with instructional strategies for supporting students in developing the knowledge and skills of scientific argumentation. They were open, however, to collaborating with the authors in a design study to investigate these issues.

Relative to teacher-researchers' views about their students, many of the teacher-researchers in the design study believed that some students were capable of understanding concepts such as those associated with argumentation, including reliability, validity, and objectivity. They were concerned, however, that many students were not mature enough to engage in this type of thinking.

Teachers' views at the conclusion of the study. By the end of the study, the teacher-researchers provided valuable information on all components of the AEI. Specifically related to claims and qualifiers, teacher-researchers, who taught classes ranging from sixth through ninth grades, thought that their students easily understood and found claims and qualifiers. One teacher-researcher indicated that, as the study progressed, students were becoming more aware of qualifiers to claims that they found in articles or infomercials outside of classes. However, teacher-researchers also indicated that some of the qualifiers that students found (particularly from outside sources) might well fit better as concerns, that is, rebuttals, counterarguments, or new questions.

By the end of the project, teacher-researchers reported that, relative to evidence (Toulmin's grounds), their students learned to distinguish between data, fact, theory, and opinion as types of evidence. An issue of concern for researchers, however, was that some teachers were still experiencing difficulty in providing clear explanations as to the difference in some evaluative components such as validity and reliability.

All in all, teachers believed that many students felt empowered in that they learned to think about a claim and were willing and able to develop questions about a claim or evidence and to organize their thinking. Thinking about the quality of evidence was particularly useful in that students thought more about reliability and bias. Students also raised issues about possible special interests or motivation of authority figures, even if those figures represented respected institutions. In this context, students raised the issue as to whether institutions might have vested interests in a claim due to grant support and funding.

Relative to the chain of reasoning (Toulmin's warrants), input from the teacher-researchers during their pilot of the AEI provided information on their beliefs about students' abilities related to higher-order reasoning that could link a claim to the evidence presented in an argument. During the development process, several teacher-researchers recommended that the intervention not use the more complex words on the AEG that were taken directly from theorists such as Toulmin. For example, they recommended using "chain of reasoning" to represent Toulmin's "warrants" and using "concerns" to represent "rebuttals." Although the researchers accepted these recommendations at the time, this ultimately raised issues regarding the wisdom of substituting some simpler synonyms for complex theoretical terms.

At the end of the study, when the teacher-researchers discussed the chain of reasoning, they believed that students seemed to understand how authority and theory served as appropriate warrants for a claim. However, the teacher-researchers believed that students had more difficulty with the complex area of logic. Relative to logic, they thought that students understood and used the term "logic" correctly in a general way, but did not understand various components of logic (as used in the AEG and in the instruction) such as analogy, correlation, causation, and generalization.

Relative to rebuttals and counterarguments, some teacher-researchers thought that these terms were difficult for students to understand, although they could more easily come up with new questions. One teacher-researcher thought that the greatest benefit came from student consideration of the last two components of the AEG: consideration of concerns and new questions, and drawing conclusions about the claims. Students, particularly in the upper grades, demonstrated some transfer of learning in that they commented on what they saw or read outside of class that contained claims. These included information found in infomercials, mailings, and various advertisements and articles.

One very important issue involves teacher-researchers' perceptions that students did not particularly enjoy the argumentation and evaluation instruction and activities. Researchers questioned whether this perception led some teachers to report that they would not use all the components of the intervention in the future. Student enjoyment is, indeed, a concern in education, but is only one consideration that must be subjected to more research.

Summary. In summary, this study challenged some of the teacher-researchers' prior beliefs about the level at which students can engage in higher-order thinking associated with argumentation. Some teacher-researchers made assumptions about the learning abilities of students in earlier grades, believing that they might not be able to acquire the higher-order thinking associated with analysis and evaluation of claims and argument. This was not borne out. In general, teacher-researchers valued the intervention and believed that it had a place in their curricula.

Implementation Issues Relative to the Argumentation and Evaluation Intervention

Background. This study is built on the work of others relative to designing curriculum materials that help teachers engage students in making explanations based on evidence, with a focus on the argumentation components put forward by Toulmin (1958). During the design study, however, teacher-researchers collaborated with the authors in attempting to modify the terminology and presentation of the Toulmin model in the AEG to adapt it to the prior knowledge, abilities, and experiences, especially of middle school students.

Implementation fidelity. Suggestions from the teacher-researchers regarding adaptations of terminology and instructional procedures raised issues related to fidelity of implementation. The strategy, graphic organizer, and instructional procedures are built on the research on Content Enhancements (Bulgren and Lenz, 1996). The work of the KU-CRL has emphasized the importance of fidelity of implementation. Rigorous research has been conducted on many of the Content Enhancement Routines, and much of this research has reinforced the importance of fidelity to the core components of research-based interventions.

The need for fidelity of implementation when using procedures that have been previously subjected to rigorous research is emphasized by Ysseldyke (2009) in his discussion of the importance of treatment integrity. He contended that when effective treatments are implemented with fidelity or integrity, the treatments have a strong effect on student outcomes, but that when that is not the case, research results can be misleading, making an observer think that a treatment is not effective when it was actually the implementation that was not effective.

Scheduling. Overall, the teacher-researchers recommended that the AEI be introduced early in the school year. They also recommended that the AEI be taught as part of the scientific process. This could help students recognize the need to look across science areas for competing claims in a variety of content materials.

Provision of examples. The teacher-researchers also recommended developing multiple types and examples of scientific and socio-scientific claims to include in a curriculum to facilitate integration of reasoning about claims and arguments across the school year. Examples suggested by the teacher-researchers included developing and/or analyzing scientific arguments that were based on published experimental studies, historical research, correlational studies (such as epidemiological studies in medicine), social-scientific claims, controversial issues (with multiple, competing claims), and student-generated results from classroom investigations.

Source of claims. Teacher-researchers used the intervention as envisioned, that is, with the written claims made by others, but also expanded its use to lab reports and provided other ways to improve the intervention. For example, teacher-researchers showed students how to write their lab reports based on the structure of the AEG as a way to clarify and present their own claim and arguments. Teacher-researchers found that the use of the AEG helped consolidate what students learned from labs. They also reported that just providing "hands-on" lab experiments did not ensure understanding by students of the critical science content related to the laboratory activity. By reviewing the AEG that students developed based on their experiments, students thought more about the experiment and how the results related to key science concepts.

Student discourse. The use of the intervention with students' own materials raised an important issue in terms of evaluation of their own results and collaborative discourse with other students. Even after students had conducted a lab experiment, they often found it hard to think of themselves as "authorities." Nonetheless, using the AEG allowed many students to challenge other students on their data analysis and accuracy leading to scientific discourse in the classroom. The teacher-researchers indicated that such a challenge might have seemed more objective (and perhaps less confrontational) when using the AEG as opposed to direct criticism. As a result, students seemed less likely to withhold comments on another student's thinking. One teacher-researcher thought that the AEG helped students most in lab reports because it provided supports for students to write summaries of the results of experiments and to write detailed justifications for their conclusions.

Terminology. Teacher-researchers suggested modifications to help students understand difficult or potentially difficult vocabulary. Therefore, an adaptation of terminology suggested by the teacher-researchers that we accepted during the study was the substitution of the word "concerns" for "rebuttals" or "counterarguments." Although well-intentioned, this suggestion by the teacher-researchers proved problematic. At the end of the study, students were interpreting the word "concerns" in a more personal way. That is, the issues they raised in that component of the analysis often included personal fears. Specifically, the problem that became obvious in the scoring sessions was that students often assumed that they should raise their own worries, rather than rebuttals, counterarguments, or new questions about a claim. This raises the issue that terms need to be clear and precise.

Sequential levels of implementation. All of the teacher-researchers provided incremental introductions to the AEG, sometimes breaking it into three parts and providing conceptual understanding of key vocabulary support at each level. These levels were as follows.

Level one argumentation emphasized the importance of initial learning activities focusing on the *big picture* of science argumentation. To do this, students were engaged in whole class and small-group discourse about their questions regarding the claim, about missing information in the argument, about rebuttals to the claim, and about students' concerns about the quality of the argument.

Level two argumentation engaged students in examining the *evidence* used to support a claim and in evaluating the quality of the evidence (which includes elements such as reliability, validity, and objectivity). Students gradually examined these elements of quality of evidence when questions arose in their discourse about the topics addressed in level one. Therefore, level two types of argumentation discourse often followed quickly and organically from level one.

Level three argumentation engaged students in examining the *chain of reasoning* (the warrant) that supports a claim. This examination includes identifying the type(s) of warrant used based on theory authority and/or logic. At this step, it is critically important that students can explain how the warrant connects the evidence to the claim, leading the reader to believe the claim. However, the teacher-researchers indicated that the use of the intervention required a great deal of time and experience for students to master these abilities.

Scaffolded materials. Other teacher-researcher suggestions related to the structure of the graphic organizer. This provided an interesting insight into the evolution of teacher thinking, perceptions, and analysis of instruction. For example, in the third year, the sixth-grade teacher-researcher indicated that using the whole AEG was overwhelming for her students. As a result, she broke the Guide into parts, put each of the parts on a separate page, and used each part, one at a time. Specifically, she prepared color-coded pages that contained the individual components of each section of the Guide. She reported that she initially believed that when younger students were able to focus on the component parts individually, they would be able to better analyze the claim with its associated qualifiers, evidence, warrants, rebuttals, and conclusions. At the end of the study, when she probed student satisfaction with the color-coded guide, however, she found that the students, in fact, did not like the color-coded approach.

Another teacher-researcher projected the article onto a white board and let students choose colored markers to highlight parts of information that supported different sections of the AEG, which proved more appealing to students and seemed to encourage class discussion. These innovative additions, and students' responses to them, raised the issue of the need for teachers to engage in ongoing discourse with students to determine relatively quickly how students perceive the usefulness of innovative procedures. What may have face validity for the teachers may not have the same appeal to the students. These findings emphasized the importance of engaging classroom teachers as teacher-researchers in development projects.

Summary. In summary, although we encouraged the teacher-researchers in this design study to take a great deal of latitude in trying adaptations and modifications, their final feedback on how they would implement the intervention raises concerns as anticipated by Ysseldyke (2009) and others (Bulgren & Lenz, 1996). At the end of the study, some teachers indicated that they would continue to use only parts of the AEI. As indicated by Ysseldyke (2009), the "pick-n-choose" approach to the use of validated interventions does not lead to optimal student learning. This may become a very important issue for this intervention if teachers pick and choose which components to use and which not to use in the future, especially if they omit the higher-order thinking related to analyzing and evaluating evidence and reasoning. Alternatively, if teachers want to modify an evidence-based intervention, there must be ways to subject those modifications to the levels of rigorous research demanded by the field. An empirical question is the current level of fidelity of implementation on the part of teachers after professional development activities and the support that some teachers need to implement research-based interventions with fidelity.

Strategy Supports and Cross-curricular Use of the Argumentation and Evaluation Intervention

Background. A contribution of this project is the use of theory about domain-specific and domain-general knowledge and dimensions of scientific reasoning processes (Klahr et al., 2001). Scientific reasoning, according to Klahr et al. (2001), is classified by domain specificity versus domain generality as well as by the type of reasoning processes involved, such as generating hypotheses, designing experiments, or evaluating evidence; each of these three processes may be explored for either domain-general or domain specific knowledge. Furthermore, Reiser, Tabak, Sandoval, Smith, Steinmuller, and Leone (2001) contended that students must develop a deep understanding of science and use general strategies in particular scientific domains if they are to approach arguments more as experts than as novices. From another point of view, Stevens, Wineburg, Herrenkohl, and Bell (2005) argued for a move from fragmented approaches to more comparative and unified approaches in instruction that could, ideally, make school a meaningful place for students to learn and analyze even more complex, overlapping issues. This raises the issue of how widely applicable strategies that are considered general can be used to support a new intervention such as the AEI, as well as how applicable an intervention such as the AEI is in cross-curricular areas.

Strategies used in support of the Argumentation and Evaluation Intervention. Supporting strategies that teacher-researchers found helpful included those associated with questioning and reading. We found that teacher-researchers spontaneously used questioning in the classroom to scaffold the learning with the AEI. This highlights the importance of teaching supporting literacy strategies, such as those requiring paraphrasing and summarizing as needed. In addition, the use of questioning in instruction has been supported by meta-analyses. For example, Rosenshine, Meister, and Chapman (1996) focused on interventions that utilize questioning techniques in some form. They reviewed the studies focusing on teaching students to generate questions as a way of improving their comprehension during or after reading or listening to a passage. In general, they found that directly teaching students to ask and answer questions yielded significant differences in favor of the experimental groups with regard to tests constructed by the researchers.

Specifically related to our study, some question types that teacher-researchers found effective were those identified by Rosenshine et al. (1996). Among these effective questioning techniques, teacher-researchers often used signal words such as "why," generic questions, and question stems. To illustrate, the sixth-grade teacher-researcher reported that challenges occurred with the components of "chain of reasoning" (warrant). The Guide and instruction included three overall types of reasoning: authority, theory, and logic. The sixth-grade teacher-researcher provided scaffolding questions (with a "fill-in-the blank" format) for the students such as the following:

"Why does _____ *authoritatively* prove the claim?"
"Why does _____ *logically* prove the claim?"
"Why does _____ *theoretically* prove the claim?"

In addition, teacher-researchers used prompting questions that students could use as they explored rebuttals and counterarguments. These were questions such as the following:

"What *more* scientific information do you need?"
"Is there an argument *against* the claim?"
"What are questions for *further* investigation?"
"What more could be done to *improve* the research?"

Relative to reading supports for students who had reading difficulties, most teacher-researchers read the articles aloud to the class when articles were the focus of the activity. They also utilized "think-alouds" and modeling as they read the article to the class—particularly in the early learning phases of using the AEG. Various learning scaffold suggestions by teacher-researchers included modeling by the teacher-researchers how and where they themselves found the types of logic in the reading or in the laboratory activity.

Cross-curricular use of argumentation. Teacher-researchers provided insights into the cross-curricular use of a strategy focusing on the analysis of claims and arguments. At the outset of the study, the science teacher-researchers saw the goal of analyzing and evaluating claims and associated argumentation as specific to the domain of science in which they were teaching.

By the end of the year, feedback from the teacher-researchers and others in the schools indicated a broader value and use of the thinking skills associated with argumentation. Specifically, the teacher-researchers received positive support from others, including reading teachers, administrators, and other content area specialists. For example, both reading teachers and administrators who observed the use of the AEI indicated that this intervention was useful as a reading support as well as a science learning support at the eighth-grade level. The reading teachers and administrators pointed out the power of the intervention to focus on conceptual understanding of words such as reliability, validity, objectivity, theory, logic, and authority. They contended that deep understanding of these words would help students succeed on a variety of assessments.

Particularly important, the sixth-grade teacher-researcher provided support for the usefulness of the AEI as a good reading strategy. She reported that by the end of the year, students understood the need to read and re-read an article for different purposes—first for an overview, second to identify the claim, third to identify the evidence, and so on. This teacher also thought that it has helped students persevere as they read and re-read an article.

In general, the teacher-researchers believed that this intervention helped students read for deep meaning and big ideas rather than just searching for facts. As a result, students referred more to the text to understand components of an argument. Interestingly, teachers in other classes such as Advanced Placement Language Arts reported teaching procedures and processes similar to Toulmin's model in their persuasive writing curriculum. Therefore, they were very supportive of using this approach in science classes, and discussed coordination of terminology and goals across subject areas.

Summary. In summary, as the third year progressed, teacher-researchers shared insights about the potential cross-curricular power of the evaluation of claims and arguments. These insights arose in two areas. First, the teacher-researchers did not believe that they could effectively teach all that was needed about the evaluation of claims and arguments in 1 year, in one science class, and in one specific domain. Second, teacher-researchers shared new information about the objectives in their districts that had cross-curricular implications. Some noted that an emphasis on evaluation of claims and arguments was becoming important not only across science grades and courses, but that it also was becoming a focus in other areas such as Language Arts. This has led to suggestions for future research on the analysis of claims and arguments across domains and content areas. Research in this area would attempt to determine the components of the AEI that may be incorporated across subjects, domains, and disciplines.

Conclusions and Recommendations

As a result of this study, the AEI was shown to help students understand and use the components of argumentation as put forth by Toulmin. However, the study also resulted in recommendations for use of the intervention and future research.

Teachers need preservice courses and in-service support as they teach students to engage in complex higher-order thinking in science. For example, we observed some teacher-researchers struggling to explain the differences in concepts such as reliability and validity. Therefore, if national standards ask that students become citizens of the world by engaging in higher-order thinking, then the field must prepare teachers to have the background to help students do this. As a result, an empirical question for future research relates to the current preservice courses taught at the undergraduate level in universities, and the level to which they incorporate adequate support for teachers to engage in the teaching of higher-order thinking such as that required to analyze and evaluate claims and arguments.

Furthermore, it is possible that teachers would benefit from ongoing collaborative meetings with other teachers, both in their content area and in other content areas. In these meetings, ongoing collaborative discourse might well support innovative instruction associated with higher-order thinking within and across content areas.

Teachers also need ways to analyze and monitor their own impressions of how students are performing in their classes. Ongoing, effective and efficient ways to analyze students' readiness to learn, their perceptions of adaptations and modifications, and their enjoyment of the instruction are needed. For example, many teachers believed that younger students, such as the sixth graders in this study, were not able to learn to engage in higher-order thinking associated with argumentation. Students at that grade level, however, out-performed many other groups of students from higher grade levels (Ellis & Bulgren, 2009). In addition, a way to effectively and efficiently monitor students' perceptions of adaptations and modifications would be useful. This would have allowed the teacher who spent a great deal of time breaking apart and color-coding components of the AEG to adjust her modifications during, rather than at the end of, the project.

Another need is to explore ways to monitor students' views and enjoyment of the new instruction. Therefore, an empirical question relates to the correlation between student enjoyment and learning, and how students report their impressions. It might be that the use of student interviews rather than, or in addition to, objective satisfaction surveys in future research studies would provide valuable insights for teachers and researchers. In addition, future research is warranted on the use of formative assessments to determine not only student progress in using an intervention, but also students' views of the instructional procedures.

Teacher-researchers also raised issues concerning the fidelity of implementation of the intervention. Teacher-researchers, in general, indicated that they would not use the entire routine in the future, presumably because of the difficulties involved in higher-order thinking associated with argumentation—the very component that is being urged by researchers and commentators. This issue needs to be addressed by the field. When research findings indicate that a package of instructional interventions serves to help students learn, the value of fidelity of implementation must become an important issue in professional development and classroom use.

Therefore, future research is needed to explore the required levels of fidelity of implementation of the component parts of a research-based procedure to assure learning outcomes at levels similar to those found in the original research. Follow-up research into the effects of using only portions of components of a research-based intervention, rather than the complete set of components, would provide much-needed information for teachers and professional developers. If teachers want to modify a research-based intervention, there must be ways to subject those modifications to standards of rigorous research.

In addition, other research and development may be needed to incorporate enjoyable ways of learning, such as learning games, into instruction. The incorporation of such games may well add not only to student enjoyment of the learning process, but also to critical learning time and collaborative engagement for students.

A related need is to determine the number of times a teacher needs to implement a new intervention for students to benefit. For example, the ninth-grade teacher-researcher implemented the intervention only six times compared to 10 times by all other teacher-researchers, and her students performed at approximately the same level as the sixth-grade students.

This study also raised the possibility that interventions such as the AEI may have benefits beyond the specific area in which the original research was conducted. This is possible because of similar content literacy and higher-order thinking demands across content areas, subjects, domains, and disciplines. For example, it was reported that a reading teacher, after observing the implementation of the instruction in a science class, indicated that the AEI was one of the best exemplars of vocabulary development and support for conceptual understanding. She suggested that its use could have a positive impact on state assessments.

In addition, teachers from other areas, such as Language Arts, indicated that the higher-order thinking was the same as they emphasized in their courses when they taught persuasive writing. Components of the AEI that were considered useful across content areas included thinking about claims and the qualifiers to the claims, analyzing and evaluating evidence and reasoning, considering other options, and coming to and defending a conclusion about the worthiness of claims. Therefore, an empirical question is whether cross-curricular use of research-based instruction, such as in AEI, would enhance learning in different content areas due to multiple exposures to higher-order thinking challenges. Future research might also address the power of such interventions to improve student performance on state assessments.

In conclusion, the valuable contribution of both qualitative and quantitative studies on a single intervention contributed to a rich understanding of the complex challenges of teaching argumentation. Furthermore, the contribution of teacher-researchers in the classrooms provided valuable insights for the study as they used the AEI in regularly scheduled science instruction.

References

American Association for the Advancement of Science (AAAS). (1993). *Benchmarks for science literacy*. New York: Oxford University Press.

Applebee, A. N., Langer, J. A., Mullis, I. V., Latham, A. S., & Gentile, C. A. (1994). *The national assessment of education progress 1992 report card*. Princeton, NJ: Educational Testing Service.

Bannan-Ritland, B. (2003). The role of design in research: The integrative learning design framework. *Educational Researcher*, *32*, 21–24.

Bransford, J., Brown, A. L., & Cocking, R. R. (Eds.). (2000). *How people learn: Brain mind, experience and school*. Washington, DC: Committee on Learning Research and Educational Practice, National Research Council.

Bulgren, J. A., Deshler, D. D., Schumaker, J. B., & Lenz, B. K. (2000). The use and effectiveness of analogical instruction in diverse secondary content classrooms. *Journal of Educational Psychology, 92*(3), 426–441.

Bulgren, J. A., & Lenz, B. K. (1996). Strategic instruction in the content areas. In D. D. Deshler, E. S. Ellis, & B. K. Lenz (Eds.), *Teaching adolescents with learning disabilities: Strategies and methods* (2nd ed., pp. 409–473). Denver, CO: Love Publishing.

Bulgren, J. A., Lenz, B. K., Schumaker, J. B., Deshler, D. D., & Marquis, J. (2002). The use and effectiveness of a comparison routine in diverse secondary content classrooms. *Journal of Educational Psychology, 94*(2), 356–371.

Bulgren, J. A., Schumaker, J. B., & Deshler, D. D. (1988). Effectiveness of a concept teaching routine in enhancing the performance of LD students in secondary-level mainstream classes. *Learning Disability Quarterly, 11*(1), 3–17.

Carver, S. (2001). Cognition and instruction: Enriching the laboratory school experience of children, teachers, parents, and undergraduates. In S. M. Carver & D. Klahr (Eds.), *Cognition and instruction: Twenty-five years of progress* (pp. 385–426). Mahwah, NJ: Lawrence Erlbaum Associates.

Davis, E. A., & Krajcik, J. S. (2005, April). Designing educative curriculum materials to promote teacher learning. *Educational Researcher, 34*(3), 3–14.

Deshler, D. D., & Schumaker, J. B. (1986). Learning strategies: An instructional alternative for low-achieving adolescents. *Exceptional Children, 52*(6), 583–590.

Driver, R., Newton, P., & Osborne, D. (2000). Establishing the norms of scientific argumentation in classrooms. *Science Education, 84*(3), 287–312.

Ellis, J., & Bulgren, J. (2009, March). *Professional development materials to develop student knowledge and skills of scientific argumentation.* (Year 3 report submitted to the National Science Foundation). Lawrence, KS: University of Kansas Center for Research on Learning.

Institute for Education Sciences—National Center for Educational Statistics. (2007). *Highlights from PISA 2006: Performance of US 15-year-old students in science and mathematics literacy in an international context.* NCES 2008016. Washington, DC: Author.

Kelly, A. E. (2004). Design research in education: Yes, but is it methodological? *The Journal of the Learning Sciences, 13*(1), 115–128.

Klahr, D., Chen, Z., & Toth, E. (2001). Cognitive development and science education: Ships that pass in the night or beacons of mutual illumination? In S. M. Carver & D. Klahr (Eds.), *Cognition and instruction: Twenty-five years of progress* (pp. 75–119). Mahwah, NJ: Lawrence Erlbaum Associates.

Klahr, D., Fay, A. L., & Dunbar, K. (1993). Heuristics for scientific experimentation: A developmental study. *Cognitive Psychology, 25*(1), 111–146.

Koslowski, B. (1996). *Theory and evidence: The development of scientific reasoning.* Cambridge, MA: MIT Press.

Koslowski, B., Okagaki, L., Lorenz, C., & Umbach, D. (1989). When covariation isn't enough: The role of causal mechanism, sampling method and sample size in causal reasoning. *Child Development, 60*, 1316–1327.

Kuhn, D. (1991). *The skills of argument.* Cambridge: Cambridge University Press.

Kuhn, D., Amsel, E., & O'Loughlin, M. (Eds.). (1988). *The development of scientific thinking skills.* San Diego, CA: Academic.

Kuhn, D., Garcia-Mila, M., Zohar, A., & Andersen, C. (1995). Strategies of knowledge acquisition. *Monographs of the Society for Research in Child Development, 60* (4, Serial No. 245).

Lawson, A. E. (2003). The nature and development of hypothetico-predictive argumentation with implications for science teaching. *International Journal of Science Education, 25*(11), 1387–1408.

Linn, M. C., Clark, D., & Slotta, J. D. (2003). WISE design for knowledge integration. *Science Education, 87*, 517–538.

Means, M. L., & Voss, J. F. (1996). Who reasons well? Two studies of informal reasoning among children of different grade, ability, and knowledge levels. *Cognition and Instruction, 14*(2), 139–178.

National Center for Education Statistics (NCES). (2004). *Trends in international mathematics and science study.* Retrieved December 9, 2004, from http://nces.ed.gov/timms/Results.asp?Results=1

National Research Council (NRC). (1996). *National science education standards.* Washington, DC: National Academies Press.

National Research Council (NRC). (2007). *Taking science to school.* Washington, DC: National Academies Press.

Penner, D., & Klahr, D. (1996). The interaction of domain-specific knowledge and domain-general discovery strategies: A study with sinking objects. *Child Development, 67*, 2709–2727.

Reiser, B., Tabak, I., Sandoval, W. A., Smith, B. K., Steinmuller, F., & Leone, A. J. (2001). BGuILE: Strategic and conceptual scaffolds for scientific inquiry in biology classrooms. In S. M. Carver & D. Klahr (Eds.), *Cognition and instruction: Twenty-five years of progress* (pp. 263–306). Mahwah, NJ: Lawrence Erlbaum Associates.

Rosenshine, G., Meister, C., & Chapman, S. (1996). Teaching students to generate questions: A review of the intervention studies. *Review of Educational Research, 66*(2), 181–221.

Schauble, L. (1996). The development of scientific reasoning in knowledge-rich contexts. *Developmental Psychology, 32*, 102–119.

Stevens, R., Wineburg, S., Herrenkohl, L. R., & Bell, P. (2005, Summer). Comparative understanding of school subjects: Past, present, and future. *Review of Educational Research, 75*(2), 125–157.

Toulmin, S. (1958). *The uses of argument.* Cambridge: Cambridge University Press.

Toulmin, S., Rieke, R., & Janik, A. (1984). *An introduction to reasoning*. New York: MacMillan.

Vygotsky, L. S., & Michael, C. (1978) *Mind in society: The development of higher psychological processes*. Cambridge, MA: Harvard University Press.

Wallace, C. S., Hand, B., & Yang, E. (2004). The science writing heuristic: Using writing as a tool for learning in the laboratory. In E. W. Saul (Ed.), *Crossing borders in literacy and science instruction* (pp. 355–368). Newark, DE: International Reading Association.

Ysseldyke, J. (2009). When politics trumps science: Generalizations from a career of research on assessment, decision making, and public policy. *Communique, 38*(4), 6–8.

Part III
Researching Argumentation in Science Education

Chapter 9
Research on Critique and Argumentation from the Technology Enhanced Learning in Science Center

Douglas B. Clark, Victor Sampson, Hsin-Yi Chang, Helen Zhang, Erika D. Tate, and Beat Schwendimann

Introduction

This chapter provides an overview and synthesizes research on critique, argument construction, and argumentation from the Technology Enhanced Learning in Science Center (TELS). TELS received funding from 2003 to 2010 by the US National Science Foundation to investigate approaches for improving learning and instruction in science classes for students in grades 6–12 with a focus on the role that information technology can play. TELS institutions included UC Berkeley, Concord Consortium, Arizona State University, Penn State, Technion, North Carolina Central University, and many others.

The work in TELS was guided by the knowledge integration (KI) framework (Linn & Eylon, 2006). This framework involves four main components: (1) eliciting current ideas, (2) introducing new ideas, (3) developing criteria for evaluating ideas, and (4) sorting and reorganizing ideas. Research and development in TELS applied and analyzed approaches and design principles based on this framework for supporting students and teachers engaging in inquiry with combined simulations, hands-on data collection, and other sources of information to make sense of complex science phenomena. Most of the curricular projects developed as part of TELS incorporated critique, argument construction, and argumentation in alignment with this framework, particularly in the context of helping students make sense of data they collected through visualizations, labs, and other evidence sources. In support of these efforts, several TELS researchers focused their research on the integration of critique, argument construction, and argumentation in TELS projects.

The first section of this chapter provides an overview of the web-based inquiry science environment (WISE), which was the principal context for much of the TELS work. The chapter then summarizes and synthesizes TELS research on critique, argument construction, and argumentation. Following our discussion of the TELS research on critique, argument construction, and collaborative argumentation, the

D.B. Clark (✉)
Vanderbilt University, Nashville, TN, USA
e-mail: doug.clark@vanderbilt.edu

M.S. Khine (ed.), *Perspectives on Scientific Argumentation*,
DOI 10.1007/978-94-007-2470-9_9,

chapter discusses the implications of these findings in terms of the overarching knowledge integration framework and future work.

Web-Based Inquiry Science Environment (WISE)

Much of the work in TELS was organized and conducted within the WISE environment. We therefore provide an overview of the WISE environment to provide context for subsequent discussion of research on critique, argument construction, and argumentation that TELS conducted in WISE. WISE is a powerful digital platform for multiple users and purposes (Fig. 9.1). It supports research innovation and teacher customization of inquiry activities in science classrooms. TELS researchers use WISE to design and develop inquiry-based online curricula. Teachers can adapt, customize, or create WISE curricula to address their local needs. Teachers use the same WISE platform to implement WISE curricula, assess their students' work, and share their experience with other WISE teachers. In addition to English,

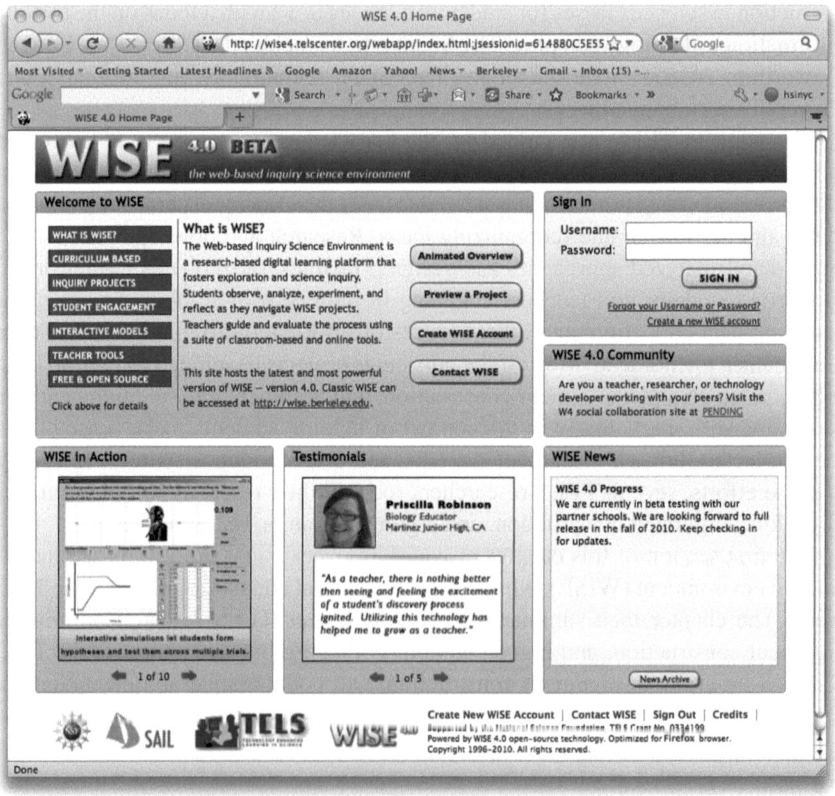

Fig. 9.1 WISE homepage

WISE includes projects in many languages such as Chinese, Dutch, Korean, and Norwegian. There are currently more than 20 developed projects in the main WISE project library on topics of physics, chemistry, earth science, biology, and physical and life science for high school or middle school students, available as open resources for teachers to use along with thousands of customized projects that various teachers and groups have created for their own contexts. WISE also supports international customization (Chang & Linn, 2010; Fig. 9.2).

The KI framework (Linn & Eylon, 2006) guides the design of WISE projects. In general, WISE projects have three main features. First, the *WISE inquiry map* reveals the structure of a WISE project and the learner's current activity and step (Fig. 9.3). The inquiry map guides students through a variety of activities and steps including visualization steps, modeling steps, reflection steps, evidence steps, and so forth. A series of steps can be aligned together to promote the KI process. For example, a series of predict–observe–explain (POE, White & Gunstone, 1992) steps can help elicit students' ideas before their observation and connect students' ideas to the new ideas after the observation step. Second, *highly interactive visualizations* enhance student learning of abstract or complex scientific concepts or phenomena that involve large-scale or unobservable levels. In the *Thermodynamics* project, for example, an interactive visualization about the molecular movement between objects with different temperature helps students learn the mechanism

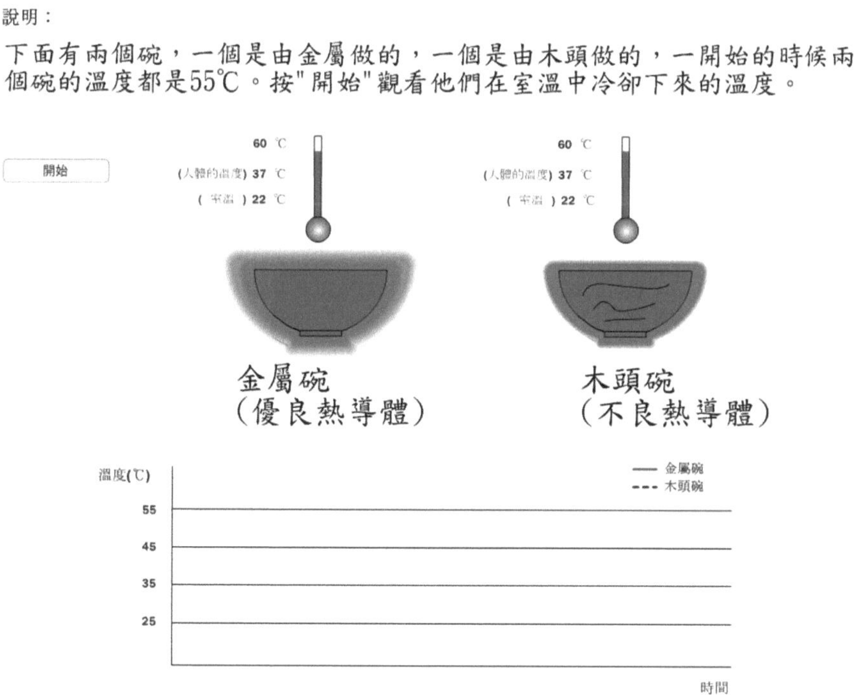

Fig. 9.2 The Chinese version of the WISE *Thermodynamics* unit

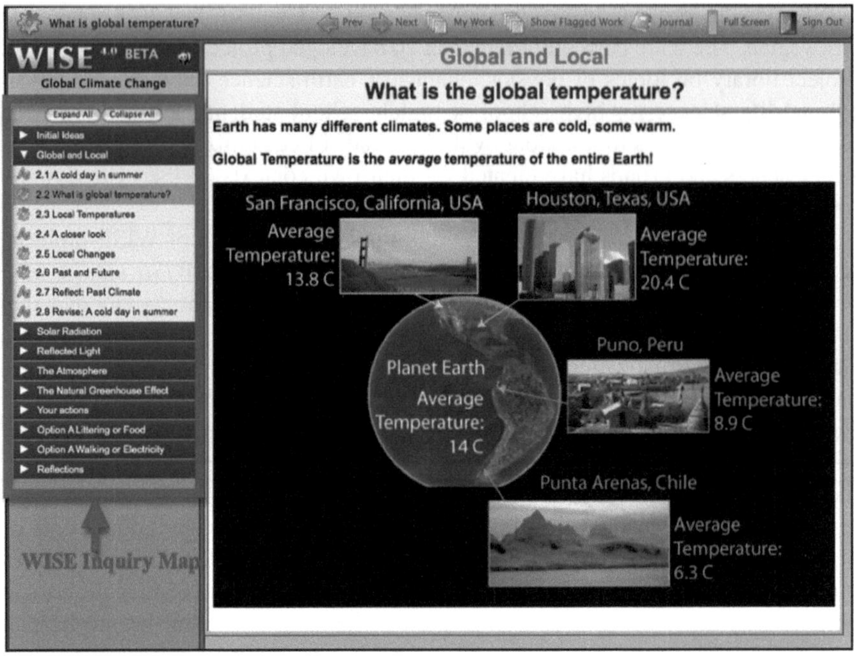

Fig. 9.3 WISE inquiry map

of heat transfer at the molecular level (Chang & Linn, 2011; Clark, 2006; Clark & Sampson, 2007, 2008) (Figs. 9.4 and 9.5). Finally, WISE projects incorporate embedded assessments to make students' thinking visible and to support students in developing conceptual understanding, decision-making, and inquiry abilities. Types of WISE embedded assessments range from multiple-choice items to open-ended textual or drawing items for curricular designers to choose from based on their needs. It is imperative for teachers to see evidence of how students are doing on the embedded assessments to help the teachers adjust their teaching and help students learn. Online feedback from the computer or the teacher helps students reconcile, reflect on, or sort their ideas.

The WISE platform supports researchers in conducting iterative design experiments (Brown, 1992; Cobb, Confrey, diSessa, Lehrer, & Schauble, 2003) and accumulating and managing research data. WISE curricula have undergone multiple iterations of designing, implementing, assessing, and refining in multiple classrooms and other educational settings. Research indicates an overall significant effect of WISE curricula over traditional instruction on students' achievements in science (Linn, Lee, Tinker, Husic, & Chiu, 2006). The study by Linn et al. (2006) reported on 12 WISE units and assessments. Each unit required about one week of class time. They compared two large time-delayed cohorts of students from schools that serve English language learners, students underrepresented in science, and students receiving free or reduced price lunches. TELS administered assessments shortly

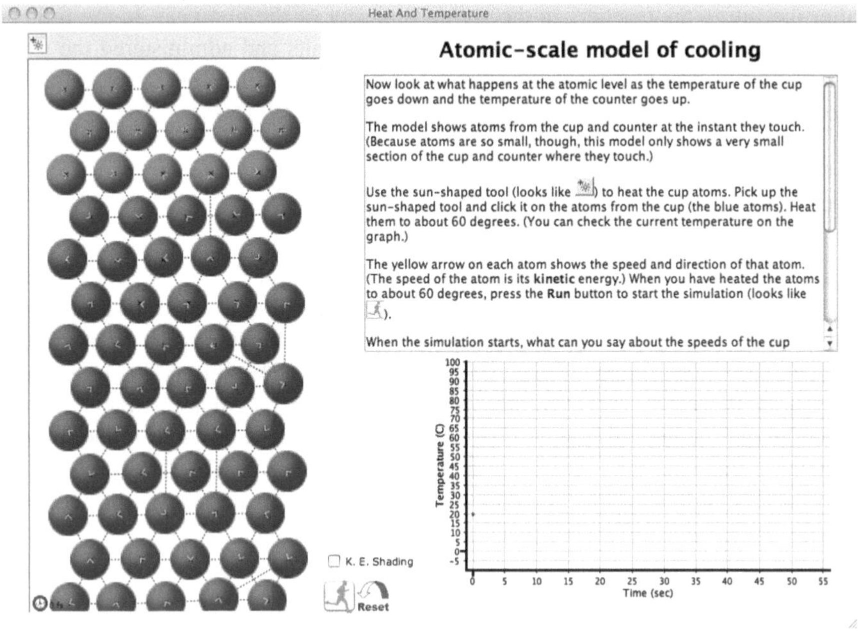

Fig. 9.4 The observation visualization

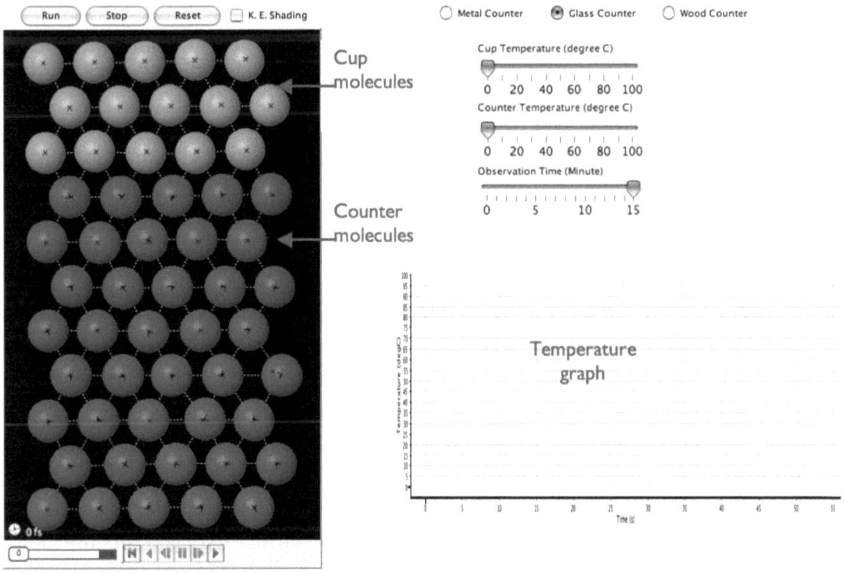

Fig. 9.5 The interactive visualization

after recruiting the teachers in the spring semester of the first year. In the following year the 25 teachers implemented the WISE units and administered the same assessments to the new cohort of students. Overall, the TELS cohort outperformed the typical cohort (effect size: 0.32, $p < 0.001$; Linn et al., 2006).

TELS Research on Critique

Scientific inquiry can be conceived as a knowledge building process where explanations are constructed to make sense of data and then presented to the broader community for critique, debate, and revision (Driver, Newton, & Osborne, 2000; Duschl, 2007; Passmore & Stewart, 2002; Sandoval & Reiser, 2004; Stewart, Cartier, & Passmore, 2005; Vellom & Anderson, 1999). Critique is thus a critical part of understanding the inquiry process and can potentially be harnessed in support of helping students make sense of complex science concepts. TELS research on critique has focused on (1) the potential of critique to support students as they conduct virtual experiments, (2) the effects of drawing and critique on enhancing student learning with dynamic visualizations, and (3) the integration of content knowledge through critique-focused concept maps.

Use of Critique to Support Students in Conducting Virtual Experiment

The first area of research on critique focuses on supporting students as they conduct virtual experiments. Interactive dynamic visualizations can engage students in conducting scientific experiments around visualizations to learn abstract scientific concepts or unobservable scientific phenomena. However, purposefully conducting virtual experiments to gain understanding in science is a challenge task for many students. For example, increasing the interactivity of a computer visualization allows students to change parameters of the visualization, but this openness may introduce extra difficulties. Students often use trial-and-error as opposed to mindful strategies (Chang & Tsai, 2010). Scaffolding can support students in efficiently conducting virtual experiments to develop adequate conceptual understanding. Research suggests coupling highly interactive visualizations with metalevel learning activities, such as self or peer evaluation (Chang, Quintana, & Krajcik, 2010; Moreno and Valdez, 2005) or critique (Chang, 2009; Chang & Linn, 2011) to help students reflect on and refine ideas (Linn, Chang, Chiu, Zhang, & McElhaney, 2011). More studies are needed to investigate how to design effective learning environments supportive of critique.

Questions. How effective are scaffolded critique activities in supporting students' understanding in science? How do students critique virtual experiments attributed to others?

Context. The TELS research by Chang (2009) and Chang and Linn (2011) used the WISE (Linn & Hsi, 2000; Linn, Davis & Bell, 2004) unit called

"Thermodynamics: Probing Your Surroundings" (Clark, 2006; Clark & Sampson, 2007, 2008). The unit begins with a thermal equilibrium hands-on experiment where students select six objects in the classroom, predict the temperatures of the objects, and then measure the temperatures using a thermal probe. Next students experiment with molecular workbench visualizations to explain heat transfer. Then students generate principles to explain patterns they observed in the temperatures of the objects and how hot or cold they felt. The final activity engages students in discussing their principle with peers and reflecting on how to revise their principles.

This one-week online inquiry project initially featured the observation version of the molecular workbench (Xie & Tinker, 2006) visualization (Fig. 9.4). Using the visualization students can observe how the molecular movement and temperature graph change when a hot cup is placed on a cold counter. Chang (2009) and Chang and Linn (2011) modified the visualization to create the interactive and critique versions while maintaining the one-week duration. In the interactive version the revised visualization (Fig. 9.5) allows students to change values of four variables to conduct virtual experiments with the visualization: (1) the counter material (metal, glass, or wood), (2) the cup temperature, (3) the counter temperature, and (4) the time of the experiment. In the critique version instead of reading guidelines about how to conduct the virtual experiments, students were guided to critique a fictitious student's, Mary's, experiment before conducting their own virtual experiments.

Methods. The study involved two science teachers and their 205 eighth-grade students in seven classes at two public middle schools in California. One teacher was able to randomly implement the critique and interactive conditions and the other teacher chose to run the observation version due to technical issues at the school. As a result, three classes used the critique version, two classes used the interactive version, and another two classes used the observation version. Data collected and analyzed included all students' responses to the pre- and posttests and embedded assessments.

Findings. How effective are the scaffolded critique activities to support students' understanding in science? The study compared student performances among the three conditions to discern the added value of critique. Effect sizes between the pre- and posttest scores for the three conditions ranged from moderate for the observation, $d = 0.57$, and interactive conditions, $d = 0.63$, to large for the critique condition, $d = 1.21$. However, a teacher or school effect might exist since the observation condition was implemented in one school while the interactive and critique conditions were implemented in a different school. On the other hand, the contexts in interactive and critique conditions were comparable. Using ANCOVA to control for differences in pretest levels, the critique condition outperformed the interactive condition on the total posttest scores [$F(1) = 6.53$, $p = 0.012$]. The results indicate that the virtual experiments were effective when coupled with the critique activity.

How do students critique virtual experiments attributed to others? The students in the critique condition showed that they were able to suggest better methods for Mary's experiment but were less successful in evaluating the interpretation of the experiment. Students demonstrated understanding of strategies such as selecting extreme values to make experimentation results significant. However, only 7% of the students related evidence from the visualization to Mary's interpretation. Moreover, rather than specific criticisms, most students responded that more experiments are better regardless of the research question.

Implications. The study provides evidence for the benefit of critique. Critique directs students to pay attention to the design of the experiment whereas conventional instruction often directs attention to producing experimental results. As students critique, they distinguish their own ideas from those attributed to Mary and develop criteria for virtual experiments.

Promoting Learning with Dynamic Visualizations: Drawing and Critique

The second area of TELS research focused on critique examined and contrasted drawing and critique as tools to support learning with dynamic visualizations. Dynamic visualizations have great potential to support science learning because they can demonstrate unseen processes (Ardac & Akaygun, 2004; Sanger, Brecheisen, & Hynek, 2001; Williamson & Abraham, 1995). Adding visualizations to instruction can increase interest and insights in science (Boo & Watson, 2001; Corliss & Spitulnik, 2008), but some researchers also warn that the impact of dynamic visualizations may not always be powerful (Tversky, Morrison, & Betrancourt, 2002). Some visualizations represent dynamic information in such an apparently simple way that learners may become convinced they understand based on superficial observations (Chiu & Linn, in press). To enhance learning with visualizations, students must observe carefully, analyze what they see, and develop criteria to decide what information to be integrated. Generating drawings has been suggested as an effective way to promote learning with visualizations (Zhang & Linn, 2008). In the present study, Zhang designed a critique activity and explored the effect of critique on enhancing student learning with visualizations.

Questions. Does critique promote student learning with dynamic visualizations? What are the effects of drawing and critique on enhancing student learning about chemical reactions with dynamic visualizations?

Context. This research was conducted during a 5-day TELS project entitled *Hydrogen Fuel Cell (HFC) Cars*. Informed by the knowledge integrate framework (Linn & Eylon, 2006), this project illustrates chemical reactions within the context of HFC cars. It starts by eliciting student ideas about gasoline powered cars and then employs different representations to introduce chemical reactions, including a video showing the burning of a hydrogen balloon, a visualization of hydrogen combustion at the molecular level (see Fig. 9.6), and a flash movie of the reaction inside HFCs. Finally, students participate in an online discussion about the advantage and disadvantage of the two cars.

Methods. Three classes of high school chemistry students participated in this study (N = 73). The classes, taught by the same teacher, were randomly assigned to one of drawing or critique groups to study HFC.

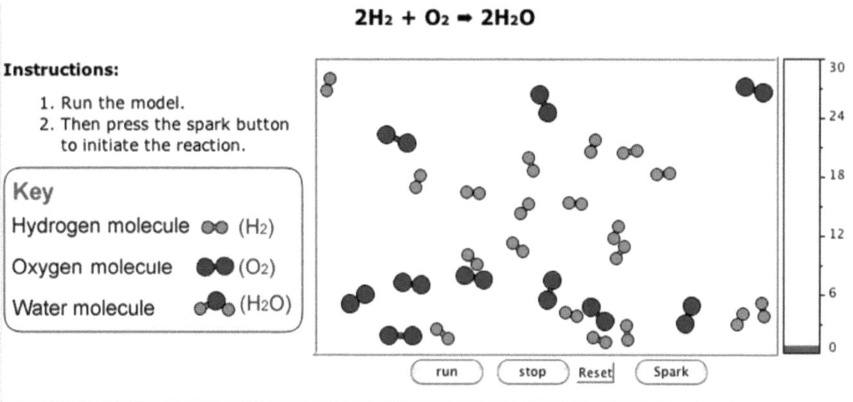

Fig. 9.6 Screenshot of the hydrogen combustion visualization

On the third day of the HFC project, students first learned chemical reactions by exploring the dynamic visualization about hydrogen combustion. Afterwards, students in the drawing group generated four pictures to represent intermediate phases during hydrogen combustion (Fig. 9.7). Students in the critique group critiqued two sets of drawings about hydrogen combustion processes (see Fig. 9.8 for one set of drawings and the critique question). Both groups completed the tasks within 40 minutes. During the remaining days of the HFC project, students in both groups worked on the same tasks of the project.

To assess student learning with the visualization, all participants were asked to complete the same tests before and after the project. The test includes three types

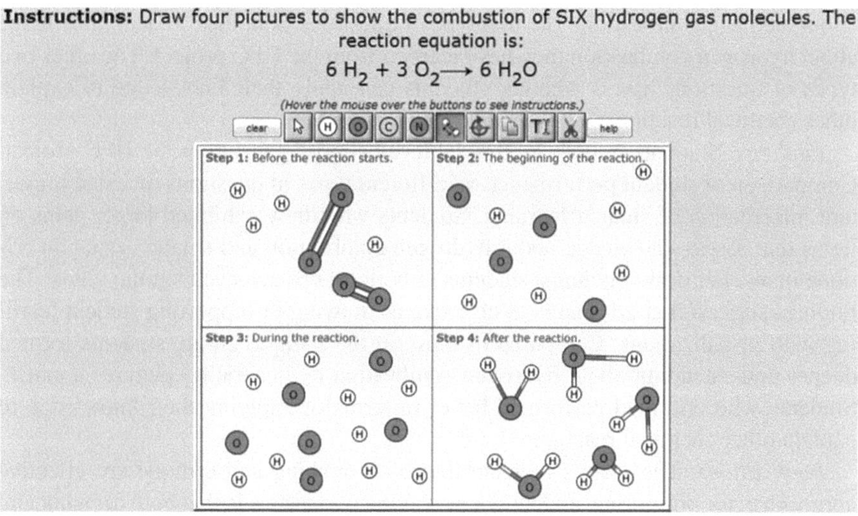

Fig. 9.7 Screenshot of the drawing activity

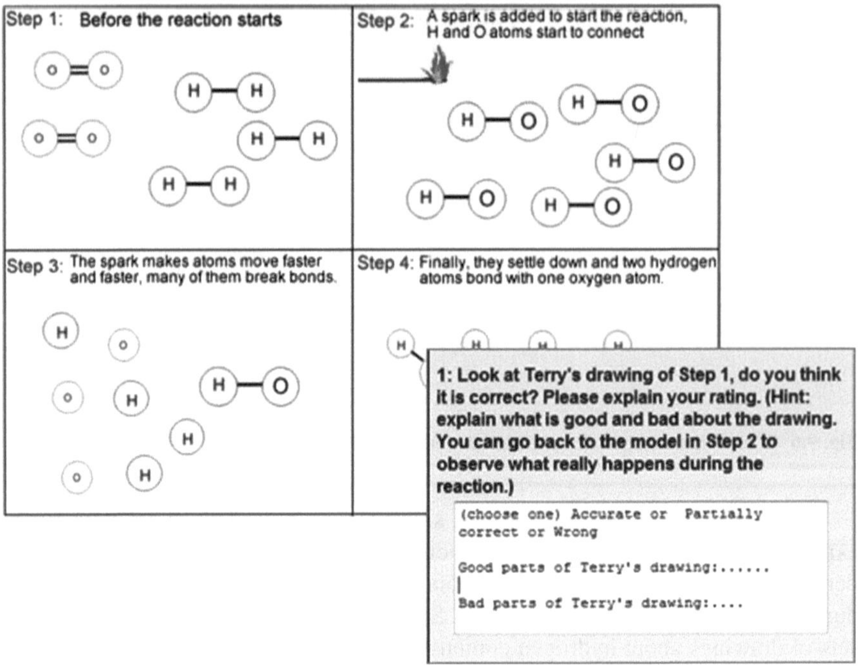

Fig. 9.8 Screenshot of the critique activity with a sample critique question

of questions: (1) items to assess content knowledge about hydrogen combustion, (2) drawing items asking students to draw how the reaction between nitrogen and hydrogen gas takes place, and (3) critique items asking students to evaluate drawings about methane combustion. The first type of questions examines student knowledge about hydrogen combustion they have learned from the HFC project. The other two types of questions assess whether students can apply their knowledge to explain other chemical reactions.

Findings. Students in both groups achieved similar gains after the HFC project. Comparison of student performance on different types of questions revealed important information of student learning. Students who drew exhibited larger gains on items that assess knowledge about hydrogen combustion and smaller gains on critique items. For drawing items, students in both groups achieved similar gains. The findings suggest that critique is as effective as drawing in supporting student learning with visualizations. Compared to those in the critique group, students formed deeper understanding about hydrogen combustion by generating pictures about it. Students who critiqued performed better in terms of applying their knowledge to explain other chemical reactions.

Implications. The results indicate that both drawing and critique are effective approaches for promoting student learning. One hypothesis is that both drawing and critique encourage students to develop criteria to distinguish among ideas. Drawing requires students to generate pictures about the details of hydrogen combustion. To accomplish this task, students need to distinguish among their own ideas and new information from the visualization to determine what to draw. They may revisit the visualization and observe carefully to help develop the criteria. Critique prompts students to evaluate some pre-made drawings. To critique, students need to analyze ideas represented in the given drawings, compare with their own ideas, and decide how to evaluate. They may also revisit the visualization to help establish criteria. The success of drawing and critique indicates that it is crucial to encourage students to develop criteria to distinguish among ideas. Further study should focus on examining what criteria are generated by students and how they are associated with learning.

Integrating Biological Knowledge Through Critique-Focused Concept Mapping

The third TELS area of research related to critique involves critique-focused concept mapping to support students in integrating biological content knowledge. Modern biology, genetics, cell biology, and evolution have been found to be conceptually difficult domains to teach and learn (Bahar, Johnstone, & Hansell 1999; Tsui & Treagust, 2003). They form a complex system with multiple interacting levels (Wilensky & Resnick, 1999). Coherent integration of such complex systems requires understanding of both the concepts and the connections between concepts. Dynamic computer-based visualizations with interactive inquiry activities allow students to explore the nature of ideas (Ainsworth, 1999). Concept maps allow making the connections between ideas within and across levels explicit (Novak, 1996).

Creating coherent concept maps is not a one-shot activity, but requires a subsequent revision step (Schwendimann, 2007). Revision activities require students to generate criteria (Chi, 2000; Linn & Eylon, 2006) that allow comparison against a benchmark. Benchmark concept maps can be generated by experts or novices.

Expert maps model expert behavior by connecting multiple levels and focus on underlying principles (Hmelo-Silver, Marathe, & Liu, 2007). On the other hand, peer generated work uses often more familiar language (Keppell, Au, Ma, & Chan, 2006) and might support deeper critical evaluation as it does not hold authoritative power over other's work. Peer evaluation can be mutually beneficial for the giver and the receiver (Topping, 2005).

Schwendimann's study compared two different critique activities: expert-generated benchmark map versus peer-generated benchmark map. Schwendimann's study used the KI (Linn & Hsi, 2000; Linn, Eylon, & Davis, 2004) in terms of focusing on connections between and distinction of a diverse repertoire of ideas.

Questions. How do expert and peer critique activities impact learning from a dynamic visualization? What connections among biology concept do students make in each condition? What criteria do students use for expert and peer critique?

Context. The week-long curriculum unit, Space Colony—Genetic Diversity and Survival, was designed in the WISE (Linn et al. , 2004). The unit consists of seven activities that emphasize connections between cell division, the underlying genetic processes, and the overarching evolution principles (Fig. 9.9). The unit includes a flash-based visualization "Evolution Lab" (biologyinmotion.com) (Fig. 9.10) that enables students to run experiments about the effects of mutations, natural selection, and evolution.

Students receive initial training in the concept mapping method. Following the visualization, students work in dyads to create a paper-based concept map from six given concepts. Students first place them in the appropriate level area (DNA, cell,

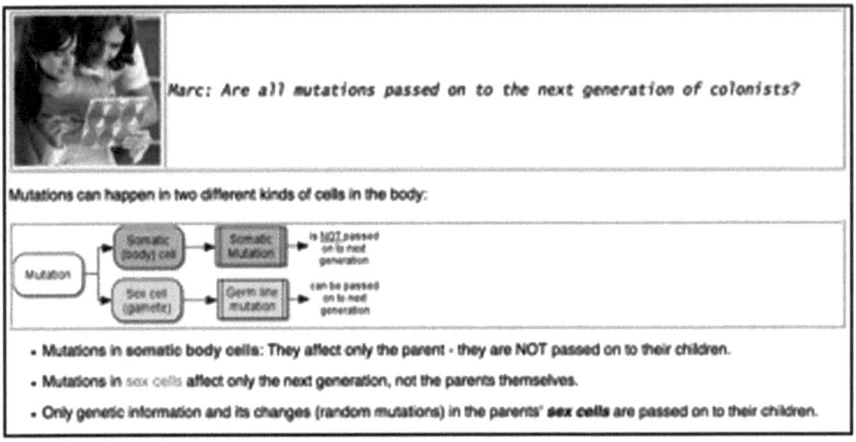

Fig. 9.9 Screenshot of WISE module space colony

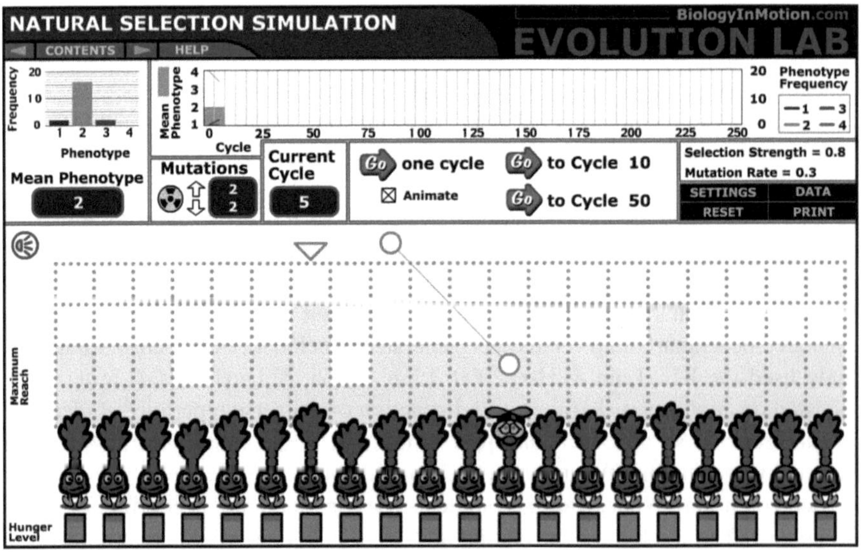

Fig. 9.10 Evolution lab visualization

or organism/population) and connect them with labeled arrows. Students then revise their map by comparing it against an expert- or peer-generated benchmark concept map. Students developed their own criteria.

Methods. The curriculum was implemented by two teachers with two ninth/tenth grade biology classes each in one public high school ($N = 81$) in the western United States. One class by each teacher was randomly selected for each treatment (expert or peer map comparison).

Pre- and posttests consisted of nine multiple choice and explanation items that assessed changes in students' connections between genetic and evolution concepts. Tests were coded using a five-scale KI rubric (Linn et al., 2006).

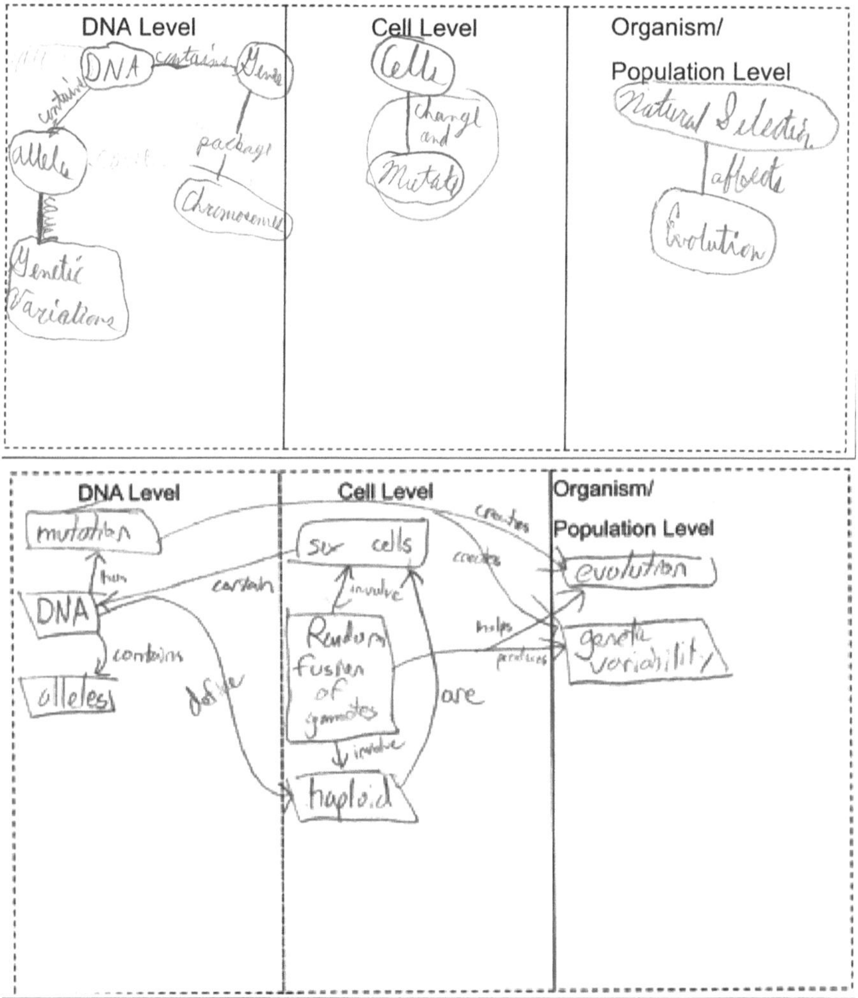

Fig. 9.11 Concept map before and after revision

Concept maps propositions were coded on a five-scale KI rubric for concept maps (Schwendimann, 2008). The rubric distinguished between link label, link direction, concept placement, and cross-links.

Findings. The results suggest that the combination of critique-focused concept mapping and a dynamic visualization helped students in both treatment groups generate novel connections across levels. Neither treatment groups differed significantly in their posttest performance. In their revised maps, the peer-review group showed more across-level connections than the expert map group. Both treatment groups significantly improved their concept maps through the critique activity [paired $t(80) = 4.13, p < 0.0001$ (two-tailed)] (Fig. 9.11).

Students in both treatment groups generated a broad variety of criteria to review and compare different aspects of concept maps. However, the groups differed from each other in the different kinds of criteria used to review their maps. This study suggests different mechanisms and criteria involved in the two critique activities. The two treatment groups differed in their use of different criteria (Fig. 9.12):

I. Students in the expert map group commented only on concept placement (61%) or missing link labels (27%). Both criteria were surface-level criteria that allowed for quick comparisons with the expert map. Critiquing other people's work is often easier than evaluating one's own work.
II. Students in the peer-map group showed a larger variety of criteria. Twenty-eight percent also criticized the misplacement of a concept and 18% pointed out a missing label, but another 28% suggested adding a missing link, and 5% analyzed the direction of an arrow. The peer-map activity engaged students to develop and use more criteria on a conceptual level, such as missing propositions

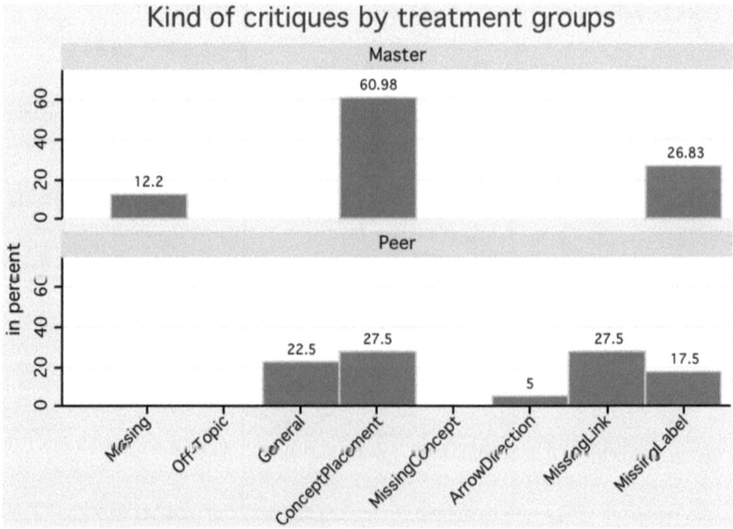

Fig. 9.12 Criteria generated by students in each treatment group

and causal directions. Comparing their own ideas against those of their peers helped students to value their own ideas while developing criteria to critically reviewing them.

Implications. Both critique methods lead to reflection through criteria generation and revision. Critical reflection supports students' self-monitoring of their learning progress. Self-monitoring is an important skill for autonomous life-long learning (Linn, Davis, & Eylon, 2004). Both surface and principle critique are important for learning. Using expert or peer benchmark work, or a combination thereof, can target specific forms of critique toward a more coherent understanding of biology.

TELS Research on Argument Construction

Generating a persuasive and convincing argument that coordinates evidence and theory in order to support or refute an explanation is an important part of the inquiry process (Driver et al., 2000; Duschl & Osborne, 2002; Jiménez-Aleixandre, Rodríguez, & Duschl, 2000; Kuhn, 1993; Kuhn, 1970; Latour, 1987; Siegel, 1989). For arguments to be considered persuasive and convincing, they must be consistent with the epistemological criteria used by the larger scientific community for "what counts" as valid and warranted scientific knowledge. Examples of central epistemological criteria in science include the importance of (a) evidentiary backing or rationales for knowledge claims and proposed tests of claims (Hogan & Maglienti, 2001), (b) coherence between theoretical frameworks and data (Passmore & Stewart, 2002), (c) establishing the credibility of evidence (Driver et al., 2000), (d) parsimony (Sandoval & Reiser, 2004), and (e) logically consistent and coherent reasoning (Zeidler, 1997). Research in TELS has focused specifically on how students warrant their claims.

How Do Students Substantiate Their Decision-Making About Community Science Issues?

Research has shown that multiple factors influence reasoning about complex science problems, such as genetic dilemmas, conservation practices, or personal and community health (Corburn, 2005; Grace & Ratcliffe, 2002; Zohar & Nemet, 2002). While science plays an important role in these issues, people often privilege other factors, such as morals and values, personal and familial gain, the uncertainty of available information, and predicted outcomes. Isolatable research methodologies identify whether students base their decisions on scientific or nonscientific knowledge (Fleming, 1986a, b; Grace & Ratcliffe, 2002; Zohar & Nemet, 2002). This study builds upon these and other integrated perspectives and investigates how students put forth multiple different perspectives, evidence that includes local knowledge, and tradeoffs to support their asthma-related decision-making.

Questions. How do students from three different local communities substantiate their decision-making?

Context. In the Asthma module, students explore the scientific dilemma of asthma by (1) constructing an integrated understanding of asthma as a community health problem and (2) practicing integrated decision-making about which asthma intervention to implement in their community. While studying the Asthma module, students are expected to engage in decision-making from multiple perspectives. In addition, students must use supporting evidence, localize the decision to specific communities, and consider tradeoffs. The Asthma project research explored how a multi-dimensional, multi-contextual methodology provides insight into students' decision-making. Table 9.1 summarizes the learning activities in Asthma module.

Methods. This study investigated 3 teachers and 108 students, across three local communities, C-town, B-Town, and R-Town. Table 9.2 summarizes the research participants and settings. The following decision prompts were embedded in three activities across the project to identify how students integrate their ideas to justify

Table 9.1 Overview of the asthma module activities and assessments

Activity	Description
1: Your asthma problem	Evidence pages and an interactive map introduce (a) the asthma problem in students' community, (b) the driving question, and (c) the diesel reduction & asthma clinic interventions
2: How does asthma affect the body?	Dynamic visualizations explain the physiology of breathing and asthma
3: What causes an asthma attack?	Static visualizations explain asthma triggers & the physiology of an allergic immune response
4: How does diesel exhaust impact your community's asthma problem?	Multiple pieces of evidence provide explanations about how diesel pollution impacts on asthma and general health
5: How can a person manage their asthma?	Multiple pieces of evidence provide explanations about how asthma management and health care can contribute to asthma-related hospitalizations rates
6: Improving your community's asthma problem	Students debate proposed solutions and generate new solutions

Table 9.2 Summary of research settings and participants

Community	School	Area	Teacher	Course	Year	# Classes	Grade	*N*
C Town	Mountain High	Urban Fringe	Sandals	Biology	2	2	9	40
B-Town	Bayview High	Urban	Pebbles	Anatomy & Physiology	2	2	10	34
R-Town	King High	Urban	Nelson	Biotechnology	3	2	11–12	34

their decisions about which program will better solve their community's asthma problem:

- Which program do you want your City Council to support? Asthma Clinic or Diesel Reduction Program
- How will this program better serve your community?
- What evidence (information) helped you make this decision?

Explain your answer.

The decision note assessment items elicited multi-dimensional responses from students. In response, KI rubrics were created for each dimension: perspectives, evidence use, tradeoffs, and localization (see Table 9.3).

Findings and Implications. Student decision-making about the asthma problem varied throughout the module and differed across communities. Table 9.4 reports the mean KI scores for each dimension. Table 9.5 summarizes the interpretation of these results with regard to decision, perspectives, evidence use, consideration of tradeoffs, and localization.

While students across all communities justified their decisions similarly, they differed in the programs they supported. Students probably varied in the programs they supported because they held preliminary ideas about asthma and their community. Also, students may have initial notions about what an effective program entails. Students switching their decisions indicates that they are grappling with ideas about the asthma problem and which solution to implement in their community. When students learn from the Asthma module, their teachers, and their peers, they have the opportunity to replace, isolate, or integrate a wide range of ideas. This restructuring of knowledge likely influenced which program they chose to support. Also, students may differ in when they change their decision because they may perceive some pieces of evidence as more compelling than others. These findings suggest the design of the Asthma module creates a rich opportunity for students to engage in KI about an authentic community science problem.

In all three local communities, students primarily supported their decisions with ideas related to risk. This was consistent with R-Town and C-Town's primary and secondary perspectives when explaining asthma as a community problem. However, B-Town students explained the asthma problem from a prevalence perspective (Tate, 2009). This inconsistency suggests that students (a) have different criteria for what constitutes a community health problem and which program best addresses that problem or (b) hold many ideas about the asthma problem and have yet to sort them out and form a coherent, integrated understanding. Future refinements to the Asthma module should include learning activities that allow students to put forth and negotiate criteria for what constitutes a community health problem and effective solutions for their community.

While students differed in the ideas they put forth to support their decisions, students in all communities exhibited similar evidence use. Most students included at least one isolated piece of evidence to justify their decisions. Seethaler and Linn (2004) and Bell (2004) argue that within an appropriately scaffolded learning

Table 9.3 Knowledge integration (KI) rubrics and example of students' responses

KI score	Perspectives	Evidence use	Tradeoffs
2: No Integration	Students use reasons or evidence to explain the problem from one perspective. "This program is better for my community because our community has many highways, railroads, public transportation, machinery/factories, and refineries." (risk)	Students simply state that a program is better for their community; no elaboration or explanation. "The program is better for my community because it will help out people with asthma much better. It will save a lot people."	Students mention no tradeoffs. "This [Asthma Clinic] program is better for my community because it will benefit the people in our community without medical insurance to be treated for asthma problems. The evidence that helped me make this decision is before the clinic, 18/80 children and teenagers were hospitalized for asthma, and 1 year after the clinic no children or teenagers were hospitalized because of the asthma clinic."
3: Partial Integration	Students explain the problem from two or more perspectives but do not link them. "This program is better for my community because it will help those in need of help with no insurance. The evidence that helped me make this decision is the number of hospitalization visits because it has given people with asthma a chance to get treated" (risk-management)	Students explain why the program is better (or worse) for the community or include evidence that is not connected to a reason. "This program is better for my community because the causes of asthma would be more directly halted if people who are directly affected by asthma could come to the clinic and receive professional medical attention. Because my community is right next to the freeway."	Students mention a tradeoff but include no supporting evidence. "The Asthma Clinic benefits are limited to people who have asthma and no health insurance, while the Diesel Reduction Program has the potential to help all people in the community."

Table 9.3 (continued)

KI score	Perspectives	Evidence use	Tradeoffs
4: Simple Integration	Students link two perspectives. "This program is better for my community because there will be less smog output reducing the number of people who will get asthma." (risk-prevalence)	Students include a reason and supporting evidence. "This program is better for my community because it provides options for people who have asthma and is not limited to just people who have asthma not caused by diesel pollution. The evidence that helped me make this decision is the chart/picture that showed that asthma is NOT only caused by irritants. This chart helped me decide that because it showed me that asthma is not caused by only one type of thing."	Students include one tradeoff that is supported with evidence. "We think our City Council should support the Diesel Reduction Program because it will help all people who have asthma, as opposed to the Asthma Clinic which would only benefit those without health insurance. Also, a reduced amount of diesel in the air will help the environment as well as the number of asthma cases."

Table 9.3 (continued)

KI score	Perspectives	Evidence use	Tradeoffs
5: Complex Integration	Students link three or more perspectives. "This program is better for my computer because most people in my community have asthma due to many triggers not just due to diesel. The clinic will benefit those without health insurance. The evidence that helped me make this decision is that over 14,500 people in [the county] do not have insurance and have asthma." (prevalence- physiology, risk-prevalence)	Students provide 2 or more instances of reasons and supporting evidence. "This program is better for my community because it will help people who have asthma and don't have health insurance. It will also help people who's asthma is not caused by irritants. Other things that will improve are number of emergency room visits and hospitalizations by non-insured patients. This program will better serve our community because it will be easier for non-insured asthma sufferers to get medicine, information, and asthma-related checkups . . . Other pieces of evidence were the pages that described the two different types asthma, irritants, and allergens. This helped us make our decision by showing us that asthma is not only caused by PM or an irritant but by also pollen or allergens."	Students include 2 or more tradeoffs, each supported with evidence. This [Diesel Reduction] program is better for my community because it helps everyone. Although the asthma clinics help those without insurance, it only helps people with asthma. The diesel reduction program can prevent lung damage caused by diesel exhaust and overall reduce CO_2 in the atmosphere, which benefits everyone. The threat of global warming affects everyone while asthma affects less than 25% in [A] County. The evidence that helped me make this decision is what diesel does to you, your body your environment. When diesel is used in vehicles, the diesel particles that contain very harmful chemicals can get into your airways and make it hard for you to breathe. Also, people who live by streets or highways have a higher chance of having diesel exhaust affect all of them. This evidence helped us to relate the issue of diesel pollution to our own personal lives."

Table 9.4 Summary of a repeated measures ANOVA for decision justifications related to students' community's asthma problem

Community	Dimension	Decision Note 1	Decision Note 2	Decision Note 3	F	p[a]	Effect size[b]
R-Town	Index	8.5 (1.46)	9.44 (1.46)	8 (1.31)	6.14	0.06	0.30
	Perspectives	3.19 (1.11)	3.57 (0.96)	2.81 (0.91)	4.01	0.03	0.35
	Evidence	3.19 (0.40)	3.34 (0.5)	2.81 (0.54)	5.34	0.10	0.53
	Tradeoffs	2 (0)	2.19 (0.4)	2.19 (0.4)	1.55	0.23	0.38
C-Town	Index	8.6 (1.24)	9.27 (1.22)	7.73 (1.27)	6.72	0.004	0.53
	Perspectives	3.13 (0.92)	3.73 (0.80)	2.87 (0.52)	4.36	0.02	0.26
	Evidence	3.27 (0.46)	3.13 (0.35)	2.87 (0.52)	4.26	0.02	0.74
	Tradeoffs	2.13 (0.35)	2.33 (0.62)	2.13 (0.35)	1.14	0.34	0
B-Town	Index	8.71 (1.42)	8.62 (2.18)	9.19 (2.87)	0.84	0.44	0.22
	Perspectives	3.19 (0.98)	3.14 (1.24)	3.29 (1.38)	0.15	0.86	0.09
	Evidence	3.19 (0.51)	3.10 (0.70)	3.24 (1.0)	0.31	0.74	0.06
	Tradeoffs	2.19 (0.40)	3.24 (0.62)	2.48 (0.87)	1.90	0.16	0.40

[a] This p value represents the significance in change in scores over time.
[b] This effect size represents the magnitude of difference between the decision notes 1 and 3.

Table 9.5 Looking across communities: Summary of findings for decision justifications

	R-town	B-town	C-town
Decision[a] AC: DRP	Undecided →[b] Asthma Clinic 1:1 → 1:1 → 2:1	Diesel Reduction Program 1:3 → 2:3 → 1:2	Undecided → Asthma Clinic 1:1 → 2:1 → 3:2
Perspectives	Multiple, isolated RISK[c]/prevalence/ management[d] →RISK/prevalence/ management/physiology →RISK/management	Multiple, isolated RISK/physiology	Multiple, isolated RISK/management →RISK/physiology →RISK/management
Evidence	Partial support	Partial support	Partial support
Tradeoffs	Limited consideration	Limited consideration	Limited, but increasing consideration
Localization	Limited localization	Limited localization	Very limited localization

[a] An estimated ratio to illustrate student program choices at each decision note.
[b] The arrow represents changes across the online decision notes in the Asthma module.
[c] The term in all caps represents the primary perspective articulated by students in their explanation of the asthma problem.
[d] The term in lowercase letters represents the secondary perspective articulated by students in explanation of the asthma problem.

environment, students can construct evidence-based justifications. The findings reported here support this claim. Students made use of the evidence provided in the module to justify their decisions. While the module was successful with regard to the availability of evidence for students to learn and include in their decision justifications, additional or improved scaffolds are needed to encourage students to generate connections among their reasons and evidence. Students also need more opportunities to learn what constitutes a well-supported and integrated decision.

In general, students in all communities provided limited localization of their decision justifications. This lack of localization can be attributed to (a) the ambiguity of the term, "community," (b) students' assumptions that others know what "community" they are referencing, or (c) students' unfamiliarity with the norms for constructing a decision justification about the asthma problem in their community. Students need more instruction from the Asthma module and teacher to explicitly localize their decision justifications. This may also include additional opportunities to negotiate and reach consensus about which aspect of the community they are addressing when they put forth and support their decisions.

Analyzing Students' Arguments

In addition to our research on students' construction of arguments, the TELS project also supported the preparation of a review of approaches used to analyze the quality of students' arguments (Sampson & Clark, 2008). The intent of this review was to provide an overview of several different analytic frameworks that science educators use to assess in terms of three focal issues: structure, justification, and content. To highlight the different foci, affordances, and constraints of these different analytic methods, the review of each framework included an analysis of the sample argument. Overall, this review highlighted how the divergent foci of the various frameworks result in different assessments of overall quality. It is therefore important for researchers to understand that analytic frameworks, such as the ones included in the review, (1) are tools created for specific tasks to investigate specific questions and (2) were originally designed for a specific context. Frameworks, as a result, are not fully interchangeable, and the foci of each framework require consideration before comparing the results of various studies.

This review also highlighted a number of overarching messages regarding the current nature of research in the field. First, the analytic frameworks available tend to focus on atomized aspects of students' arguments. While this type of emphasis has proven fruitful, future research will need to also include more holistic considerations of the quality of the arguments that students produce as part of the

Directions: The first three questions are designed to determine what you think counts as a good *scientific* argument. In each question you will be given a claim. Following the claim are 6 different justifications. Your job is to rank the justifications in order using the following scale (**For each question, you can only use each ranking once**):

1 = This is the **most** convincing justification
2 = This is the **2nd most** convincing justification
3 = This is the **3rd most** convincing justification
4 = This is the **4th most** convincing justification
5 = This is the **5th most** convincing justification
6 = This is the **least** convincing justification

Question #1. Your task is to rank these 6 different justifications in terms of how convincing you think they are. Remember that you can only rank one justification as 1, one justification as 2, one justification as 3, and so on.

Claim: Objects that are in the same room are the same temperature even though they feel different because…	**Your Ranking**
…when we measured the temperature of the table, it was 23.4°C, the metal chair leg was 23.1°C, and the computer keyboard was 23.6°C.	________
…good conductors feel different than poor conductors even though they are the same temperature.	________
…objects that are in the same environment gain or lose heat energy until everything is the same temperature. Our data from the lab proves that point: the mouse pad and plastic desk were both 23°C.	________
…objects will release and hold different amounts of heat energy depending on how good of an insulator or conductor it is.	________
…our textbook says that all objects in the same room will eventually reach the same temperature.	________
…we measured the temperature of the wooden table and the chair leg and they were both 23°C even though the metal chair leg feels colder. If the metal chair leg was actually colder it would have been a lower temperature when we compared it to the temperature of the table.	________

Fig. 9.13 An example of an ASRT item

inquiry process. This work, however, will require new approaches that examine the structural, conceptual, epistemic, and social aspects of argument generation in a more synergistic fashion rather than looking at each of these aspects independently. Second, the review of the available literature suggests that much research on argument in science education has thus far focused on the identification of patterns and themes in students' arguments (e.g., "students tend to produce arguments that lack sufficient justification" or "students tend to produce arguments that have a simplistic structure") rather than focusing on the underlying reasons for these patterns. Studies that explore the causes of these patterns and themes will prove valuable in developing new curricular materials, instructional approaches, and technology-enhanced learning environments to promote and support more productive argumentation inside the classroom.

Clark and Sampson (2008) also developed the *Argumentation in Science Rating Task (ASRT)* in order to assess the criteria used by students for evaluating the quality of arguments and the quality of challenges to arguments. The ASRT consists of six items, three that focus on the quality of argument that can be used to justify a claim and three that focus on the quality of a challenge to an argument. For each item, individuals are asked to rank six arguments or six challenges to an argument in terms of quality. An example of an ASRT item is shown in Fig. 9.13.

TELS Research on Collaborative Argumentation

Much of the work in TELS adopts the view of dialogical argumentation as a process where "different perspectives are being examined and the purpose is to reach agreement on acceptable claims or course of actions" (Driver et al., 2000, p. 291). Much of this work therefore views dialogical argumentation as a social and collaborative process that is employed "to solve problems and advance knowledge" (Duschl & Osborne, 2002, p. 41) rather to "justify or refute a particular standpoint" (van Eemeren, Grootendorst, & Henkemans, 2002, p. 38). This view of argumentation emphasizes collaboration over competition and suggests that activities that promote dialogical argumentation can enable individuals to use each others' ideas to construct and negotiate a shared understanding of a particular phenomenon in light of existing data and new evidence (Abell, Anderson, & Chezem, 2000; Andriessen, Baker, & Suthers, 2003; Boulter & Gilbert, 1995; deVries, Lund, & Baker, 2002; Veerman, 2003). Thus, in practice, TELS work conceptualizes dialogical argumentation as a process of proposing, supporting, evaluating, and refining ideas to make sense of complex or ill-defined problems or phenomena. In the following sections, we provide an overview of studies investigating optimal grouping and seeding of online discussions for argumentation, the relative affordances and processes involved in collaborative versus individual engagement in argument construction, and the development and consideration of approaches for analyzing argumentation.

Optimal Grouping and Seeding of Online Discussions for Argumentation

The design of many online learning environments can be thought of in terms of "scripts" that orchestrate and control students' interactions with each other and the environments (Hesse, 2007; King, 2007; Weinberger, Stegmann, Fischer, & Mandl, 2007). One particular class of scripts focuses on grouping students together with other students who have expressed differing perspectives or stances. This general scripting approach can be referred to as a "conflict schema" (Dillenbourg & Jermann, 2007, p. 292). Yet, there are many ways to group students under the broad category of a "conflict schema" and there is little research available that explicitly examines the efficacy of different approaches.

Questions. How can the grouping of students for argumentation be informed by the content of their ideas? How should these discussions be seeded with initial ideas?

Context. This research investigated the efficacy of a conflict schema approach and also on optimal approaches to seeding the resulting online discussions with initial comments for discussion (Clark, 2004; Clark, D'Angelo, & Menekse, 2009; Clark & Sampson, 2005, 2007, 2008; Cuthbert, Clark, & Linn, 2002). The context of the research study was the *Thermodynamics: Probing Your Surroundings* project discussed earlier in this chapter. The version of the project for this research included eight activities. During the first five activities, students make predictions and collect real time data about the temperatures of objects found inside the classroom and explore interactive simulations dealing with such ideas as heat transfer, thermal conductivity, and thermal sensation. The sixth activity then scaffolds students in creating an explanation to explain patterns they notice within the data they have collected. This step involves a series of pull-down menus with sentence fragments (Fig. 9.14). The software underlying this interface then sorts the students into discussion groups in a manner determined by the researchers for each research condition. The seventh activity engages students in discussions where they critique a set of provided explanations, outline evidence for and against each explanation,

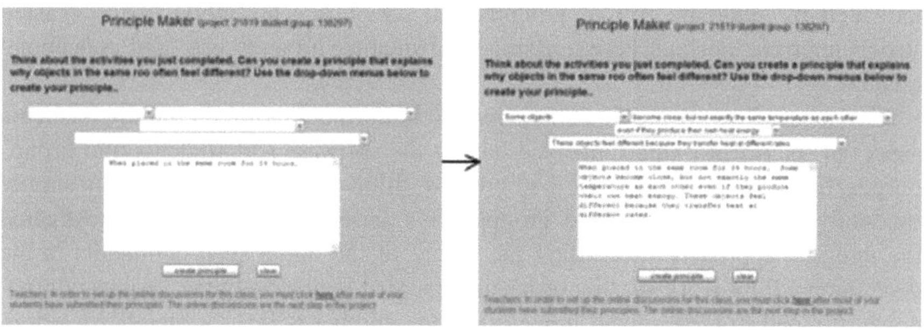

Fig. 9.14 Interface providing pull-down menus of sentence fragments that students use to construct their explanations for patterns they notice in the data

and engage in collaborative argumentation to work toward consensus. The nature of these provided initial explanations also varies by research condition. The eighth activity then allows students to construct a revised explanation after the discussions and to reflect on how their ideas have changed.

Methods. Some early studies did not compare across conditions and instead simply focused on the nature of learning supported within the discussion environment using various approaches to grouping students and to choosing the initial explanations to be discussed. Later studies involved random assignment of students to conditions within any given classroom in the study. Conditions in each study varied the nature of how students were grouped with one another and the nature of the initial explanations that they discussed in their online forum. Sample sizes for any given study generally included approximately 100 students. Studies were conducted in diverse public middle school and high school classrooms in California and Arizona.

Findings. Early work measured the structural quality of argumentation and participation in the ensuing discussion showed that the original personally seeded script that (1) sorted students into groups with students who had created explanations that were different from their own and (2) inserted the students' own explanations as the starting seed comments for the discussions was superior to standard online discussions that involved (1) no preexploration of the explanation fragments that constitute the preset explanations, (2) random group assignment, and (3) preset explanations as seeds (Clark, 2004; Clark & Sampson, 2005; Cuthbert et al., 2002). Subsequent research (Clark & Sampson, 2007) elaborated on these findings showing that carefully structured online environments integrating the personally seeded discussion approach can effectively scaffold high quality scientific argumentation in the classroom as measured from a structural perspective, particularly in light of the low levels of argumentation that typically take place within classrooms.

The research then proceeded to extend beyond structural perspectives in terms of students' discourse moves to also consider students' use of grounds and the conceptual quality of students' contributions (Clark & Sampson, 2008). This work suggested that personally seeded discussions are an effective way to encourage students to justify their ideas and challenge the ideas of others as indicated by students' use of grounds and rebuttals. This work also suggested strong interrelationships between structural quality, grounds use, and conceptual quality.

The next series of studies compared the contributions of the conflict-schema aspect of the script (which involved having the software sort students into groups purposefully with students who had created different explanations) versus simply randomly sorting students into groups. This series of studies also investigated the multiple approaches regarding the nature of the initial seed comments in the discussions. Pilot work for this series underscored how closely the various components of a pedagogical model hinge upon one another (Clark, Schleigh, Menekse, D'Angelo, & Sampson, 2008) and also suggested that students in discussions with their own comments participated more but also contributed more comments involving social pressure for others to "pick" their explanations and not those of others. These findings were reinforced by a subsequent study (Clark, D'Angelo, & Menekse, 2009). Comparisons between the trials in terms of modified gain scores in Clark, D'Angelo, & Menekse (2009) also showed that students in the conflict schema condition (where the software grouped students with other students who had created different explanations) outperformed students in the nonconflict schema trials (where students were assigned to partners randomly). In terms of conditions regarding the nature of the initial seed comments, students in the augmented-preset condition (where initial comments were preselected to represent an optimized range of possible student conceptions) demonstrate significant gains on their explanations compared to students in the personally seeded conditions (where the students' own explanations were included as the initial seed comments). This was true overall but particularly strong when combined with the conflict schema approach to group creation. Furthermore, the actual discussions of the students in the augmented-preset groups generally demonstrate the same or better overall argumentation quality in terms of structure, discourse moves, and grounds quality. Their participation levels were slightly lower, but the overall outcomes favored the augmented-preset condition in terms of the discussions themselves.

One hypothesis explaining the advantages of the augmented-preset performance is that the sets of seed comments for the personally seeded groups (which were their own explanations) often did not include the same diversity of ideas as the sets of preset seed comments in the augmented-preset groups. The average standard deviation for the sets of seed comment scores in the augmented-preset groups was higher than the average standard deviation of the seed comment scores in the personally seeded groups. This hypothesis was supported by the fact that the augmented-preset groups (1) showed a higher average gain and normalized gain than the personally seeded groups and also (2) included a higher proportion of group members who improved their scores than the personally seeded groups. The augmented-preset condition thus potentially results in more productive learning than the personally seeded condition by exposing students to a wider range of ideas on average. Another possible explanation is that students in the augmented-preset condition are guaranteed to have a fully normative explanation as one of the seed comments in their group while students in the personally seeded condition have their own explanations as seed comments for their group and thus may or may not have a fully normative explanation included depending on their group.

Implications. This series of studies suggests that supporting productive argumentation in online discussions is greatly facilitated by attention to group composition and to the initial structuring of the discussions in terms of their initial seed comments. In particular, the conflict schema approach to purposefully organizing students into groups with other students who have expressed different perspectives on the topic is valuable. The work also suggests that optimizing a range of initial seed-comments in terms of potential student perspectives is ultimately more valuable than including the students' own initial explanations even though the latter approach results in potentially higher levels of engagement.

Collaborative Versus Individual Argument Construction and Argumentation

Many researchers (Abell, Anderson, & Chezem, 2000; Bell & Linn, 2000; Kuhn & Reiser, 2005; McNeill, Lizotte, Krajcik, & Marx, 2006; Schwarz & Glassner, 2003) have encouraged students to work in collaborative groups when they engage in scientific argumentation. The work of these authors suggests that opportunities to collaborate with others can lead to more productive scientific argumentation and improved learning outcomes because groups can pool knowledge and take advantage of different cognitive or monitoring resources. Few studies, however, have explicitly compared individual and group performance on tasks that require students to engage in argumentation or examined the benefits of collaboration during an episode of argumentation for individual learning in the context of science education. Given this gap in the literature, the overall objectives of this study were (a) to evaluate the benefits of collaboration on argumentation outcomes and for individual learning and (b) to identify potential reasons for variation in group-level performance.

Questions. Do students who engage in argumentation in groups craft better arguments and learn more than students who engage in argument construction on their own? Do individuals adopt and internalize the group outcome? What are the characteristics of high versus low performing groups as they engage in collaborative argumentation?

Context. This research was conducted as a foundation for a proposed WISE project that did not ultimately reach completion focusing on issues of conductivity. Participants were asked to complete a complex task that required them to engage in argumentation in order to make sense of a discrepant event. This task, which is called the *ice melting blocks problem*, required them to determine which explanation, of six plausible alternatives, was the most valid or acceptable way to explain why ice placed on an aluminum block melts faster than ice placed on a plastic block even though the aluminum block feels much colder. Once the participants had determined which explanation was the best way to make sense of the phenomenon, they were asked to create a written argument that articulated and justified this explanation with appropriate evidence and reasoning. This study took place in a large suburban public high school located in the southwest United States.

Methods. The 168 participants in this study, who were enrolled in five different sections of chemistry at the same high school, were randomly assigned (within each classroom using a matched-pairs design) to one of two conditions to complete this task. Students assigned to the individual argumentation condition completed this task alone, while students assigned to the collaborative argumentation condition worked in a same-gender group of three (triads). In order to assess student understanding of the phenomenon in question, all of the participants were asked to complete the *ice melting blocks problem* for a second time. For this administration of the problem, each student was required to generate his or her own written argument for the *ice melting blocks problem*. To assess the participants' ability to apply what they have learned in a different context, individuals completed a conceptually identical task, the *why do objects feel different problem*. This problem was used because the discrepant event in this problem has the same underlying cause as the ice melting blocks problem. As before, these tasks required each student to produce a written argument that articulates and justifies an explanation for the event in question.

An in-depth qualitative analysis of the argumentation that took place within two more successful triads and two less successful triads was also conducted in order to identify major contrasting dimensions in group interaction that can be linked to differences in group outcomes. The four groups were selected based on differences in the quality of their written solutions to the *ice melting blocks problem* and because their interactions seemed representative of the kinds of interactions that took place in the more and less successful groups.

Findings. The results of this study indicate that, although groups of students did not produce substantially better products than the students who worked alone, students in the collaborative condition performed better on the mastery and application problems with moderate effect sizes. There was also a great deal of variation in the quality of the arguments produced by the triads. The qualitative analysis of the two more and two less successful groups suggests that the numbers of ideas students introduce into a discussion, how individuals respond to these ideas, the willingness of participants to challenge the ideas of others, the criteria individuals use to distinguish between ideas, and how students use data as they work seemed to influence on the overall quality of their final argument.

Implications. These findings indicate that collaboration was beneficial for individual learning but not for initial performance on the task. This result was unexpected given the extensive literature that suggests that collaborative effort can and should result in a product that exceeds what is possible by an individual working alone (Andriessen et al., 2003; Mason, 1998; Rochelle, 1992; Scardamalia & Bereiter, 1994). It seems that the ability to engage in productive argumentation with others is not something that comes easily to many students. These findings also suggest that students may need to learn how to engage in argumentation with others in a more productive way before individuals can reap all of the potential benefits of collaboration. Finally, the five differences in the ways more and less successful groups engaged in collaborative argumentation will help lay the groundwork for future studies that examine how individuals and their interactions influence group understanding and outcomes and why some groups are so much more productive than others.

Analyzing Argumentation

In addition to research on supporting argumentation, TELS also supported the development of a framework for analyzing argumentation (Clark & Sampson, 2005, 2007, 2008) and a review of approaches to analyzing argumentation (Clark, Sampson, Weinberger, & Erkens, 2007). Essentially, the framework developed by Clark and

Sampson focuses on the relationships between levels of opposition found within a discourse episode, the types of comments student make, the grounds quality included in those comments, and the conceptual quality of their ideas. By focusing on the relationships between these aspects of argumentation, the framework offers researchers a specific analytic tool to examine possible connections between argumentation and subject matter learning. Analysis grounds and conceptual quality is supported by flowcharts involving a series of binary decisions on the part of the coder to increase reliability of coding (Fig. 9.15). Clark and Sampson (2008) provide the most detailed account of the framework, including these flowcharts, detailed explanation of episode segmenting protocol, and many other issues, including description of the Cochran–Mantel–Haenzel χ^2 analyses based on table scores to determine the significance of the relationship between the *discourse move* of a comment and the *grounds quality* or *conceptual quality* of that comment. Jeong, Clark, Sampson, and Mushin (2011) explore the potential of expanding analysis with the framework by incorporating sequential analysis.

The review by Clark, Sampson, Weinberger, and Erkens (2007) examines five categories of analytic frameworks for measuring participant interactions within these environments focusing on (1) formal argumentation structure, (2) conceptual quality, (3) nature and function of contributions within the dialogue, (4) epistemic nature of reasoning, and (5) argumentation sequences and interaction patterns.

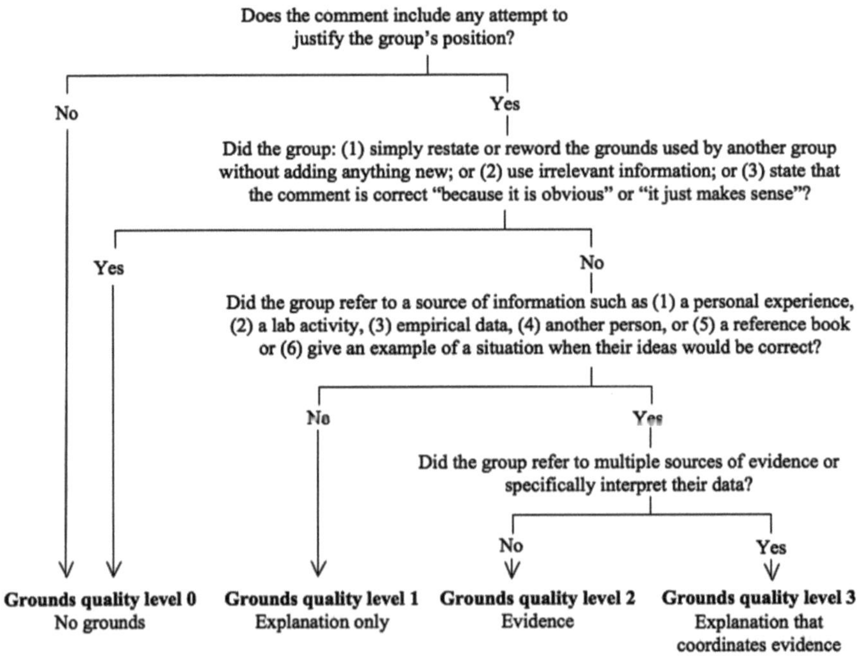

Fig. 9.15 Flowchart for analyzing grounds

Reviewed frameworks include Toulmin (1958), Erduran, Simon, and Osborne (2004), Clark and Sampson (2008), Kuhn and Udell (2003), deVries, Lund, and Baker (2002), Janssen, Erkens, Jaspers, and Kanselaar (2006), Baker, Andriessen, Lund, van Amelsvoort, and Quignard (2007), Jiménez-Aleixandre, Rodríguez, and Duschl (2000), Duschl (2008), Leitão (2000), Hogan, Nastasi, and Pressley (2000), Baker (2003), and Weinberger and Fischer (2006). The review highlights the diversity of theoretical perspectives represented in approaches to analyzing argumentation, the importance of clearly specifying theoretical and environmental commitments throughout the process of developing or adopting an analytic framework, and the role of analytic frameworks in the development of learning environments for argumentation.

Discussion: Implications for the KI Framework and Next Steps

As outlined in the Introduction to this chapter the design of projects in TELS was guided by the KI framework (Linn & Eylon, 2006). We now discuss the implications and next steps of the research described above for the four components of the KI framework as approaches for supporting learning: (1) eliciting current ideas, (2) introducing new idea, (3) developing criteria for evaluating criteria, and (4) sorting and reorganizing ideas.

Eliciting Current Ideas

Constructivist perspectives on learning assert that students learn by building upon their existing ideas. Some researchers (Inhelder & Piaget, 1969; Strike & Posner, 1985) suggest that eliciting students' current ideas helps them identify contradictions between their current ideas and the phenomena under investigation. Other researchers suggest that eliciting students' ideas supports students in building or refining connections to these ideas across contexts (Bransford, Brown, & Cocking, 1999; Linn & Hsi, 2000; Collins, Brown, & Holum, 1991; Brown & Campione, 1994). This high level goal of eliciting students' current ideas can be pursued through multiple avenues. TELS work on critique, argument construction, and argumentation provides insight into several of these avenues.

Structured Concept Maps. Research suggests that making connections between ideas explicit through concept maps can help students reflect on them. Structuring concept maps into domain-specific areas makes existing and missing connections within and across domains explicit. The spatial arrangement of concepts into domain-specific areas can indicate students' ontologies. Making connections between ideas visually explicit can be beneficial for collaborative learning. Concept maps can make changes in students' connections between ideas explicit. Expressing ideas and their connections into visuo-spatial forms can support students' reflection on their repertoire of ideas.

Asthma Module. A key theme of community science instruction is that learning materials build upon students' prior knowledge about science and the community. To elicit these ideas, the Asthma module and assessments prompt students to express their ideas about science and their personal world. The design of the decision justification assessment items prompted students to predict and iteratively refine their reasons and evidence in support of a particular intervention. Research reports that students significantly changed their explanations and justifications throughout their study of the Asthma module. This suggests that students not only expressed and reviewed their initial ideas and connections, but they were able to see how their understanding about the problem transformed over time.

Critiquing Virtual Experiments. TELS research on critiquing a fictitious student's (Mary's) virtual experiments (Chang, 2009; Chang & Linn, 2011) elicited students' idea and reflection on what counts a good experiment. The critique activity guided students to examine Mary's research question, method, and conclusions, consistent with critique activities proposed in other design principle research (Linn & Hsi, 2000; Linn & Eylon, 2006). Moreover, Mary's virtual experiment modeled the process of conducting experiments with a visualization. Modeling the process may help students who are confused by the visualization make sense of it (Betrancourt, 2005; Lowe, 2003, 2004). After critiquing Mary's experiment students had a clearer understanding of the visualization than they did in the other treatments without critique, as indicated by student performances on the embedded assessments.

Introducing New Ideas

A central goal of science education involves introducing new ideas to students. While most students manage to add ideas introduced during instruction, they face significant challenges in integrating these ideas to each other and to their prior knowledge (Clark, 2006). Traditional approaches to science instruction, such as lecture and textbook-based exercises, introduce ideas in ways that result in brittle decontextualized knowledge that is difficult to apply effectively (AAAS, 1993; Bjork, 1994; Bransford et al., 1999; NRC, 1996). Instruction that builds on students' normative ideas as well as their misconceptions can help students to add ideas that build from their prior understandings and promote durable and relevant scientific knowledge (Clement, 1993; Linn & Eylon, 2006). Effective science instruction should introduce new ideas in ways that allow students to generate connections among them. Although multiple approaches can be effective, specific approaches must be selected at the appropriate level of complexity (Feynman, Leighton, & Sands, 1995). TELS research on critique, argument construction, and argumentation has made contributions in clarifying several such approaches.

Seeded Discussions. Research on seeded discussions suggests that grouping students with other students who expressed different ideas than their own is more effective when the group is provided an optimized range of ideas that includes the scientifically normative idea than when only including the students' own ideas.

This may be because the range of ideas is optimized to represent a broad range of common misconceptions or because it is guaranteed to include the scientifically normative ideas. Future research will further explore and clarify the optimal ways to introduce new ideas in group contexts.

Asthma Module. The Asthma study presents results related to students' understanding of asthma as a community problem. The improved KI scores on the asthma explanation item from pre- to posttest demonstrate that students acquired ideas from the Asthma module, specifically the regional- or county-level community ideas. In addition, analysis of the embedded decision justifications provided evidence that the ideas contained in student responses often reflected the most immediate instruction. This study provides sufficient evidence that the Asthma module effectively added ideas to students' repertoire. Future design and research should focus on the develop criteria and sort ideas phases of the KI process.

Drawing and Critique. The drawing and critique study suggests how new ideas can be introduced through dynamic visualizations. In the HFC project, students first watched a video that shows the explosion of a hydrogen balloon. Then they interacted with a dynamic visualization demonstrating how chemical bonds change during hydrogen combustion. The visualization is built upon students' prior knowledge or experience. First, the representations of hydrogen and oxygen particles resemble the ball-and-stick physical models commonly used in science classrooms. Second, to relate to student personal experience about ignition (e.g., setting up a campfire), this visualization includes a "spark" button so that students can control how much energy is provided to ignite the reaction. Third, this visualization includes a dynamic temperature bar. Students can observe synchronous temperature change during the reaction and relate this to the explosion they observed in the video. With these features, the visualization introduced new ideas effectively and supported students to better integrate new ideas with prior experience.

Developing Criteria for Evaluating Ideas

Learners need to develop coherent ways to evaluate the scientific ideas they encounter as they add, refine, connect, promote, and demote ideas within their repertoires. Developing and understanding these criteria is not merely of philosophical or historical importance. Students maintain rich conceptual ecologies involving many prior ideas about many topics (Clark, 2006). As students encounter new ideas during instruction, a goal of science education involves helping them connect these new ideas in normative ways. Students must thus evaluate new and old ideas as they promote, demote, and refine ideas and connections between ideas. The criteria that students need to adopt in making these decisions normatively from the perspective of science as a discipline are not necessarily the ones that students bring with them from everyday life. While "compromising" and agreeing that "everyone is sort of right" may provide productive approaches for resolving social conflicts, for example, students need to understand the epistemological criteria of science if they are to engage productively in KI in science (Duschl & Osborne, 2002; Newton, Driver, &

Osborne, 1999; Keller, 1993; Longino, 1994). TELS research on critique, argument construction, and argumentation provides insight in terms of ways to support these goals.

Asthma Module. Students need several opportunities to develop and refine criteria for what counts as an effective community intervention to address their community's asthma problem. Students' low KI scores for the tradeoffs dimension suggest that students did not consider the positive and negative aspects of each program. Students may also need more instruction and practice constructing decision justifications that make explicit the criteria that inform their decision for which program will best address the asthma problem as they understand it.

Drawing and Critique. How can we encourage students to develop criteria to evaluate new and old ideas and to refine connections between ideas? The drawing and critique comparison study suggests two promising approaches. In the drawing approach, students need to evaluate among prior conceptions and new ideas they learned from the visualization to determine what to draw. In the critique activity, students need to evaluate among their knowledge and ideas represented in the given drawings to decide what to critique. This study provides evidence for the success of both approaches. It also suggests that by developing criteria to evaluate various ideas and conceptions, students are prompted to refine their knowledge and develop complicated links among ideas.

Structured Concept Maps. Research on structured concept mapping suggests that it can help students developing a wide range of criteria to critique connections between ideas. Structuring concept maps into different domain-specific areas makes connections within and across domains explicit, which fosters collaborative critique activities. Research suggests that critiquing existing connections might be easier for students with low prior knowledge as it provides them with starting points for their critical reasoning. Students with more prior knowledge might prefer creating their own concept maps which allows them to follow their own train of thought.

Critiquing Virtual Experiments. TELS research on critiquing virtual experiments encouraged students to develop criteria for virtual experiments when they distinguish ideas. This type of activity could promote metacognitive skills or metavisualization abilities by encouraging students to monitor their own reasoning (Gilbert, 2008; Hegarty 2004, 2005). For the experimentation items during the posttests, the critique group outperformed the other groups without critique activities, consistent with the argument that critique promoted developing criteria and distinguishing of ideas.

Sorting and Reorganizing Ideas

The fourth and final component of the KI framework builds on the first three by supporting students in developing, reorganizing, and refining connections among ideas. As part of this reorganization process, students apply their criteria to their new and preexisting ideas as they sort through potential contradictions, promote and demote ideas within their conceptual ecologies, revise and reprioritize connections between ideas, and identify situations where more information is needed (Bransford et al., 1999; Clark, 2001, 2006; diSessa, 1993; diSessa, Gillespie, & Esterly, 2004; diSessa & Wagner, 2005; Dufresne, Mestre, Thaden-Koch, Gerace, & Leonard, 2005; Linn & Hsi, 2000; Scardamalia & Bereiter, 1999). This process benefits from metacognitive skills and scaffolding to focus students' efforts most effectively (Bielaczyc, Pirolli, & Brown, 1995; Lin & Schwartz, 2003). Unfortunately, many students default to rote memorization (Songer & Linn, 1992), which results in brittle knowledge that is compartmentalized, difficult to apply or transfer, and quickly forgotten (AAAS, 1993; Bjork, 1994; Bransford et al., 1999; NRC, 1996). Students instead need significant support in engaging actively, consciously, and strategically in refining and restructuring their understandings (Clark, 2006). TELS research on critique, argument construction, and argumentation has also provided insights into these goals.

Seeded Discussions. The seeded discussions research focused primarily on this component of the KI framework. How can we encourage students to reflect on their ideas, compare these ideas to other ideas, and make informed decisions as they sort through evidence and arguments in terms of these ideas? This research suggests strongly the value of conflict schema approaches where students are grouped with other students who have expressed ideas different than their own to facilitate the sorting, evaluation, and reorganization of explanations and ideas for challenging science phenomena. Essentially, this approach "crowd sources" some of the cognitive load of integrating and contrasting ideas for the students while also potentially leveraging social motivations for engagement with the process.

Asthma Module. Students need more support linking different perspectives as they justify their decision about which community intervention to implement. Even though students' justifications included multiple perspectives, students struggled to connect the perspectives so they form a broad, integrated view of the asthma problem. Specifically, students isolated community and physiological perspectives. Since the module successfully adds ideas about physiology and the social implications of asthma, future revisions to the module should focus on the design of learning activities that promote connections among them. For example, students could role play and answer questions from a variety of assigned perspectives. This may prompt them to evaluate their existing ideas and generate connections they would not otherwise consider.

The research reported in the Asthma study also indicates that students struggled to demonstrate integrated evidence use. Analysis of the evidence use dimension revealed that the evidence presented in the Asthma module was added to students' repertoire, but not linked to reasons or other evidence. This finding was consistent across all communities, indicating that additional scaffolded instruction from the module and teacher would benefit a wide range of students. The revisions to the module should emphasize the develop criteria and sort ideas phases of knowledge integration. In particular, students need more explicit instruction on what constitutes an integrated explanation and decision justification and several opportunities to apply these criteria, such as the peer critique and the debate activities.

Structured Concept Maps. Research on structured concept mapping found that making the connections between domain-specific areas explicit can help students sorting out and reorganizing ideas. Concept maps constrain learners to decide on only one relationship between two concepts. This constraint requires students to apply criteria to select one connection and use supporting evidence when collaboratively working on a concept map. Research suggests that initial concept maps need refinement through critical revision. Presenting students with flawed concept maps can effectively support students' criteria generation and application. Findings show that students can apply criteria to their peers' work as well as their own work. Teacher-guided classroom discussions can support students' metacognitive understanding of different forms of criteria. Future research will extend critique-focused concept mapping to other science domains.

Critiquing Virtual Experiments. TELS research on critiquing virtual experiments engaged students in distinguishing their own ideas from those attributed to Mary. Neither the interaction condition nor the observation condition required students to distinguish among ideas (Chang, 2009; Chang & Linn, 2011). Observation and interaction may have encouraged adding ideas but not integrating ideas (Linn et al., 2004). In contrast, when students critique Mary's virtual experiment they need to sort and reorganize their and Mary's ideas in order to make a claim, link to the evidence, and provide arguments as they critique Mary's experiment.

Final Thoughts

The TELS research presented in this chapter represents a diverse set of science learning experiences that feature various scaffolding strategies, designed content, and supported modes of participation. Taken together, they strongly endorse the inclusion of critique and argumentation learning activities as an effective way to improve students' understanding of core science concepts. Furthermore, these research studies offer the field several analytic approaches to the assessment of science learning among a wide range of students.

While this chapter celebrates diversity in science education research, it also underscores the importance of a solid theoretical foundation. The KI framework guides the design and research of each study reported in this chapter as researchers work toward a collective goal—a deep, integrated understanding of science among learners. The principled approach of the KI framework affords designers and researchers flexibility in implementation and focus. This supports the creation of varied projects that are not limited or repetitive in their format, and also promote the high level of cumulative learning demonstrated in TELS overarching studies across multiple projects (Linn et al., 2006).

Consistent with the framework's emphasis on connections, each study reports on how students linked their prior and new ideas to the science content, their peers' dialogue, or information about their community. For example, the Space Colony study supports students' use of critique to promote connections between key genetic and evolutionary ideas. The Heat and Temperature study demonstrates how critique can guide students' identification of valid experiments, a key scientific process. Studies also promoted science learning in the form of engagement in professional and personal scientific practices. The HFC Cars study shows that students can be supported to use visualizations to inform their scientific understanding more effectively when they engage in critique. Related to argumentation, the Asthma study discusses how scaffolding the use of evidence and the consideration of tradeoffs can enhance students' engagement in integrated decision-making about community science. Finally, seeded discussion identifies scaffolding strategies that foster substantiated dialogue, a skill set that can be applied to numerous, relevant topics through learners' lifetimes.

These studies not only highlight successful approaches for engaging students in critique and argumentation, they also highlight room for improving these approaches. This creates fertile ground for continued research that focuses on the implementation and investigation of critique and argumentation as tools to promote deep science learning. In addition, these studies demonstrate how a shared overarching conceptual framework can encourage the integration of creative and diverse approaches to design, assessment, and analytic ideas from multiple theoretical perspectives, which in turn has contributed to an increasingly powerful and expansive database of design principles (Kali, 2006; Kali & Linn, 2008; Kali, Linn, & Roseman, 2008) for leveraging the KI framework to promote deep, complex science learning.

References

Abell, S. K., Anderson, G., & Chezem, J. (2000). Science as argument and explanation: Exploring concepts of sound in third grade. In J. Minstrell & E. H. Van Zee (Eds.), *Inquiry into inquiry learning and teaching in science* (pp. 100–119). Washington, DC: American Association for the Advancement of Science.

Ainsworth, S. E. (1999). A functional taxonomy of multiple representations. *Computers and Education,* 33(2/3), 131–152.

American Association for the Advancement of Science (AAAS). (1993). *Benchmarks for science literacy*. Washington, DC: Author.

Andriessen, J., Baker, M., & Suthers, D. (2003). Argumentation, computer support, and the educational contexts of confronting cognitions. In J. Andriessen, M. Baker, & D. Suthers (Eds.), *Arguing to learn: Confronting cognitions in computer-supported collaborative learning environments* (pp. 1–25). Dordrecht: Kluwer.

Ardac, D., & Akaygun, S. (2004). Effectiveness of multimedia-based instruction that emphasizes representations on students' understanding of chemical change. *Journal of Research in Science Teaching, 41*(4), 317–337.

Bahar, M., Johnstone, A. H., & Hansell, M. H. (1999). Revisiting learning difficulties in biology. *Journal of Biological Education, 33*(2), 84–86.

Baker, M. (2003). Computer-mediated argumentative interactions for the co-elaboration of scientific notions. In J. Andriessen, M. Baker, & D. Suthers (Eds.), *Arguing to learn: Confronting cognitions in computer-supported collaborative learning environments* (pp. 47–78). The Netherlands: Kluwer.

Baker, M., Andriessen, J., Lund, K., van Amelsvoort, M., & Quignard, M. (2007). Rainbow: A framework for analyzing computer-mediated pedagogical debates. *International Journal of Computer Supported Collaborative Learning, 2*(2), 315–357.

Bell, P. (2004). The educational opportunities of contemporary controversies in science. In M. Linn, E. Davis, & P. Bell (Eds.), *Internet environments for science education* (pp. 233–260). Mahwah, NJ: Lawrence Erlbaum Associates.
Bell, P., & Linn, M. C. (2000). Scientific arguments as learning artifacts: Designing for learning from the web with KIE. *International Journal of Science Education, 22*(8), 797–818.
Betrancourt, M. (2005). The animation and interactivity principles in multimedia learning. In R. E. Mayer (Ed.), *The Cambridge handbook of multimedia learning* (pp. 287–296). New York: Cambridge University Press.
Bielaczyc, K., Pirolli, P., & Brown, A. L. (1995). Training in selfexplanation and self-regulation strategies: Investigating the effects of knowledge acquisition activities on problem solving. *Cognition and Instruction, 13*(2), 221–252.
Bjork, R. A. (1994). Memory and metamemory considerations in the training of human beings. In J. Metcalfe & A. Shimamura (Eds.), *Metacognition: Knowing about knowing* (pp. 185–205). Cambridge, MA: MIT Press.
Boo, H. K., & Watson, J. R. (2001). Progression in high school students' (aged 16–18) conceptualizations about chemical reactions in solution. *Science Education, 85*(5), 568–585.
Boulter, C. J., & Gilbert, J. K. (1995). Argument and science education. In P. J. M. Costello & S. Mitchell (Eds.), *Competing and consensual voices: The theory and practices of argument.* Clevedon: Multilingual Matters.
Bransford, J., Brown, A., & Cocking, R. (1999). *How people learn: Brain, mind, experience and school.* Washington, DC: National Academy of Science Press.
Brown, A. (1992). Design experiments: Theoretical and methodological challenges in creating complex interventions in classroom settings. *The Journal of Learning Sciences, 2*(2), 141–178.
Brown, A. L., & Campione, J. (1994). Guided discovery in a community of learners. In K. McGilly (Ed.), *Classroom lessons: Integrating cognitive theory and classroom practice* (pp. 229–270). Cambridge, MA: MIT Press/Bradford Books.
Chang, H.-Y. (2009). *Use of critique to enhance learning with an interactive molecular visualization of thermal conductivity.* Poster presented at the annual meeting of National Association for Research in Science Teaching (NARST) 2009, Garden Grove, CA.
Chang, H.-Y., & Linn, M. C. (2010, January). *Transition to inquiry: Instructional practice of inquiry-based online science curricula in Taiwan.* Paper presented at the Sixth International Conference on Science, Mathematics and Technology Education, Hualien, Taiwan.
Chang, & Linn. (2011). Learning from a molecular visualization: Observe, interact or critique? Manuscript submitted for publication.
Chang, H.-Y., & Tsai, K. C. (2010, June). *Investigating the role of physical and virtual experiments in developing integrated understanding of thermal conductivity and equilibrium.* Presented in the symposium "Using visualization to link abstract science and everyday experience", the International Conference for the Learning Sciences 2010.
Chang, H.-Y., Quintana, C., & Krajcik, J. (2010). The impact of designing and evaluating molecular animations on how well middle school students understand the particulate nature of matter. *Science Education, 94*(1), 73–94.
Chi, M. T. H. (2000). Self-explaining: The dual processes of generating inference and repairing mental models. In R. Glaser (Ed.), *Advances in instructional psychology: Educational design and cognitive science* (Vol. 5, pp. 161–238). Mahwah, NJ: Lawrence Erlbaum Associates.
Chiu, J., & Linn, M. C. (in press). The role of self-monitoring in learning chemistry with dynamic visualization. In A. Zohar, & Y. J. Dori (Eds.), *Metacognition and science education: Trends in current research.* London: Springer.
Clark, D. B. (2001). *New representations of student knowledge integration in CLP: Theories or repertoires of ideas?* Paper presented at the AERA, Seattle, WA.
Clark, D. B. (2004). Hands-on investigation in Internet environments: Teaching thermal equilibrium. In M. C. Linn, E. A. Davis, & P. Bell (Eds.), *Internet environments for science education* (pp. 175–200). Mahwah, NJ: Lawrence Erlbaum Associates.

Clark, D. B. (2006). Longitudinal conceptual change in students' understanding of thermal equilibrium: An examination of the process of conceptual restructuring. *Cognition and Instruction, 24*(4), 467–563.

Clark, D. B., D'Angelo, C. M., & Menekse, M. (2009). Initial structuring of online discussions to improve learning and argumentation: Incorporating students' own explanations as seed comments versus an augmented-preset approach to seeding discussions. *Journal of Science Education and Technology, 18*(4), 321–333.

Clark, D. B., & Sampson, V. (2005, June). *Analyzing the quality of argumentation supported by personally-seeded discussions.* Paper presented at the annual meeting of the Computer Supported Collaborative Learning (CSCL) Conference, Taipei, Taiwan.

Clark, D. B., & Sampson, V. (2008). Assessing dialogic argumentation in online environments to relate structure, grounds, and conceptual quality. *Journal of Research in Science Teaching, 45*(3), 6.

Clark, D. B., & Sampson, V. D. (2007). Personally-seeded discussions to scaffold online argumentation. *International Journal of Science Education, 29*(3), 253–277.

Clark, D., Sampson, V., Weinberger, A., & Erkens, G. (2007). Analytic frameworks for assessing dialogic argumentation in online learning environments. *Educational Psychology Review, 19*(3), 343–374.

Clark, D. B., Schleigh, S. P., Menekse, M., D'Angelo, C. M., & Sampson, V. (2008). *Improving the quality of student argumentation through the initial structuring of online discussions.* Paper presented at the American Educational Research Association 2008 meeting. New York.

Clement, J. (1993). Using bridging analogies and anchoring intuitions to deal with students' preconceptions in physics. *Journal of Research in Science Teaching, 30*(10), 1241–1257.

Cobb, P., Confrey, J., diSessa, A. A., Lehrer, R., & Schauble, L. (2003). Design experiments in educational research. *Educational Researcher, 32*(1), 9–13.

Collins, A., Brown, J. S., & Holum, A. (1991). Cognitive apprenticeship: Making thinking visible. *American Educator, 15*, 6–11 & 38–46.

Corburn, J. (Ed.). (2005) Street science: Characterizing local knowledge. In His *Street science: Community knowledge and environmental health justice* (pp. 47–77). Cambridge, MA: MIT Press.

Corliss, S., & Spitulnik, M. (2008). Student and teacher regulation of learning in technology-enhanced science instruction. In *International perspectives in the learning sciences: Creating a learning world. Proceedings of the 8th international conference of the learning sciences* (Vol. 1, pp. 167–174). Utrecht, the Netherlands: International Society of the Learning Sciences.

Cuthbert, A. J., Clark, D. B., & Linn, M. C. (2002). WISE learning communities: Design considerations. In K. A. Renninger & W. Shumar (Eds.), *Building virtual communities: learning and change in cyberspace* (pp. 215–246). Cambridge: Cambridge University Press.

deVries, E., Lund, K., & Baker, M. (2002). Computer-mediated epistemic dialogue: Explanation and argumentation as vehicles for understanding scientific notions. *Journal of the Learning Sciences, 11*(1), 63–103.

Dillenbourg, P., & Jermann, P. (2007). Designing integrative scripts. In F. Fischer, H. Mandl, J. Haake & I. Kollar (Eds.), *Scripting computer-supported communication of knowledge—cognitive, computational and educational perspectives* (pp. 275–301). New York: Springer.

diSessa, A. A. (1993). Toward an epistemology of physics. *Cognition and Instruction, 10*(2 & 3), 105–225.

diSessa, A. A., Gillespie, N., & Esterly, J. (2004). Coherence versus fragmentation in the development of the concept of force. *Cognitive Science, 28*, 843–900.

diSessa, A. A., & Wagner, J. F. (2005). What coordination has to say about transfer. In J. Mestre (Ed.), *Transfer of learning from a modern multi-disciplinary perspective* (pp. 121–154). Greenwich, CT: Information Age.

Driver, R., Newton, P., & Osborne, J. (2000). Establishing the norms of scientific argumentation in classrooms. *Science Education, 84*(3), 287–313.

Dufresne, R., Mestre, J., Thaden-Koch, T., Gerace, W., & Leonard, W. (2005). Knowledge representation and coordination in the transfer process. In J. Mestre (Ed.), *Transfer of learning from a modern multi-disciplinary perspective* (pp. 155–215). Greenwich, CT: Information Age.

Duschl, R. (2007). Quality argumentation and epistemic criteria. In S. Erduran & M. Jimenez-Aleixandre (Eds.), *Argumentation in science education: Perspectives from classroom-based research* (pp. 159–175). Dordreht: Springer Academic Publishers.

Duschl, R. (2008). Science education in three-part harmony: Balancing conceptual, epistemic, and social learning goals. *Review of Research in Education, 32*, 268–291.

Duschl, R. A., & Osborne, J. (2002). Supporting and promoting argumentation discourse in science education. *Studies in Science Education, 38*, 39–72.

Erduran, S., Simon, S., & Osborne, J. (2004). TAPping into argumentation: Developments in the application of Toulmin's argument pattern for studying science discourse. *Science Education, 88*, 915–933.

Feynman, R. P., Leighton, R. B., & Sands, M. L. (1995). *Six easy pieces: Essentials of physics, explained by its most brilliant teacher*. Reading, MA: Addison-Wesley.

Fleming, R. (1986a). Adolescent reasoning in socio-scientific issues. Part I. Social cognition. *Journal of Research in Science Teaching, 23*, 677–687.

Fleming, R. (1986b). Adolescent reasoning in socio-scientific issues. Part II. Nonsocial cognition. *Journal of Research in Science Teaching, 23*, 689–698.

Gilbert, J. K. (2008). Visualization: An emergent field of practice and enquiry in science education. In J. K. Gilbert (Ed.), *Visualization: Theory and practice in science education* (pp. 3–24). Dordrecht, The Netherlands: Springer.

Grace, M., & Ratcliffe, M. (2002). The science and values that young people draw upon to make decisions about biological conservation issues. *International Journal of Science Education, 24*(11), 1157–1169.

Hegarty, M. (2004). Dynamic visualizations and learning: getting to the difficult questions. *Learning and Instruction, 14*, 343–351.

Hegarty, M. (2005). Multimedia learning about physical systems. In R. E. Mayer (Ed.), *The Cambridge handbook of multimedia learning* (pp. 447–465). New York: Cambridge University Press.

Hesse, F. W. (2007). Being told to so something or just being aware of something? An alternative approach to scripting in CSCL. In F. Fischer, H. Mandl, J. Haake, & I. Kollar (Eds.), *Scripting computer-supported communication of knowledge – cognitive, computational and educational perspectives* (pp. 275–301). New York: Springer.

Hmelo-Silver, C. E., Marathe, S., & Liu, L. (2007). Fish swim, rocks sit, and lungs breathe: Expert–novice understanding of complex systems. *Journal of the Learning Sciences, 16*(3), 307–331.

Hogan, K., & Maglienti, M. (2001). Comparing the epistemological underpinnings of students' and scientists' reasoning about conclusions. *Journal of Research in Science Teaching, 38*(6), 663–687.

Hogan, K., Nastasi, B., & Pressley, M. (2000). Discourse patterns and collaborative scientific reasoning in peer and teacher-guided discussions. *Cognition and Instruction, 17*(4), 379–432.

Inhelder, B., & Piaget, J. (1969). *The early growth of logic in the child*. New York: Norton.

Janssen, J., Erkens, G., Jaspers, J., & Kanselaar, G. (2006, June/July). *Visualizing participation to facilitate argumentation*. Paper presented at the 7th International Conference of the Learning Sciences, Bloomington, IN.

Jeong, A., Clark, D., Sampson, V., & Mushin, M. (2011). Assessing and comparing dialogical scientific argumentation across asynchronous online discussion environments with sequential analysis. In S. Puntambekar, C. Hmelo-Silver, & G. Erkens (Eds.), *Analyzing interactions in CSCL: Methodology, approaches, and issues* (pp. 207–233). New York: Springer.

Jiménez-Aleixandre, M. P., Rodríguez, A. B., & Duschl, R. (2000). Doing the lesson or doing science: Argument in high school genetics. *Science Education, 84*, 757–792.

Kali, Y., (2006). Collaborative knowledge-building using the design principles database. *International Journal of Computer Support for Collaborative Learning, 1*(2), 187–201.

Kali, Y., & Linn, M. C. (2008). Technology-Enhanced Support Strategies for Inquiry Learning. In J. M. Spector, M. D. Merrill, J. J. G. Van Merriënboer, & M. P. Driscoll (Eds.), *Handbook of research on educational communications and technology* (3rd ed., pp. 145–161). New York: Lawrence Erlbaum Associates.

Kali, Y., Linn, M. C., & Roseman, J. E. (Eds.). (2008). *Designing coherent science education*. New York: Teachers College Press.

Keller, E. F. (1993). *A feeling for the organism: The life and work of Barbara McClintock* (10th Anniversary Ed.). New York: W.H. Freeman.

Keppell, M., Au, E., Ma, A., & Chan, C. (2006). Peer learning and learning-oriented assessment in technology-enhanced environments. *Assessment & Evaluation in Higher Education* (Special Issue: Learning-Oriented Assessment: Principles and Practice), *31*(4), 453–464.

King, A. (2007). Scripting collaborative learning processes: A cognitive perspective. In F. Fischer, H. Mandl, J. Haake, & I. Kollar (Eds.), *Scripting computer-supported communication of knowledge—cognitive, computational and educational perspectives* (pp. 275–301). New York: Springer.

Kuhn, T. S. (1970). *The structure of scientific revolutions* (2nd ed.). Chicago: University of Chicago Press.

Kuhn, D. (1993). Science as argument: Implications for teaching and learning scientific thinking. *Science Education, 77*(3), 319–337.

Kuhn, L., & Reiser, B. (2005). *Students constructing and defending evidence-based scientific explanations*. Paper presented at the annual meeting of the National Association for Research in Science Teaching, Dallas, TX.

Kuhn, D., & Udell, W. (2003). The development of argument skills. *Child Development, 74*(5), 1245–1260.

Latour, B. (1987). *Science in action: How to follow scientists and engineers through society*. Cambridge: Harvard University Press.

Leitão, S. (2000). The potential of argument in knowledge building. *Human Development, 43*, 332–360.

Lin, X. D., & Schwartz, D. (2003). Reflection at the crossroad of cultures. *Mind, Culture & Activities, 10*(1), 9–25.

Linn, M. C., Chang, H.-Y., Chiu, J., Zhang, H., & McElhaney, K. (2011). Can desirable difficulties overcome deceptive clarity in scientific visualizations? In A. Benjamin (Ed.), *Successful remembering and successful forgetting: A Festschrift in honor of Robert A. Bjork* (pp. 235–258). New York: Psychology Press.

Linn, M. C., & Eylon, B. S. (2006). Science education: Integrating views of learning and instruction. In P. A. Alexander & P. H. Winne (Eds.), *Handbook of educational psychology* (2nd ed., pp. 511–544). Mahwah, NJ: Lawrence Erlbaum Associates.

Linn, M. C., & Hsi, S. (2000). *Computers, teachers, peers: Science learning partners*. Mahwah, NJ: Lawrence Erlbaum Associates.

Linn, M. C., Davis, E. A., & Bell, P. (2004). *Internet environments for science education*. Mahwah, NJ: Lawrence Erlbaum Associates.

Linn, M. C., Davis, E. A., & Eylon, B. S. (2004). The scaffolded knowledge integration framework for instruction. *Internet environments for science education*. (pp. 47–72). (1) Graduate School of Education, University of California, Berkeley, CA; (2) U Michigan, MI; (3) Weizmann Institute of Science, Israel Mahwah, NJ: Lawrence Erlbaum Associates.

Linn, M. C., Eylon, B. S., & Davis, E. A. (2004). The knowledge integration perspective on learning. *Internet environments for science education*. (pp. 29–46). (1) Graduate School of Education, University of California, Berkeley, CA, US; (2) Weizmann Institute of Science, Israel; (3) University of Michigan, MI, US Mahwah, NJ: Lawrence Erlbaum Associates Publishers.

Linn, M. C., Lee, H. S., Tinker, R., Husic, F., & Chiu, J. L. (2006). Teaching and assessing knowledge integration. *Science, 313*(5790), 1049–1050.

Longino, H. (1994). The fate of knowledge in social theories of science. In F. F. Schmidt (Ed.), *Socializing epistemology: The social dimension of knowledge*. Lanham, MD: Rowman & Littlefield.

Lowe, R. (2003). Animation and learning: Selective processing of information in dynamic graphics. *Learning and Instruction, 13*, 157–176.

Lowe, R. (2004). Interrogation of a dynamic visualization during learning. *Learning and Instruction, 14*, 257–274.

Mason, L. (1998). Sharing cognition to construct scientific knowledge in school context: The role of oral and written discourse. *Instructional Science, 26*, 359–389.

McNeill, K. L., Lizotte, D. J., Krajcik, J., & Marx, R. W. (2006). Supporting students' construction of scientific explanations by fading scaffolds in instructional materials. *The Journal of the Learning Sciences, 15*(2), 153–191.

Moreno, R., & Valdez, A. (2005). Cognitive load and learning effects of having students organize pictures and words in multimedia environments: The role of student interactivity and feedback. *Educational Technology Research and Development, 53*(3), 35–45.

National Research Council (NRC). (1996). *National science education standards*. Washington, DC: National Academy Press.

Newton, P., Driver, R., & Osborne, J. (1999). The place of argumentation in the pedagogy of school science. *International Journal of Science Education, 21*(5), 553–576.

Novak, J. D. (1996). Concept mapping: A tool for improving science teaching and learning. In D. F. Treagust, R. Duit, & B. J. Fraser (Eds.), *Improving teaching and learning in science and mathematics* (pp. 32–43). New York: Teachers College Press.

Passmore, C., & Stewart, J. (2002). A modeling approach to teaching evolutionary biology in high schools. *Journal of Research in Science Teaching, 39*(3), 185–204.

Rochelle, J. (1992). Learning by collaborating: Convergent conceptual change. *The Journal of the Learning Sciences, 2*, 235–276.

Sandoval, W. A., & Reiser, B. J. (2004). Explanation-driven inquiry: Integrating conceptual and epistemic supports for scientific inquiry. *Science Education, 88*, 345–372.

Sanger, M. J., Brecheisen, D. M., & Hynek, B. M. (2001). Can computer animations affect college biology students' conceptions about diffusion and osmosis? *American Biology Teacher, 63*(2), 104–109.

Sampson, V., & Clark, D. (2008). Assessment of the ways students generate arguments in science education: Current perspectives and recommendations for future directions. *Science Education, 92*(3), 447–472.

Scardamalia, M., & Bereiter, C. (1994). Computer support for knowledge-building communities. *Journal of the Learning Sciences, 3*(3), 265–283.

Scardamalia, M., & Bereiter, C. (1999). Schools as knowledge building organizations. In D. K. C. Hertzman (Ed.), *Today's children, tomorrow's society: The developmental health and wealth of nations* (pp. 274–289). New York: Guilford.

Schwarz, B., & Glassner, A. (2003). The blind and the paralytic: Supporting argumentation in everyday and scientific issues. In J. Andriessen, M. Baker, & D. Suthers (Eds.), *Arguing to learn: Confronting cognitions in computer-supported collaborative learning environments* (pp. 227–260). The Netherlands: Kluwer.

Schwendimann, B. (2007). *Integrating interactive genetics visualizations into high school biology*. TELS AERA Meeting.

Schwendimann, B. (2008). Scaffolding an interactive dynamic model to promote coherent connections in high school biology. In *Annual meeting of the American education research association (AERA)*. New York.

Seethaler, S., & Linn, M.C. (2004). Genetically modified food in perspective: An inquiry based curriculum to help middle school students make sense of tradeoffs. *International Journal of Science Education, 26*(14), 1765–1785.

Siegel, H. (1989). The rationality of science, critical thinking and science education. *Synthese, 80*(1), 9–42.

Songer, N. B., & Linn, M. C. (1992). How do students' views of science influence knowledge integration? In M. K. Pearsall (Ed.), *Scope, sequence and coordination of secondary school science: Vol. 1. Relevant research* (pp. 197–219). Washington, DC: National Science Teachers Association.

Stewart, J., Cartier, J. L., & Passmore, C. (2005). Developing understanding through model-based inquiry. In S. Donovan & J. D. Bransford (Eds.), *How students learn science in the classroom* (pp. 147–198). Washington, D.C.: The National Academies Press.

Strike, K., & Posner, G. (1985). A conceptual change view of learning and understanding. In L. H. T. West & A. L. Pines (Eds.), *Cognitive structure and conceptual change* (pp. 189–210). New York: Academic.

Tate, E. D. (2009). *Asthma in the community: Designing instruction to help students explore scientific dilemmas that impact their lives*. Unpublished dissertation, University of California, Berkeley, CA.

Topping, K. J. (2005). Trends in peer learning. *Educational Psychology* (Special Issue: Developments in Educational Psychology: How Far Have We Come in 25 Years?) *25*(6), 631–645.

Toulmin, S. (1958). *The uses of argument*. Cambridge: Cambridge University Press.

Tsui, C. Y. & Treagust, D. F. (2003). Genetics reasoning with multiple external representations. *Research in Science Education, 33*(1), 111–135.

Tversky, B., Morrison, J. B., & Betrancourt, M. (2002). Animation: Can it facilitate? *International Journal of Human Computer Studies, 57*, 247–262.

van Eemeren, F., Grootendorst, R., & Henkemans, A. F. (2002). *Argumentation: Analysis, evaluation, presentation*. Mahwah, NJ & London, England: Lawrence Erlbaum Associates.

Veerman, A. (2003). Constructive discussions through electronic dialogue. In J. Andriessen, M. Baker, & D. Suthers (Eds.), *Arguing to learn: Confronting cognitions in computer-supported collaborative learning environments* (pp. 117–143). The Netherlands: Kluwer.

Vellom, R. P., & Anderson, C. W. (1999). Reasoning about data in middle school science. *Journal of Research in Science teaching, 36*(2), 179–199.

Weinberger, A., & Fischer, F. (2006). A framework to analyze argumentative knowledge construction in computer-supported collaborative learning. *Computers & Education, 46*, 71–95.

Weinberger, A., Stegmann, K., Fischer, F., & Mandl, H. (2007). Scripting argumentative knowledge construction in computer-supported learning environments. In F. Fischer, H. Mandl, J. Haake, & I. Kollar (Eds.), *Scripting computer-supported communication of knowledge – cognitive, computational and educational perspectives* (pp. 191–211). New York: Springer.

White, R., & Gunstone, R. F. (1992). Prediction-observation-explanation. In R. White & R. Gunstone (Eds.), *Probing understanding* (pp. 44–64). London: The Falmer Press.

Wilensky, U., & Resnick, M. (1999). Thinking in levels: A dynamic systems approach to making sense of the world. *Journal of Science Education and Technology, 8*(1), 3–19.

Williamson, V. M., & Abraham, M. R. (1995). The effects of computer animation on the particulate mental models of college chemistry students. *Journal of Research in Science Teaching, 32*(5), 521–534.

Xie, Q., & Tinker, R. (2006). Molecular dynamics simulations of chemical reactions for use in education, *Journal of Chemical Education, 83*, 77–83.

Zeidler, D. L. (1997). The central role of fallacious thinking in science education. *Science Education, 81*, 483–496.

Zhang, Z., & Linn, M. C. (2008). Using drawings to support learning from dynamic visualizations. In *Proceedings of the 8th international Conference on Learning Sciences* (Utrecht, The Netherlands, June 24–28, 2008). International Conference on Learning Sciences. International Society of the Learning Sciences.

Zohar, A., & Nemet, F. (2002). Fostering students' knowledge and argumentation skills through dilemmas in human genetics. *Journal of Research in Science Teaching, 39*(1), 35–62.

Chapter 10
Evaluating Arguments About Climate Change

Adam Corner

Introduction – Communicating Climate Change

Anthropogenic climate change – the impact of human activity on the climate – has become a global political priority. Delegations from 192 countries and more than 60 Heads of State gathered for the United Nations Conference of Parties in Copenhagen in December 2009, and individual nations have now begun the formidable task of decarbonising their economies. In the United Kingdom, for example, the government has set a target of an 80% reduction in levels of greenhouse gases by 2050 (HM Government, 2008) and published a plan that sets out some of the major structural changes that will be required to achieve this goal.

Despite these major and significant developments in national and international policy, a considerable amount of uncertainty remains in public attitudes about the reality and seriousness of climate change. In fact, a range of public opinion polling data suggests that scepticism about anthropogenic climate change has recently *increased* (e.g., BBC, 2010; Pew Research Centre, 2009). The increase in uncertainty about climate change has been most marked in the United States, where a significant proportion of the public do not accept that climate change is caused by human activity (Pew Research Centre, 2009). There is also evidence that an increasing number of people believe that claims about human impacts on the climate have been exaggerated (Dunlap & McCright, 2008; Whitmarsh, 2011). It is sobering to contrast these data on public opinion with a survey of active and publishing climate scientists. Among this group, Doran and Zimmerman (2009) found that 97% agreed that human activity was contributing to climate change.

Despite the fact that climate is a statistical phenomenon – the pattern of weather over a particular time period or geographical location – the communication of

A. Corner (✉)
School of Psychology, Cardiff University, Cardiff, Wales CF10 3AT, UK
e-mail: corneraj@cardiff.ac.uk

M.S. Khine (ed.), *Perspectives on Scientific Argumentation*,
DOI 10.1007/978-94-007-2470-9_10,

climate change is not usually achieved using probabilistic data or numerical risk information. Rather, *arguments* are constructed and transmitted, from scientists to politicians, from politicians to the media and from all of these groups to the general public. One such group is the Intergovernmental Panel on Climate Change (IPCC) – the body of independent scientists charged with providing periodic assessments of climate science. In their most recent assessment report, they stated that:

> Most of the observed increase in globally-averaged temperatures since the mid-20th century is very likely due to the observed increase in anthropogenic greenhouse gas concentrations. . .(these are) expected to have mostly adverse effects on natural and human systems. (Pachauri & Reisinger, 2007)

The statement seems clear and unequivocal – yet, a considerable number of people are not persuaded of its truth. Why is it that a significant proportion of international public opinion has not been convinced by arguments about anthropogenic climate change?

There is growing interest in answering this question, in developing more effective ways of communicating about climate change and in engaging the public more successfully (e.g., American Psychological Association, 2009; Spence, Pidgeon, & Uzzell, 2009). While there are some well-documented small-scale projects that have successfully communicated climate change messages to the general public (e.g., McKenzie-Mohr & Smith, 1999; WRAP, 2008), for the most part the story of climate change communication is one of failure. While general awareness about climate change is growing, there is little evidence that behavioural engagement has shifted significantly over the past few years (Upham, Whitmarsh, Poortinga, Purdam, Darnton, McLachlan & Devine-Wright, 2009).

A recent example of an emphatically unsuccessful climate change communication campaign was the 'Bedtime Story' advertisement commissioned by the British government's Department for Energy and Climate Change in 2009. The advert was designed to communicate the seriousness and urgency of climate change by depicting a young child being read a bedtime story about climate change. As a narrative about the destructive future effects of climate change unfolds, scary music plays in the background and vivid graphical representations of 'evil carbon' cause floods to rise around the house of the child. The advert ends with the message that it is up to the viewer how the 'story' of climate change ends – that it is not too late to avert the negative consequences of unmitigated climate change. The advert was intended to make climate change more personally relevant to British viewers (by depicting an 'average' neighbourhood becoming flooded). However, following a number of complaints from viewers that the advert was unnecessarily scary and several critical analyses from climate change communication experts, the advert was abandoned.

Why is it that the process of communicating climate change to the public is not straightforward? In this chapter, I will outline some possible answers to this question. With reference to analyses of popular climate change media narratives and empirical data on climate change argument evaluation, I will examine the way that people evaluate arguments, evidence and messages about climate change.

How do People Evaluate Arguments About Climate Change?

Knowledge About Climate Change

Is it the case that people simply do not know or understand enough about climate change? Would educating or teaching members of the public more about climate change make them more likely to accept arguments about the human impact on the climate?

Until fairly recently, it was often assumed by scholars and science communicators that if a particular scientific or technological development was unpopular among the general public, that the public must have a 'deficit' of knowledge that needed addressing through information provision. Increasingly, however, this view has been challenged, as studies have consistently shown that people's perception of science and technology is not straightforwardly attributable to their level of knowledge about it (e.g.,Kahan, Braman, Slovic, Gastil, & Cohen, 2009; Malka, Krosnick, & Langer, 2009). Rather, disagreement over scientific and technological developments may be due to divergent values, distrust in risk communicators or differing attitudes towards risk management and regulatory systems. Furthermore, assuming a 'deficit' of knowledge is not conducive to establishing a genuinely participatory interaction between science communicators and the broader public (Irwin & Wynne, 1996). It is now widely recognised that differences in opinion between science and industry and members of the public cannot simply be put down to a lack of knowledge or awareness. Correspondingly, behavioural research has found that information alone is insufficient to promote pro-environmental behaviour. Even if informational campaigns are successful in influencing attitudes, there is often a significant gap between people's attitudes and their behaviour (Maio, Verplanken, Manstead, Stroebe, Abraham, Sheeran, & Connor, 2007).

Adams (1999) examined how college students, scientists and policy analysts evaluated 'questionable' scientific claims about climate change. Participants were interviewed as they evaluated the claims made in a particular article, were asked whether they agreed with the article, what they thought of the article and how reliable they thought it was. The qualitative responses they gave indicated that despite being the least knowledgeable of the three groups about climate change, the college students seemed able to apply a 'generic' evaluative criteria to the reports, asking questions about the source's validity and the degree to which it was appropriate to generalise or extrapolate given the available evidence. The results of this study suggest that being knowledgeable about climate change is not a prerequisite for evaluating the merit of climate change arguments.

However, while the relationship between lay knowledge and attitudes towards climate change is not straightforward, there is some evidence to suggest that people who demonstrate a greater understanding of climate change are more likely to support arguments about government action to reduce greenhouse gases (Bord, O'Connor, & Fischer, 2000) and that learning about climate change through structured educational programmes promotes a heightened sense of agency around climate change – that is, an increased acceptance of the human impact on climate and the ability of human behaviour to mitigate it (Hogg & Shah, 2010). A comprehensive review of the role of psychology in addressing climate change by the American Psychological Association (APA, 2009) suggested that a lack of knowledge about appropriate behavioural responses to climate change (and their impact – Stern, 2000) was a major barrier to public engagement with climate change.

In addition, as Weber (2010) has argued, even a good technical understanding of the causes and effects of climate change cannot prevent experiential learning from personal experience of everyday weather from interfering. Climate is a statistical phenomenon, comprised of patterns of weather over a period of time for a particular region. Confusingly, however, weather is not necessarily a good guide to climate – while an increase in greenhouse gases is expected to increase the frequency and severity of extreme weather events, no single weather event can be unambiguously attributed to 'climate change'. This means that climate change is perceptually vague, abstract and difficult to visualise, while most people's experience of weather is concrete and visceral. Where there is a conflict between the weather (e.g., a particularly cold winter) and climatic predictions (e.g., warming over a 50-year period), the fast and automatic associative processes that drive learning from personal experience are likely to trump the cognitive effort required for learning from statistical descriptions or written reports (Weber, 2010).

In an attempt to make climate change more tangible and more relevant to people's lives, the American think-tank EcoAmerica played people recordings of actors delivering speeches about climate change (Western Strategies & Lake Research Partners, 2009). The version that people were most positive towards talked about 'air pollution' rather than 'climate change' – because pollution is something visible that they could relate to, with strong connotations of dirtiness and poor health. Research comparing the climate change attitudes of flood victims and ordinary citizens has also identified a positive association between air pollution and concern about climate change (Whitmarsh, 2008). While flood victims were no more likely than other people to be concerned about climate change (perhaps, because they did not associate their personal experience with the global phenomenon of climate change), people who reported direct experience of air pollution affecting their health were more concerned about climate change (see also Spence, Poortinga, Butler, and Pidgeon, 2011, for a more recent discussion of the links between flooding experience and attitudes towards climate change). Weber (2010) has suggested that the concretisation of future climate-related events may hold promise as a method of increasing awareness and concern about climate change. One reason that arguments about climate change so often seem to fail is that they are arguments about abstract concepts, intangible effects and psychologically distant consequences.

The link between knowledge and attitudes towards climate change is complex and learning about climate change – either through structured educational programmes or by using techniques to overcome the challenges that climate change poses to our perceptual and cognitive systems – is one determinant of how climate change messages will impact on members of the public. But what does it mean to say that someone is 'engaged' by climate change? Lorenzoni, Nicholson-Cole and Whitmarsh (2007) presented an analysis of what they considered constituted 'engagement' with climate change. They identified engagement as an individual's state, comprised of cognitive, affective and behavioural elements, suggesting that

> (I)t is not enough for people to know about climate change in order to be fully engaged; they also need to care about it, be motivated and be able to take action. (Lorenzoni et al., 2007, p. 445)

Lorenzoni et al. (2007) asked members of the public about their perceptions of and responses to climate change and identified two broad classes of barriers that people perceived to engaging with climate change – individual and social. Individual barriers included a lack of knowledge about where to find relevant information – but also a perceived *overload* in the amount of information available; confusion about conflicting scientific evidence and a lack of trust in the sources delivering messages about climate change (e.g., politicians/environmental campaigners/the media). Social barriers included a lack of perceived political action on climate change, social norms and expectations to live (or aspire to) high consuming lifestyles, and concern about 'free riders' who might avoid taking action on climate change (leaving an unfair burden on those willing to change).

This research makes clear that it is not simply a lack of knowledge about climate change that acts as a barrier to the communication of arguments about climate change. In the next section, I will present a more detailed analysis of one of these barriers – uncertainty – and describe some forthcoming research (Corner, Whitmarsh, & Xenias, in press) that sheds light on how uncertainty impacts on the evaluation of arguments about climate change.

Uncertainty

Despite the overwhelming body of evidence showing that human activity is altering the global climate, debates about climate change are characterised by an enormous amount of *uncertainty* (Hulme, 2009; Zehr, 2000). Uncertainty is a multi-facetted and complex phenomenon, which is present in almost every debate about science and society (Friedman, Dunwoody, & Rogers, 1999). Some of the uncertainty about climate change stems from the science itself: important questions about the extent and impact of climatic changes remain unanswered. Many of the predicted effects of climate change are quantifiable but uncertain, and are only accurately expressed as probability distributions or ranges (see, e.g., UK Climate Impact Programme, 2009). As in economic forecasts, medical diagnoses and policy making, uncertainty is a fundamental feature of climate science. Yet, more uncertainty arises from policy debates about what constitutes 'dangerous' climate change (Lorenzoni, Pidgeon, & O'Connor, 2005; Oppenheimer, 2005) and which mitigation and adaptation measures will be required to prevent it. However, among ordinary members of the public, a substantial amount of uncertainty remains about the reality or seriousness of human-caused climate change.

While the everyday meaning of uncertainty is negative, as it is commonly equated with ignorance (Shome & Marx, 2009), uncertainty is not an enemy of science that must be conquered. Rather, it is a stimulus that drives science forward. Pollack (2005) has suggested that there is a tendency for the media and non-scientists in general to infer from the fact that scientists do not know *everything* about a topic, that they do not know *anything* about it. This means that uncertainty can be problematic when people seek to evaluate arguments, evidence and media reports on climate change.

Some concerted attempts have been made at quantifying and communicating the uncertainties around climate science. In their most recent assessment report, the IPCC used specific terms to indicate the confidence with which particular conclusions were held (Pachauri & Reisinger, 2007). The term 'very likely' was used, for example, to indicate 90% confidence in a statement. By using numerically defined terms – Bayesian expressions of belief in a hypothesis based on scientific evidence – the IPCC hoped to quantify uncertainty in a meaningful way. However, the way that people interpret evidence about climate change is impacted by well-documented biases that influence judgments about numerical and non-numerical risk information (Pidgeon, Kasperson, & Slovic, 2003; Weber, 2006; Weber, 2010) and several studies have shown that the average person's interpretation of the verbal labels used by the IPCC does not match their intended meaning.

Budescu, Broomwell and Por (2009) found that there were significant discrepancies between the meaning intended by the IPCC in their risk statements and the numerical values that people assigned to them – even when people were provided with the numerical definitions of the verbal terms. Patt and Schrag (2003) proposed that the use of specific language to describe probability ranges in climate change risks – the strategy employed by the IPCC – tended to result in miscommunication. In a study designed to examine the relationship between the severity of an environmental risk and the numerical probability people assigned to a verbal description of it, Harris and Corner (2011) found that severe events such as volcanoes elicited higher probability judgments than more neutral events (even when the language used to describe the likelihood of these events occurring was held constant), suggesting that more severe outcomes are easier to 'simulate' in the imagination (Risen & Gilovich, 2007).

The communication of risk and uncertainty is a major challenge for the IPCC. However, the overwhelming majority of risk information that people receive about climate change comes not through formal IPCC reports (which are designed for policy makers), but through arguments summarising risk information presented by the media. Norris, Phillips, and Korpan (2003) studied university students' evaluations of brief scientific stories, found evidence that they overestimated the certainty with which they could make conclusions based on the data reported in the report and seemed to display an inflated view of how well they could understand the report.

Several analyses of media coverage of climate change have concluded that a discourse of uncertainty is unsuited to the typically adversarial style of English language journalism (e.g., Boykoff, 2007). Radio, television and newspaper reports have been criticised for interpreting too simplistically the notion of providing a 'balanced' set of views, which can lead to competing points of view on a scientific issue being presented as equal when in fact they are not (Zehr, 2000). While there is evidence that this is changing (in the United Kingdom at least – Boykoff, 2007), Butler and Pidgeon (2009) have shown that people view the media as offering a range of viewpoints on climate change, creating the impression that the causes of climate change are more controversial than they in fact are. Corbett and Durfee (2004) have emphasised that the word 'uncertainty' need not be present in an article in order for the science to be portrayed as uncertain – all that is necessary is that 'duelling experts' are presented without any sense of how the weight of evidence is distributed.

Difficulties in interpreting scientific uncertainty can be overcome through a more structured process of evaluation. Ratcliffe (1999) studied the ability of 11–14-year-old pupils, 16–18-year-old college students and university graduates to evaluate the content of short articles taken from the *New Scientist* magazine. Participants were required to note any areas about which they felt uncertainty and to make a list of questions they would like to ask about the article. The graduates showed more advanced evaluative skills than the students and the students more advanced skills than the pupils (as measured by the taxonomy developed in Korpan, Bisanz, Bisanz, & Henderson, 1997); however, all age groups demonstrated relevant evaluation of the claim to some extent, acknowledging the role of uncertainty and recognising that the reports also contained established facts. Corbett and Dufree (2004) have argued that providing more 'context' for claims about climate change (i.e., general information about climatic trends alongside specific claims about individual phenomena) is an effective way of reducing the unintended communication of uncertainty.

However, while presentational devices may mitigate unintended uncertainty to some extent, uncertainty about climate change is often attributable to more deep-rooted and psychological differences. Studies in the United States that have suggested that scepticism about climate change is increasing (Pew Research Centre, 2009) have also demonstrated that public opinion about climate change in the United States is dividing along ideological lines. Supporters of the Republican Party are far more likely to express scepticism about anthropogenic climate change than the Democrats. In the United Kingdom, Whitmarsh (2011) found that between 2003 and 2008 public uncertainty about climate change remained constant in most respects. However, the belief that claims about climate change has been exaggerated almost doubled over that period from 15 to 29%. In addition, beliefs about climate change were strongly influenced by stated political affiliation, with conservative voters the most sceptical about the human causes of climate change.

A growing body of research by Dan Kahan and his colleagues at the Yale Centre for Cultural Cognition suggests that there may be an even more important factor than an individual's political preferences for predicting their attitude towards scientific risks such as climate change. Drawing on the long-standing anthropological work of Douglas and Wildavsky (1982), Kahan and his colleagues have demonstrated that people with opposing 'cultural worldviews' tend to polarise in their perception of the risks posed by climate change, as well as other areas of science and technology (Kahan et al., 2009). According to Kahan et al., people's cultural worldviews (their beliefs about the relationship between nature and society and their attitudes towards risk and regulation) lead them to assimilate and integrate new information about science and technology in a biased way, such that following exposure to 'balanced' information about climate change, people's attitudes divide along cultural lines.

In fact, there is a long history of research in social psychology (Lord, Ross, & Lepper, 1979; Miller, McHoskey, Bane, & Dowd, 1993; Munro & Ditto, 1997) demonstrating that people with opposing views on controversial topics sometimes polarise when they receive new information. For example, when presented with balanced/mixed evidence for and against a hypothesis (e.g., the desirability of capital punishment), pro-capital punishment people become *more* convinced of their beliefs, while anti-capital punishment people become *more* convinced of theirs (Lord et al., 1979). Despite viewing the very same evidence, people report that their beliefs move in different directions.

Corner, Whitmarsh, and Xenias (in press) conducted an experiment with undergraduate students at Cardiff University to establish whether individuals who expressed different attitudes about climate change would evaluate uncertain evidence about climate change differentially. Drawing on the typology of uncertainty presented in Tannert, Elvers, and Jandrig (2007; see also Patt (2007), for a distinction between model-based and conflict-based uncertainty), Corner et al. presented participants with two newspaper 'editorials' that offered opposing arguments about climate change (the editorials were constructed for the purpose of the study). In one condition of the experiment, the two opposing articles focussed on climate science (one headline read 'We are as certain about climate change as we are about anything', while the other read 'If we can't predict the weather, how can we predict the climate?') and was designed to generate data-based or *epistemological* uncertainty. In the other condition, the two opposing articles focussed on *moral* uncertainty – one headline read 'US politicians are committing treason against the planet', while the other read 'Why are environmentalists exaggerating claims about climate change?'

Corner et al. asked participants to indicate how convincing and how reliable they found the two editorials to be, and reported that participants' evaluations of the editorials depended on their prior attitudes towards climate change – in particular, the extent to which they perceived climate change to be uncertain. For people who expressed lower levels of uncertainty about climate change, the pro-climate change editorials were rated as both more convincing and more reliable than the anti-climate change editorials. However, the opposite pattern was observed for individuals who expressed higher levels of uncertainty about climate change. These findings suggest that when presented with arguments about climate change, prior beliefs and attitudes towards climate change are likely to play an important role in how compelling these arguments will be. Arguments that seem compelling to those who are already persuaded of the reality or importance of climate change may not be as effective for people who are uncertain about climate change in the first place. Despite the fact that uncertainty is an inescapable part of any complex scientific topic, perceived uncertainty may play a critical role in determining the extent to which people accept arguments about climate change.

An irony of the debate about the uncertainty associated with predictions about climate change is that climate models sketch out possible, rather than inevitable futures. One crucial uncertainty that cannot be captured in any climate model is the extent to which action is taken to cut the emissions of greenhouse gases – something directly contingent on the public acceptance of arguments about climate change. The IPCC was initiated as a body that could assess the predictions made by climate models in order to give policy makers and the public some idea of what lies ahead. Climate models, replete with their inherent uncertainties about impacts and effects, provide policy makers with a critical opportunity to change course.

Preventing the negative consequences of climate change is of course central to the growing level of interest in how to better communicate about climate change. In the next section, I present quantitative data from an experiment with college students in South Wales that examined the way in which *consequentialist* arguments about climate change were evaluated.

The Consequences of Climate Change

Many scientific arguments about climate change are based on the *consequences* that our current actions will have for future generations. A dissuasive consequentialist argument (or deterrent) warns against a particular course of action on the grounds that it will lead to an undesirable outcome or consequence (Bonnefon & Hilton, 2004). We may be warned, for example, that if the global climate continues to increase in temperature, glacial ice will melt at an accelerated rate, sea levels will rise and low lying homes will be flooded. This is certainly a negative consequence, but avoiding it might require personal sacrifices that many consider unacceptable. For example, the aviation industry is one of the fastest growing sources of carbon dioxide emissions in the world (Bows, Upham, & Anderson, 2005). It may be the case that the use of aeroplanes will be curbed or restricted in some way in the future, although this is a sacrifice that few are currently willing to make as a method of reducing carbon dioxide emissions (Defra, 2008). Presumably, this is because people find the prospect of international travel less appealing if large distances cannot be covered quickly using an aeroplane. The negativity of the outcome (i.e., the adverse effects of climate change) must be balanced against the personal cost involved in avoiding it.

In 2007, 64 students aged 16–18 from three schools in South Wales took part in an experiment where they were required to evaluate the strength of consequentialist arguments. The experiment was part of a project called 'Evaluating Scientific Arguments', which was an initiative designed to engage young people in a scientific activity. The experiment followed the design of the 'Consequentialist Arguments Task' reported in Corner and Hahn (2009) and was designed to replicate the results of this study with a different sample. Two features of the consequentialist arguments were varied – the negativity of the outcome and the sacrifice required to avoid the outcome in the experiment – creating four experimental conditions:

1. Very negative outcome/small sacrifice required
2. Less negative outcome/small sacrifice required
3. Very negative outcome/big sacrifice required
4. Less negative outcome/big sacrifice required

Participants were required to evaluate one consequentialist argument about a scientific topic (flooding caused by climate change) and one consequentialist argument about a non-scientific topic (sleeping through an alarm clock). Each individual participant contributed data to two (randomly selected) conditions of the experiment – one for each argument topic. Participants were asked to indicate how convincing they found the arguments, on a scale from 0 (very unconvincing) to 10 (very convincing).

The four variations of the climate change argument were as follows:

1. "If global warming continues at the current rate, it will cause the sea levels to rise and 10,000 people in Britain will lose their homes within 5 years. To prevent this, we must switch all the light bulbs in our houses to energy efficient ones." (Very negative outcome/small sacrifice)
2. "If global warming continues at the current rate, it will cause the sea levels to rise and 1,000 people in Bangladesh will lose their homes in 50 years time. To prevent this, we must all never use an aeroplane to go on holiday ever again." (Less negative outcome/big sacrifice)
3. "If global warming continues at the current rate, it will cause the sea levels to rise and 10,000 people in Britain will lose their homes within 5 years. To prevent this, we must all never use an aeroplane to go on holiday ever again." (Very negative outcome/big sacrifice)
4. "If global warming continues at the current rate, it will cause the sea levels to rise and 1,000 people in Bangladesh will lose their homes in 50 years time. To prevent this, we must switch all the light bulbs in our houses to energy efficient ones." (Less negative outcome/small sacrifice)[1]

The non-scientific argument followed a similar format, but involved walking a short or long distance (level of sacrifice) in order to buy batteries for an alarm clock

[1] That the experiment contrasted the prospect of people in Bangladesh losing their homes with people in the United Kingdom losing their homes does not indicate that the housing security of British citizens is of greater value than that of Bangladeshi citizens. Rather, it was an attempt to render the negative outcome not only *more negative* (in the sense that 10,000, rather than 1,000 people's homes were at risk and in 5, rather than 50 years time), but also *more relevant* (based on the assumption that a typical 16–18-year-old British citizen has more empathy with the security of houses in their own country within the next 5 years than the security of houses in a foreign country within the next 50 years). Bangladesh was selected as a comparison country simply because as a geographically low-lying nation, it faces very real threats from rising sea levels attributable to human-caused climate change.

that was necessary to wake up in time for an important exam or an ordinary 'non-work' day (negativity of outcome).

Corner and Hahn (2009) used the framework of Bayesian decision theory to make predictions about the strength of consequentialist arguments (Edwards, 1961; Keeney & Raiffa, 1976; Savage, 1954). Applying decision theory to consequentialist arguments, the more (subjective) negative utility there is associated with a consequence, the stronger that consequentialist argument should be (Corner, Hahn, & Oaksford, 2006; Hahn & Oaksford, 2006, 2007). As the perceived negativity of the outcome and the level of sacrifice required to avoid it both contribute to the subjective utility of a consequentialist argument, Corner and Hahn (2009) predicted and found that both these factors influenced the strength of scientific and non-scientific consequentialist arguments. Arguments containing more negative outcomes were rated as significantly stronger than arguments containing less negative outcomes, while arguments requiring a smaller sacrifice were rated as significantly stronger than arguments requiring a bigger sacrifice. Figure 10.1 displays the ratings of argument strength obtained from participants in each condition of the current experiment.

An analysis of variance (ANOVA) was conducted with outcome negativity (very negative vs. less negative), level of sacrifice (big sacrifice vs. small sacrifice) and topic (scientific vs. non-scientific) as independent variables. Only level of sacrifice had a significant effect on baseline ratings of argument strength (F (1, 124) = 18.83, $p < 0.001$), meaning that neither the negativity of the outcome nor the topic of the argument had a statistically significant effect on participants' ratings of argument strength.

The results suggest that participants were highly sensitive to the level of sacrifice in the arguments. Where a high level of sacrifice was required, people assigned a

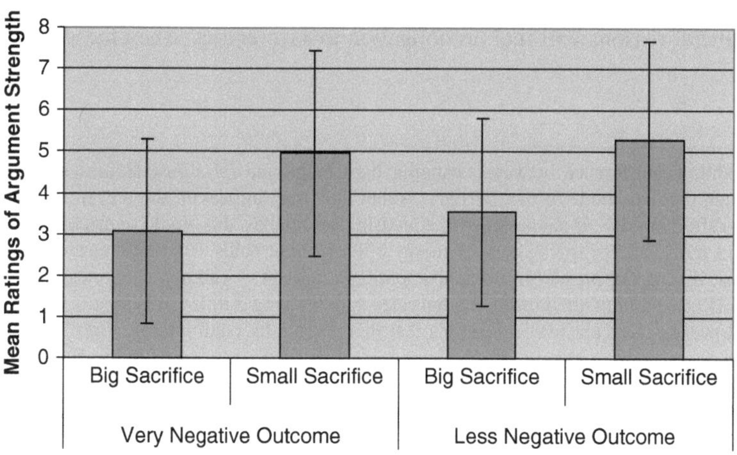

Fig. 10.1 Mean baseline ratings of argument strength (scientific and non-scientific topics combined). Error bars represent one standard deviation

significantly lower rating of strength to the argument.[2] Could this be part of the reason that many people remain unconvinced by the evidence for climate change?

Research on the use of fear appeals in persuasive communication suggests that there is a danger of inducing defensive reactions if the severity of the message is too high (de Vries, Ruiter, & Leegwater, 2002), and that simply increasing severity does not necessarily add to the persuasive impact of a message (Hoog, Stroebe, & de Wit, 2005). This kind of defensive reaction is all the more likely where the behaviour targeted is highly valued, pleasurable or central to one's identity – as in the case of many energy-intensive activities (Randall, 2009). Most people in the United Kingdom do not feel personally threatened by climate change (Lorenzoni et al., 2007; O'Neill & Nicholson-Cole, 2009) which means that doomsday scenarios and apocalyptic language are unlikely to be effective communication strategies. Research also suggests that if climate change risks are perceived as temporally or geographically distant, then they are likely to be psychologically discounted (APA, 2009; Spence et al., 2011; Uzzell, 2000).

The results of this experiment suggest that there is an additional factor to consider in constructing arguments about climate change that contain negative consequences – that the convincingness of these arguments will be partly dependent on the *sacrifice* that is required to avoid the negative outcome. Promoting environmental behaviour may be more effectively achieved by emphasising the positive effects of pro-environmental behaviours (e.g., the health benefits of cycling rather than using a car), although constructing an artificially positive 'spin' on messages that ultimately require a certain level of sacrifice may be problematic in the longer term (Randall, 2009). For example, while the prospect of saving money may motivate energy saving behaviours in the home, Thogersen and Crompton (2009) have argued that it does not trigger critical psychological mechanisms that make the performance of *other* pro-environmental behaviours more likely. In general, people like to avoid feeling hypocritical and will take steps to avoid any *dissonance* between their actions (Cooper & Fazio, 1984; Festinger, 1957). However, people saving energy for financial reasons will feel no obligation to save energy when these reasons are

[2] An additional difference between changing light bulbs and refraining from using aeroplanes (other than the magnitude of the sacrifice) is that they may impact on the prevention of the outcome in different ways. If people *were* to stop using aeroplanes, this would reduce greenhouse gas emissions more than swapping over to energy efficient light bulbs. This difference in the *efficacy* of the sacrifice is not present in the non-scientific argument – walking 2 minutes to the shop is no less effective as a method of buying batteries than walking 3 miles; it is simply more of a sacrifice. However, there are two indications that this potentially confounding effect does not seem to have influenced the outcome of the experiment. Firstly, if participants in the experiment were paying attention to this difference in efficacy, the arguments containing big sacrifices should have been rated as *more* compelling than the arguments containing small sacrifices. However, this was not the case. Secondly, no differences were observed in the impact of the level of sacrifice variable between the scientific and non-scientific arguments. It would seem, therefore, that participants treated the arguments as representing greater and lesser sacrifices, rather than more or less effective methods of avoiding the negative outcome.

absent – they might switch off appliances at home (where they pay the bills), for example, but leave them on at work (where they do not).

The picture that is starting to be built up around the efficacy of employing different types of arguments to communicate climate change is complex. However, the results of this experiment suggest that there is a compromise to be struck between being honest about the negativity of the predicted effects of climate change and avoiding disengaging people through ill-considered fear appeals. Supporting the findings reported in Corner & Hahn (2009), paying attention to the perceived sacrifice contained in an argument about the consequences of climate change is an important determinant of its strength.

Of course, consequentialist arguments about the behaviour required to avoid a particular outcome are about far more than just climate change science – they are normative statements about behaviour and policy. In the next section, I use a recent controversy over climate change communication to illustrate why the indeterminate lines between climate science, climate change communication and climate change advocacy may act as a barrier to communicating arguments about climate science itself.

Trust in the Communicators of Climate Change Arguments

Until fairly recently, science was typically viewed as value free and apolitical – a lack of trust in science and scientists was not something of concern. This view was challenged by some high profile scientific controversies (such as the debate over agricultural biotechnology – Walls, Rogers-Hayden, Mohr, & O'Riordan, 2005) and social scientists who highlighted the role of social and cultural influences in science (e.g., Irwin & Wynne, 1996; Collins & Evans, 2007). The picture of science that emerged was one that stressed the importance of the scientific community as the location and source of legitimation of scientific norms, judgements and knowledge – but not one that viewed science as *value free*.

Climate change provides perhaps the most compelling example of the ways in which political interests, personal involvement or corporate allegiances can colour the interpretation of scientific evidence. In October 2007, the IPCC and the American politician Al Gore won the Nobel Peace Prize. The IPCC was recognised for

> Efforts to build up and disseminate greater knowledge about man-made climate change, and to lay the foundations for the measures that are needed to counteract such change. (www.ipcc.ch)

Al Gore was recognised for his film, *An Inconvenient Truth*, which attempts to translate the science of the IPCC and make it accessible to the widest possible audience. As a piece of science communication, it is a tour de force – the content of the film is dominated by scientific evidence, graphs and statistics. Evidence of the uncontroversial nature of the film and its message can be found in the British Government's decision to send the film to every secondary school in the United

Kingdom, as an educational tool for teaching about the effects of climate change. However, some of the arguments in the film are also clearly political: urgent action is required if the worst predicted effects of human-caused climate change are to be avoided.

In the same week as Gore received his Nobel Peace Prize, a British High Court judge ruled that there were nine errors in *An Inconvenient Truth* and that the film should be accompanied by guidance for schools, pointing these errors out. Although the judge was careful to emphasise that the film's main arguments were based on uncontroversial scientific data, the effect of the ruling was to create some doubt about the reliability of the evidence presented in the film.

Viewers of the film may have had their reservations moderated by the knowledge that Stewart Dimmock, the claimant who brought the case against the government, was funded by a Scottish quarrying magnate who has a track record of lobbying against environmentalist groups. However, the example highlights many of the issues associated with the public evaluation of arguments about climate change. To the extent that science was being debated, the participants were not scientists but a politician, a lobbyist, lawyers and a judge, while the public learnt about the debate indirectly through its reporting in the news and other media. The lines between the science of climate change, the communication of climate change science and normative policy or behavioural recommendations based on climate science are indeterminate and blurry.

In *Why We Disagree About Climate Change*, Hulme (2009) proposed that while 'climate change' has a physical meaning – literally the changes in patterns of weather over a period of time – it also has a *social* meaning and many competing and overlapping narratives have developed around the *idea* of climate change. According to Hulme, people use climate change to tell stories about human progress and disagree about how to respond to climate change because it (socially) means very different things to different people. Thus, when evaluating arguments about seemingly 'objective' facts about temperature variation, ocean acidification or biodiversity loss, an awful lot of ideological baggage may be weighing people's judgements behind the scenes.

Because climate change has different social meanings, messages about climate change are 'framed' in very different ways – and how a message is framed is an important determinant of how persuasive or effective it is (Corner & Hahn, 2010), particularly for different audiences (Nisbett, 2009). According to Nisbett, frames are "interpretive storylines that set a specific train of thought in motion, communicating why an issue might be a problem, who or what might be responsible for it, and what should be done about it" (Nisbett, 2009). Certain words activate certain frames – for example, the words 'protecting the environment' trigger a conceptual framing of the environment as external to human behaviour – something distinct to 'be protected' rather than inextricably linked to human behaviour (Lakoff, 2010). Arguments about international climate change policy – for example, the provision of finance for an adaptation fund – tend to focus on *fairness* and *justice*. Concerns about domestic energy policy are often couched in terms of *threat* or *security*. At the individual level, the most frequent frame for climate change messages in the United Kingdom is the *financial gain* or *personal benefit* that pro-environmental behaviour may bring (see, e.g., Defra, 2008). Different frames tend to be utilised by different information sources – and the trustworthiness of these sources is central to the persuasiveness of their arguments.

There is empirical evidence that suggests that trust in science (and scientists) as reliable sources of information is essential for arguments about climate change to be perceived as convincing (Corner & Hahn, 2009; Hahn, Harris, & Corner, 2009). In an experiment with undergraduate students, Corner and Hahn (2009) found that participants were particularly sensitive to the reliability of sources of scientific information. While sources presenting strong scientific arguments were rated as *more* reliable than sources presenting non-scientific arguments, when scientific sources presented weak arguments they were perceived as *less* reliable than non-scientific sources presenting weak arguments. Corner and Hahn (2009) concluded that weak scientific arguments reflect badly on the perceived reliability of scientific sources and suggested that contradictory scientific evidence may impact badly on perceptions of scientists themselves – which will then feed into the evaluation of scientific messages.

A study conducted by Malka et al. (2009) supports this conclusion. Malka et al. conducted a telephone survey with participants in the United States to establish whether people who trusted scientists responded differently to information about climate change compared to people who did not trust scientists. The study found that increased knowledge about climate change was positively correlated with increased concern about climate change – but only among those individuals who reported trusting scientists. Among those individuals who expressed distrust towards scientists, additional knowledge about climate change had no impact on their level of concern – that is, the effect of the argument was moderated by the trust that participants expressed towards scientists.

Another recent event highlighted just how important public trust in climate science is. In November 2009, less than a month before the highly publicised climate change negotiations at the United Nations Conference of Parties in Copenhagen, a series of private email exchanges between members of the Climate Research Unit (CRU) at the University of East Anglia (UEA) in the United Kingdom and external collaborators were illegally published on the internet. Initial reports focussed on a handful of emails involving Professor Phil Jones, the head of CRU, which were said to contain evidence of climate scientists seeking to suppress inconvenient data and subvert the peer-review process. In the months following the release of the emails, a House of Commons Science and Technology Committee inquiry found no evidence of wrongdoing by Phil Jones or his colleagues. A separate independent investigation – a Science Assessment Panel comprised of climate scientists from other institutions – found no evidence of dishonesty or impropriety. However, the event received media attention around the world, generating a significant amount of controversy, and the incident became widely known as 'climategate'.

The British Broadcasting Corporation (BBC) opinion polls conducted around the time of the incident found that the percentage of the public agreeing that climate change is 'largely man-made' fell from 50 to 34% between November 2009 and February 2010 (BBC, 2010). The BBC data were supported by evidence from the United States, where several polls had indicated a substantial increase in scepticism towards climate change in the second half of 2009 (e.g., Pew Research Centre, 2009). Media commentators began to speculate about the impact of the UEA emails on public opinion. However, data collected by Corner et al. (in press) from university students found that most of the students had heard little or nothing about the email controversy, despite a significant amount of coverage in the UK press. Interestingly, among the students who had heard about the event, only 25% reported becoming more sceptical about climate change because of it.

While 'climategate' did not seem to alter their belief in the reality of climate change, Corner et al. found that the incident had a negative impact on participants' perceptions of the trustworthiness of climate scientists. This finding fits with the conclusions of Corner and Hahn (2009) – arguments containing 'mixed messages' produced lower judgments of source reliability – an effect that was strongest for arguments about scientific topics. Scientific controversies may lead to sources of scientific arguments being perceived as less reliable, undermining their ability to construct compelling and trustworthy arguments in the future. A critical factor in the acceptability of arguments about climate change is the extent to which the general public trust the communicators of the arguments.

Conclusion

In this chapter, I have sought to examine four challenges in the communication of climate change that all have a direct bearing on the way that people evaluate arguments about climate change – knowledge about climate change, uncertainty about climate change, the negative consequences of climate change and trust in the communicators of climate change arguments. The four challenges I have focussed on are by no means the only barriers to the successful communication of climate change. The 'social' barriers to engagement with climate change identified by Lorenzoni et al. (2007) include such diverse factors as a lack of pro-environmental social norms and concern about 'free riders' who might reap the rewards of others making behavioural sacrifices as a response to messages about climate change.

However, the analyses in this chapter do shed some light on the sorts of concerns that anyone interested in communicating about climate change must consider. In particular, and in keeping with Hulme (2009), this chapter demonstrates that the relationship between climate change (as a scientific phenomenon) and the construction of climate change (as a social phenomenon) is complex and challenging. But a consideration of the ways in which people evaluate arguments about climate change does at least highlight the considerable potential for social science researchers to contribute towards the goal of making climate change communication more effective.

The research presented in this chapter has some practical implications for climate change communicators. While structured education programmes are likely to enhance engagement with climate change, simply bombarding individuals with more and more information about climate change is unlikely to be an effective communication tactic. Similarly, a consideration of the way in which people evaluate consequentialist arguments about climate change suggests that a seemingly persuasive argument that revolves around the negative consequences of climate change will be mediated by the level of sacrifice required by the recipient of the argument. Uncertainty is an unavoidable aspect of communicating climate change, but paying more attention to the prior attitudes and beliefs of the recipients of climate change arguments might help to predict how they will respond to uncertain evidence. Trustworthy communicators are essential for arguments about climate change to seem compelling – and people seem especially sensitive to the reliability of sources presenting scientific messages.

A major factor influencing the extent to which the negative impacts of climate change will be mitigated is the level of public engagement with climate change and an international preparedness to support and promote sustainability policies and behaviours. The way that people evaluate arguments about climate change is an important determinant of their level of engagement. This means that there is a pressing need for argumentation scholars (and social researchers more broadly) to conduct and communicate research on the way that ordinary people respond to arguments about climate change. This chapter represents a small contribution to that goal.

Acknowledgements I would like to thank Ulrike Hahn, who co-designed the experiment reported in this chapter, and Lorraine Whitmarsh and Rob Evans for helpful comments on earlier versions of this chapter.

References

Adams, S. (1999). Critiquing claims about global warming from the world wide web: A comparison of high school students and specialists. *Bulletin of Science, Technology & Society, 19*(6), 539–543.

American Psychological Association Task Force on the Interface Between Psychology and Global Climate Change. (2009). *Psychology and global climate change: Addressing a multi-faceted phenomenon and set of challenges*. Washington, DC: American Psychological Association.

Bonnefon, J. F., & Hilton, D. J. (2004). Consequential conditionals: Invited and suppressed inferences from valued outcomes. *Journal of Experimental Psychology: Learning, Memory & Cognition, 30*(1), 28–39.

Bord, R. J., O'Connor, R. E., & Fischer, A. (2000). In what sense does the public need to understand global climate change? *Public Understanding of Science, 9*, 205–218.

Bows, A., Upham, P., & Anderson, K. (2005). *Growth scenarios for EU & UK aviation: Contradictions with climate policy*. Report for Friends of the Earth Trust Ltd.

Boykoff, M. (2007). Flogging a dead norm? Media coverage of anthropogenic climate change in United States and United Kingdom, 2003–2006. *Area, 39*(4), 470–481.

British Broadcasting Corporation. (2010). *BBC Climate Change Poll – February 2010.* Retrieved May 1, 2010, from http://news.bbc.co.uk/nol/shared/bsp/hi/pdfs/05_02_10climatechange.pdf

Budescu, D. V., Broomell, S., & Por, H. (2009). Improving communication of uncertainty in the reports of the intergovernmental panel on climate change. *Psychological Science, 20*, 299–308.

Butler, C., & Pidgeon, N. (2009). Media communications and public understanding of change – reporting scientific consensus on anthropogenic climate change. In T. Boyce & J. Lewis (Eds.), *Climate change and the media* (pp. 43–58). New York: Peter Lang.
Collins, H. M., & Evans, R. (2007). *Rethinking expertise*. Chicago: University of Chicago Press.
Cooper, J., & Fazio, R. H. (1984). A new look at dissonance theory. In L. Berkowitz (Ed.), *Advances in experimental social psychology* (Vol. 17, pp. 229–266). New York: Academic.
Corbett, J. B. & Durfee, J. L. (2004). Testing public (un)certainty of science: Media representations of global warming. *Science Communication, 26*, 129–151.
Corner, A., & Hahn, U. (2009). Evaluating scientific arguments: Evidence, uncertainty & argument strength. *Journal of Experimental Psychology: Applied, 15*(3), 199–212.
Corner, A., & Hahn, U. (2010). Message framing, fallacy & normative advocacy. *Argumentation, 24*(2), 153–163.
Corner, A., Hahn, U., & Oaksford, M. (2006). The slippery slope argument: Probability, utility and category boundary re-appraisal. In *Proceedings of the 28th annual conference of the cognitive science society* (pp. 1145–1151). Vancouver: Cognitive Science Society.
Corner, A., Whitmarsh, L., & Xenias, D. (in press). Trust, uncertainty and attitudes towards climate change. *Climatic Change*.
Defra. (2008). *A framework for pro-environmental behaviours*. Retrieved January 8, 2010, from http://www.defra.gov.uk/evidence/social/behaviour/documents/behaviours-jan08-report.pdf
de Vries, N., Ruiter, R., & Leegwater, Y. (2002). Fear appeals in persuasive communication. In: G. Bartels & W. Nelissen (Eds.), *Marketing for sustainability: Towards transactional policy making* (pp. 96–104). Amsterdam, The Netherlands: IOS Press.
Doran, P. T., & Zimmerman, M. K. (2009). Examining the scientific consensus on climate change. *EOS, Transactions American Geophysical Union, 90*(3), 22–23.
Douglas, M., & Wildavsky, A. (1982). *Risk and culture: An essay on the selection of technological and environmental dangers*. Berkley, CA: University of California Press.
Dunlap, R., & McCright, A. M. (2008). A widening gap: Republican and democratic views on climate change. *Environment, 50*(5), 26–35.
Edwards, W. (1961). Behavioural decision theory. *Annual Review of Psychology, 12*, 473–498.
Festinger, L. (1957). *A theory of cognitive dissonance*. Stanford, CA: Stanford University Press.
Friedman, S. M., Dunwoody, S., & Rogers, C. L. (Eds.). (1999). *Communicating uncertainty: Media coverage of new and controversial science*. Hillsdale, NJ: Erlbaum.
Hahn, U., Harris, A. J. L., & Corner, A. (2009). Argument content and argument source – and exploration. *Informal Logic, 29*(4), 337–367.
Hahn, U., & Oaksford, M. (2006). A Bayesian approach to informal fallacies. *Synthese, 152*, 207–237.
Hahn, U., & Oaksford, M. (2007). The rationality of informal argumentation: A Bayesian approach to reasoning fallacies. *Psychological Review, 114*, 704–732.
Harris, A. J. L., & Corner, A. (2011). Communicating environmental risks: Clarifying the severity effect in interpretations of verbal probability expressions. *Journal of Experimental Psychology: Learning, Memory, and Cognition*. Advance online publication. doi: 10.1037/a0024195
HM Government. (2008). *Climate change bill*. London. Retrieved January 2010, from http://www.publications.parliament.uk/pa/ld200708/ldbills/087/2008087.pdf
Hogg, M., & Shah, H. (2010). *The impact of global learning on public attitudes and behaviours towards international development and sustainability*. Development Education Authority, UK.
Hoog, N., Stroebe, W., & de Wit, J. B. F. (2005). The impact of fear appeals on processing and acceptance of action recommendations. *Personality & Social Psychology Bulletin, 31*, 24–33.
Hulme, M. (2009). *Why we disagree about climate change: Understanding controversy, inaction and opportunity*. Cambridge: Cambridge University Press.
Irwin, A., & Wynne, B. (Eds.). (1996). *Misunderstanding science? The public reconstruction of science and technology*. Cambridge, MA: Cambridge University Press.
Kahan, D. M., Braman, D., Slovic, P., Gastil, J., & Cohen, G. (2009). Cultural cognition of the risks and benefits of nanotechnology. *Nature Nanotechnology, 4*(2), 87–90.

Keeney, R. L., & Raiffa, H. (1976). *Decisions with multiple objectives: Preferences and value tradeoffs*. New York: Wiley.
Korpan, C. A., Bisanz, G. L., Bisanz, J., & Henderson, J. M. (1997). Assessing literacy in science: Evaluation of scientific news briefs. *Science Education, 81*, 515–532.
Lakoff, G. (2010). We are the polar bears: What's wrong with the way the environment is understood. In *From Hot Air to Happy Endings: How to inspire public support for a low carbon economy*. London: Green Alliance.
Lord, C. G., Ross, L., & Lepper, M. R. (1979). Biased assimilation and attitude polarization: The effects of prior theories on subsequently considered evidence. *Journal of Personality and Social Psychology, 37*(11), 2098–2109.
Lorenzoni, I., Nicholson-Cole, S., & Whitmarsh, L. (2007). Barriers perceived to engaging with climate change among the UK public and their policy implications. *Global Environmental Change, 17*, 445–459.
Lorenzoni, I., Pidgeon, N. F., & O'Connor, R. E. (2005). Dangerous climate change: The role for risk research. *Risk Analysis, 25*(6), 1287–1398.
Maio, G., Verplanken, B., Manstead, T., Stroebe, W., Abraham, C., Sheeran, P., & Conner, M. (2007). Social psychological factors in lifestyle change and the relevance to policy. *Social Issues & Policy Review, 1*(1), 99–137.
Malka, A., Krosnick, J. A. & Langer, G. (2009). The association of knowledge with concern about global warming: Trusted information sources shape public thinking. *Risk Analysis, 29*, 633–647.
McKenzie-Mohr, D., & Smith, W. (1999). *Fostering sustainable behavior: An introduction to community-based social marketing*. Canada: New Society Publishers.
Miller, A. G., McHoskey, J. W., Bane, C. M., & Dowd, T. G. (1993). The attitude polarization phenomenon: Role of response measure, attitude extremity, and behavioural consequences of reported attitude change. *Journal of Personality and Social Psychology, 64*(4), 561–574.
Munro, G. D., & Ditto, P. H. (1997). Biased assimilation, attitude polarization, and affect in reactions to stereotype-relevant scientific information. *Personality & Social Psychology Bulletin, 23*(6), 636–653.
Nisbett, M. (2009). Communicating climate change: Why frames matter for public engagement. *Environment, 51*(2), 514–518.
Norris, S. P., Phillips, L. M., & Korpan, C. A. (2003). University students' interpretation of media reports of science and its relationship to background knowledge, interest and reading difficulty. *Public Understanding of Science, 12*, 123–145.
O'Neill, S., & Nicholson-Cole, S. (2009). "Fear won't do it": Promoting positive engagement with climate change through visual and iconic representations. *Science Communication, 30*, 355–379.
Oppenheimer, M. (2005). Defining dangerous anthropogenic interference: The role of science, the limits of science. *Risk Analysis, 25*(6), 1399–1407.
Pachauri, R. K., & Reisinger, A. (Eds). (2007). *Climate change 2007: Synthesis report*. Geneva: IPCC.
Patt, A. (2007). Assessing model-based and conflict-based uncertainty. *Global Environmental Change, 17*, 37–46.
Patt, A. G., & Schrag, D. P. (2003). Using specific language to describe risk and probability. *Climatic Change, 61*, 17–30.
Pew Research Centre for the People and the Press. (2009). *Fewer Americans see solid evidence of global warming*. Retrieved January 2010, from http://people-press.org/report/556/global-warming
Pidgeon, N. F., Kasperson, R. K., & Slovic, P. (2003). *The social amplification of risk*. Cambridge: Cambridge University Press.
Pollack, H. N. (2005). *Uncertain science . . . uncertain world*. Cambridge, U K: Cambridge University Press.

Randall, R. (2009). Loss and climate change: The cost of parallel narratives. *Ecopsychology, 1*(3), 118–129.
Ratcliffe, M. (1999). Evaluation of abilities in interpreting media reports of scientific research. *International Journal of Science Education, 21*, 1085–1099.
Risen, J. L., & Gilovich, T. (2007). Another look at why people are reluctant to exchange lottery tickets. *Journal of Personality and Social Psychology, 93*, 12–22.
Savage, L. J. (1954). *The foundations of statistics*. New York: Wiley.
Shome, D., & Marx, S. (2009). *The psychology of climate change communication: A guide for scientists, journalists, educators, political aides and the interested public*. Columbia: Centre for Research on Environmental Decisions.
Spence, A., Pidgeon, N., & Uzzell, D. (2009). Climate change – psychology's contribution. *The Psychologist, 22*, 108–111.
Spence, A., Poortinga, W., Butler, C., & Pidgeon, N. (2011). Perceptions of climate change and willingness to save energy related to flood experience. *Nature Climate Change, 1*(1). doi:10.1038/NCLIMATE1059
Stern, P. C. (2000). Towards a coherent theory of environmentally significant behavior. *Journal of Social Issues, 56*(3), 407–424.
Tannert, C., Elvers, H. D., & Jandrig, D. (2007). The ethics of uncertainty: In the light of possible dangers, research becomes a moral duty. *European Molecular Biology Organisation Reports, 8*(10), 892–896.
Thøgersen, J., & Crompton, T. (2009). Simple and painless? The limitations of spillover in environmental campaigning. *Journal of Consumer Policy, 32*, 141–163.
UK Climate Impacts Programme. (2009). *Adapting to climate change: UK climate projections*. Retrieved November 2009, from http://ukclimateprojections.defra.gov.uk/content/view/824/517/
Upham, P., Whitmarsh, L., Poortinga, W., Purdam, K., Darnton, A., McLachlan, C., et al. (2009) *Public attitudes to environmental change: A selective review of theory and practice*. A research synthesis for the Living with Environmental Change Programme, Research Councils UK.
Uzzell, D. L. (2000). The psycho-spatial dimensions of global environmental problems. *Journal of Environmental Psychology, 20*(4), 307–318.
Walls, J., Rogers-Hayden, T., Mohr, A., & O'Riordan, T. (2005). Seeking citizens' views on GM Crops – Experiences from the United Kingdom, Australia, and New Zealand. *Environment, 47*(7), 22–36.
Weber, E. U. (2006). Evidence-based and description-based perceptions of long-term risk: Why global warming does not scare us (yet). *Climatic Change, 77*, 103–120.
Weber, E. U. (2010). What shapes our perceptions of climate change? *Wiley Interdisciplinary Reviews: Climate Change*. doi:10.1002/wcc.41
Western Strategies & Lake Research Partners. (2009). *Climate and energy truths: Our common future*. Washington, DC: EcoAmerica.
Whitmarsh, L. (2008). Are flood victims more concerned about climate change than other people? The role of direct experience in risk perception and behavioural response. *Journal of Risk Research, 11*(3), 351–374.
Whitmarsh, L. (2011). Scepticism and uncertainty about climate change: Dimensions, determinants and change over time. *Global Environmental Change, 21*, 690–700.
WRAP/WI. (2008). *Love food champions report*. Retrieved June 2009, from http://www.wrap.org.uk/retail/case_studies_research/report_love_food.html
Zehr, S. (2000). Public representations of scientific uncertainty about global climate change. *Public Understanding of Science, 9*, 85–103.

Chapter 11
The Effects of University Students' Argumentation on Socio-Scientific Issues via On-Line Discussion in Their Informal Reasoning Regarding This Issue

Ying-Tien Wu and Chin-Chung Tsai

Introduction

Contemporary development in science and technology, such as genetic engineering and nuclear power usage, often brings about social dilemmas. These social dilemmas have conceptual or technological associations with science, and, in these issues, science and society represent interdependent entities, and both the social and scientific factors play the central roles (Sadler, 2004). As a result, these social dilemmas are termed "socio-scientific issues." In this 21st century, learners, as the citizens in democratic society, have more and more opportunities to encounter a variety of socio-scientific issues. Therefore, improving learners' ability in dealing with socio-scientific issues has been regarded as one of the important goal for modern science education (e.g., American Association for the Advancement of Science (AAAS), 1989, 1993).

Recently, more and more educators have acknowledged that students' informal reasoning and argumentation ability play an important role when dealing with socio-scientific issues (e.g., Osborne, Erduran, & Simon, 2004; Sadler, 2004). However, the findings derived from previous studies have suggested that how to improve learners' reasoning quality or argumentation level when they deal with socio-scientific issues should be highlighted (e.g., Osborne et al., 2004; Wu & Tsai, 2007). Jonassen (1996, 2000) has argued that the Internet can be utilized as a *cognitive tool*. However, still not much research has addressed how Internet-based learning environments affect the development of students' informal reasoning quality. Therefore, this study was conducted to explore the effect of the use of anonymous on-line discussion forums on university students' informal reasoning outcomes regarding a socio-scientific issue.

Y.-T. Wu (✉)
Graduate Institute of Network Learning Technology, National Central University, JhongLi, Taiwan
e-mail: ytwu@mail.ntust.edu.tw

M.S. Khine (ed.), *Perspectives on Scientific Argumentation*,
DOI 10.1007/978-94-007-2470-9_11,

Theoretical Framework

Informal Reasoning and Socio-Scientific Issues

In general, science educators often refer reasoning (or scientific reasoning) to formal reasoning, which is characterized by rules of logic and mathematics. In fact, as formal reasoning, informal reasoning is also recognized as a rational process of constructing and evaluating arguments (Kuhn, 1993). However, unlike formal reasoning, the problems of informal reasoning are ill-structured (Sadler, 2004). In particular, the premises are often not explicitly stated in informal reasoning tasks. As a result, the conclusions of the arguments in informal reasoning may not be demarcated. In general, informal reasoning is often used in situations where reasons exist both supporting and against the conclusion, such as making decisions about what to believe or what actions to be taken (Shaw, 1996). Therefore, an individual's informal reasoning ability plays an important role in dealing with socio-scientific issues (Sadler, 2004).

Until today, the nature of learners' informal reasoning regarding socio-scientific issues is not well-understood. The dual-process theories for explaining human's thinking and some recent findings in psychology (e.g., Evans, 2002, 2003; Sloman, 1996) may provide some insights for the aforementioned issue. The dual-process theories posit the existence of two distinct cognitive systems: one essentially pragmatic and the other capable of deduction and hypothetical thought (e.g., Evans, 2002, 2003; Sloman, 1996). These two distinct systems are often termed as "System 1" and "System 2." The utilization of System 1 is unconscious, pragmatic, and contextualized, while the operation of System 2 is conscious and involves logical and abstract thinking. Moreover, System 1 processes are rapid, parallel, and automatic in nature; while System 2 thinking is slow and sequential in nature and makes use of the central working system. Besides, Evans (1996) has suggested that people decide first and think afterward in order to justify choices that are unconsciously determined. Evans and Curtis-Holmes (2005) further argued that a central phenomenon in dual-process accounts of reasoning is that of "belief bias," the tendency to evaluate the validity of an argument on the basis of whether it agrees with the conclusion, rather than on whether it follows logically from the premises. Nevertheless, System 2 reasoning can override pragmatic influence and lead to normative correct solutions (Evans, 2002).

According to the dual-process theories and the perspectives mentioned above, it seems that, when encountering a socio-scientific issue, an individual will unconsciously evoke System 1 reasoning, and the usage of System1 will trigger him/her to make an initial decision first. Then, he (she) may utilize System 2 reasoning to justify his/her initial decision. Moreover, it has also been proposed that experts can make the "right" decisions quickly because they have complex models that allow them to see underlying causes (e.g., Randel, Pugh, & Reed, 1996), indicating that differences between expert and novice on their informal reasoning regarding a socio-scientific issue may exist. When encountering a socio-scientific issue, experts' complex mental models regarding a socio-scientific issue can help them intuitively

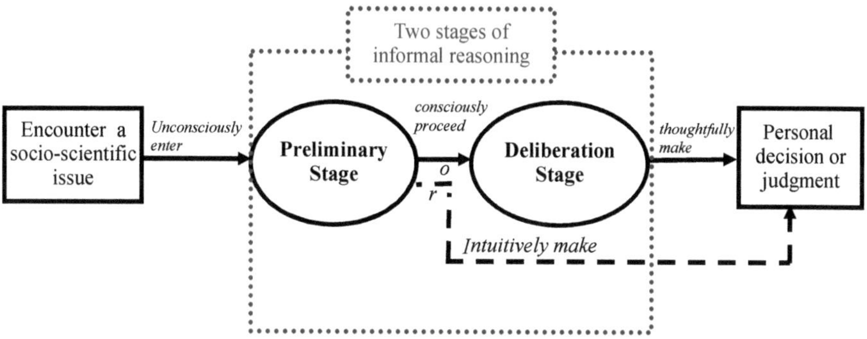

Fig. 11.1 The process of learners' informal reasoning regarding a socio-scientific issue

make the "right" decisions and justify their decisions quickly and efficiently. The same as experts, novices (such as high school students) also make intuitive decision; however, unlike experts' quick justification on their decisions, they will spend more time in conducting logical and abstract thinking (i.e., conduct System 2 reasoning) to justify their initial decisions. Even novices may not use System 2 reasoning and make only intuitive decisions as their final decisions. The process of learners' informal reasoning on a socio-scientific issue may be further illustrated as Fig. 11.1.

As revealed in Fig. 11.1, when an individual learner has to make a decision or judgment on a socio-scientific issue, there are two stages of his/her informal reasoning regarding this issue: *the Preliminary Stage* and *the Deliberation Stage* (Wu & Tsai, 2010). In the Preliminary Stage, on the basis of his/her past experiences (including prior knowledge and personal beliefs instantaneously retrieved from long-term memory), System 1 will trigger an individual learner immediately to make an initial decision on the socio-scientific issue accordingly. It should be noticed that an individual learner might (and frequently do) only experience the Preliminary Stage. In other words, he (or she) may make only an intuitive decision on this issue at the end. However, he (or she) may, then, proceed to reason in the Deliberation Stage. In the Deliberation stage, System 2 will help him (or her) form some evaluative criteria, justify his (or her) initial decision accordingly, and, then, make his (or her) final decision until a conclusion is reached. That is, in the Preliminary Stage, a learner may use System 1 to make an intuitive decision regarding a socio-scientific issue, which may be also his (her) final decision, or he (or she) will further employ System 2 to elaborate his/her thinking and then make a final decision on this issue in the Deliberation Stage.

Socio-Scientific Argumentation and Anonymous On-Line Discussion Forums

Argumentation is concerned with how individuals make and justify claims and conclusions (Driver, Newton, & Osborne, 2000). In general, argumentation is recognized as a socially situated activity that enables scientific discourse among learners

(Erduran, Simon, & Osborne, 2004; Sadler & Fowler, 2006). For a long time, argumentation has been regarded as one of the key components of contemporary science education (AAAS, 1993; National Research Council (NRC), 1996). Educators and researchers in science education have also advocated that socio-scientific issues can be used to provide students with opportunities for argumentation (e.g., Sadler & Donelly, 2006; Zeidler, Sadler, Simmons, & Howes, 2005). The term "socio-scientific argumentation" has been proposed to refer to learners' argumentation practices in which socio-scientific issues are used as the contexts for argumentation (e.g., Sadler, 2004).

In recent years, Toulmin's definition of argument has been widely applied as a methodological tool for the analysis of argumentation in science education (e.g., Erduran et al., 2004; Jiménez-Aleixandre, Rodríguez, & Duschl, 2000; Zohar & Nemet, 2002). For Toulmin, the essential elements of argument are climes, data, warrants, backings, and rebuttals (Erduran et al., 2004). The Toulmin's Argument Pattern (TAP) illustrates the structure of an argument including an interconnected sets of a claim, data (supporting the claim), warrants (providing a link between the data and the claim), backings (strengthening the warrants), and rebuttals (pointing to the circumstances under which the claim would not hold true) (Erduran et al., 2004). Based on TAP, many studies in science education assessed the quality of students' argumentation (e.g.,Jiménez-Aleixandre et al., 2000; Osborne et al., 2004). However, it should be noticed that the use of TAP in these studies has mainly focused on the description of small-group discussions among students (e.g., Osborne et al., 2004). Rather than attempting to sort unique elements of an individual's argument according to TAP, Sadler and Fowler (2006) modified the framework in Osborne et al. (2004) and designed a rubric to score argumentation quality. The rubric developed in Sadler and Fowler (2006) was used in analyzing face-to-face discussion, and designed to distinguish quality of arguments based on how the claims were further justified.

Previous studies have revealed students' difficulties with argumentation (Driver et al., 2000). Therefore, how to enhance the quality of students' argumentation has been highlighted by researchers and educators in science education (e.g., Osborne et al., 2004; Zohar & Nemet, 2002). Some studies have conducted to address the aforementioned issue. However, diverse findings were revealed in relevant studies. For example, Zohar and Nemets (2002) revealed the positive result with an argumentation intervention in the context of dilemmas in human genetics, and it suggested that instruction can be beneficial in the promotion of argumentation skills. Nevertheless, Kortland's (1996) unsuccessful intervention showed that not all instruction for argumentation works well. Recently, on-line learning environments were designed to support learners' or teachers' argumentation regarding socio-scientific issues and the usefulness of on-line learning environments was reported (e.g., Clark & Sampson, 2007; Zembal-Saul, Munford, Ceawford, Friedrichsen, & Land, 2002). However, in most of these studies, on-line systems were designed to scaffold users' construction in light of TAP argument. Still not much research focused on students' self-directed argumentation regarding a socio-scientific issue, such as argumentation regarding a socio-scientific issue in anonymous on-line discussion forums.

Argumentation is often viewed as a central part of informal reasoning (Zohar & Nemet, 2002). Although informal reasoning is a somewhat fuzzy construct, its core includes skills in argument generation and evaluation (Means & Voss, 1996). In computer-supported collaborative learning (CSCL), such as anonymous on-line discussion forums, learners are supposed to engage in an argumentative discourse (Weinberger & Fischer, 2006). In anonymous on-line discussion forums, students may work together to construct and critique arguments. It seems that anonymous on-line discussion forums may serve as a tool to support learners' argumentation practices regarding a socio-scientific issue.

According to the perspectives above, it seems that students' informal reasoning ability in dealing with socio-scientific issues may be improved with the practice of socio-scientific argumentation. Therefore, this study aimed to investigate the effects of anonymous on-line discussion (i.e., anonymous on-line argumentation) on a group of university students' informal reasoning regarding a socio-scientific issue. In addition, the effects of anonymous on-line discussion on students with different reasoning ability were further explored.

Methodology

Participants and the Socio-Scientific Issue

The participants of this study were 37 university students (consisting of 16 males and 21 females) from a national university in Taiwan. They were all non-science majors, and took the "Natural, Technology and Society" course. Among these students, 17 students were sophomores, 18 students were juniors, and 2 students were seniors.

In the "Natural, Technology and Society" course, socio-scientific issues, such as nuclear power usage, global warning, and genetic engineering, were introduced. During the 2-hour class period per week, a socio-scientific issue was introduced by instructors every week. Then, the students were asked to read a relevant report and discuss on this issue in groups. In this study, "xenotransplantation" (i.e., the use of genetic engineering in transplanted animal organs to human) was used as the socio-scientific issue for exploring learners' informal reasoning and on-line argumentation. Before the conduct of this study, the participants had already learned about genetic engineering in the "Natural, Technology, and Society" course.

General Research Design

To explore the effects of on-line discussion on students' informal reasoning regarding a socio-scientific issue, this study was conducted by using a single group quasi-experimental research design. Before the conduct of this study, the participants' personal positions and informal reasoning regarding genetic engineering (xenotransplantation) were assessed (i.e., the pre-test). The research treatment of this study was on-line discussion task. In this study, an on-line discussion forum was used. The participants' personal positions revealed in the pre-test were used to assign them into on-line discussion groups. Each discussion group had four students, consisting of two students supporting genetic engineering and two students with the opposite position. The students were asked to discuss on "xenotransplantation" anonymously in the on-line discussion forum in groups during the period of a week (7 days). After the conduct of on-line discussion task, the participants' informal reasoning regarding genetic engineering was assessed again (i.e., the post-test).

Instruments, Data Collection and Data Analyses

In this study, data regarding the students' reasoning regarding genetic engineering as well as data regarding their on-line argumentation regarding this issue were collected. A detailed description about the data collection and data analyses of these variables is as follows.

Assessing Students' Informal Reasoning Regarding Genetic Engineering (Xenotransplantation)

Wu and Tsai (2007) have developed an open-ended questionnaire for assessing learners' informal reasoning on nuclear power usage. Similar to this, another open-ended questionnaire was developed in this study to collect the data regarding students' informal reasoning regarding genetic engineering.

In the questionnaire developed in this study, the participants were asked to write down their answers about the following questions:

(1) Do you agree with xenotransplantation? Why? (assessing students' position on this issue)
(2) If you want to convince your friend with your position, what arguments you will propose to convince him/her? (evaluating students' ability to generate supportive arguments)
(3) If someone holds an opposite position with you on this issue, what arguments he/she may have? (assessing students' counter-argument construction)
(4) According to the arguments you have proposed in question 3, can you write down your opposing ideas to justify your position? (evaluating students' rebuttal construction)

After data collection, each of an individual student' arguments shown in his/her responses to the aforementioned open-ended questions were analyzed both qualitatively and quantitatively with a framework similar to that of Wu and Tsai (2010) (for a detailed description, please refer to Wu and Tsai, 2007). The framework used in this study includes several qualitative indicators and quantitative measures described as follows:

(1) *Qualitative indicators*: Three qualitative indicators were used for assessing students' informal reasoning regarding genetic engineering, including the following:

a. *Reasoning level*: Kuhn (1993) has argued that rebuttals are critical because they complete the structure of argument, integrating argument and counter-argument. In this study, according to the informal reasoning outcomes revealed in the pre-test, a student's reasoning quality was categorized as "lower level" if he (or she) made only simple claims (supportive arguments) or counter-arguments, while his (or her) reasoning quality was categorized as "higher level" if he (or she) generated not only simple claims (supportive arguments) and counter-arguments, but also rebuttals.
b. *Reasoning mode revealed in an argument*: Learners may generate their arguments from different aspects. For example, they may propose arguments from "religious, social, or moral"; "human right to live"; "risk-oriented"; "scientific or medical development"; "nature-oriented"; or "equity-based" perspectives. In this study, each argument proposed by the participants was categorized into one of the aforementioned reasoning modes.

(2) *Quantitative measures*: After qualitative analyses, the following quantitative measures were also obtained for representing students' informal reasoning regarding xenotransplantation:

a. *Number of supportive arguments*: The amount of supportive arguments a learner constructs. The more the supportive arguments a learner proposed, the more he (or she) was able to provide supportive evidences for his/her position.
b. *Number of counter-arguments*: The amount of counter-arguments a learner proposes. This measure assessed the ability of a learner to reason from the counter-position.
c. *Number of rebuttals*: The amount of rebuttals a learner generated. The more the rebuttals a learner constructed, the more he (or she) was able to justify for his position.
d. *Total number of arguments*: The total amount of the three kinds of arguments (i.e., supportive arguments, counter-arguments, and rebuttals). This measure evaluated an individual learner's ability to make arguments regarding a socio-scientific issue.
e. *Total number of reasoning modes*: The total number of reasoning modes an individual utilized in his/her informal reasoning. Similar to Wu and Tsai (2007, 2010), each argument generated by an individual student was categorized into one of the following six reasoning modes: "religious, social, or moral"; "human right to live"; "risk-oriented"; "scientific or medical development"; "nature-oriented"; and "equity-based." Then, the total number of reasoning modes the individual used was calculated. The more the total number of reasoning modes an individual learner used, the more he (or she) was oriented to reason from multiple perspectives.

To assess the reliability of the aforementioned analyses, another researcher was asked to analyze 20 participants' responses on open-end questionnaire independently. Their inter-coder agreements for these analyses were assessed, and the discrepancies were discussed to achieve final agreements. In this study, all the inter-coder agreements for the analyses were greater than 0.80, indicating that the qualitative analyses of students' informal reasoning in this study were sufficiently reliable.

Evaluating Students' On-Line Argumentation

The students' argumentation regarding xenotransplantation (i.e., the use of genetic engineering in the transplantation of animal organs to human) in anonymous on-line discussion was analyzed. Weinberger and Fischer (2006) proposed a multi-dimensional approach to analyzed argumentative knowledge construction in CSCL. Four different dimensions are identified: participation, epistemic, argumentative, and social mode. This study modified Weinberger and Fischer's (2006) framework, and the modified framework for evaluating students' argumentation in on-line discussion forums included the following three aspects:

(1) *Participation*: Discusses to what extent an individual student participated in on-line discussion. The number of the posts that an individual student proposed was counted in this study as the indicator for his/her participation in on-line discussion forums.
(2) *Purposes of the posts*: Students may propose posts in on-line discussion forums for different purposes, such as to propose personal positions in a post ("proposing personal position"), to reply to others' posts ("replying"), to explain or elaborate personal ideas or position ("explaining or elaborating"), or to question other participants ("questioning"). Besides, the participants also proposed some irrelevant posts ("irrelevant posts"). The purpose for each post revealed in on-line discussion forums was analyzed and categorized into one of the aforementioned types of purposes.
(3) *Argumentation quality*: The argumentation quality revealed in each post was analyzed in this study. Similar to Sadler and Fowler's (2006) framework, a rubric was developed in this study to analyze the argumentation quality revealed in each post. The rubric was shown in Table 11.1. Based on the rubric, the

Table 11.1 A rubric for assessing students' argumentation quality for each post

Argumentation level	Score	Description
0	0	Irrelevant posts
1	1	No justification and only position
2	2	Position with explanation or questioning
3	3	Position with supportive data
4	4	Position with both supportive data and questioning
5	5	Position with rebuttals

argumentation quality revealed in each post was also analyzed and scored. Then, the "achieved argumentation quality" was further analyzed. The "achieved argumentation quality" means the highest score for a post that an individual student obtained among all of his/her posts. Moreover, the "average argumentation quality" for each student was also calculated. The "average argumentation quality" is equal to "the total score of the posts / number of posts".

Results

The Effects of the Use of Anonymous On-Line Discussion Forums on Students' Informal Reasoning Outcome

In this study, the effects of the use of anonymous on-line discussion forums on students' informal reasoning outcome were examined. As shown in Table 11.2, after completing the anonymous on-line discussion task, the students proposed significantly more arguments than they did in the pre-test ($p < 0.05$). More importantly, the students also generated significantly more counter-arguments and rebuttals in the post-test ($p < 0.05$). In this study, students' rebuttal construction was viewed as the indicator for their informal reasoning quality. It seemed that, in general, the anonymous on-line discussion task in this study did improve students' informal reasoning quality regarding a socio-scientific issue. Besides, the results of Table 11.2 also revealed that the anonymous on-line discussion task helped the students reason from significantly more perspectives ($p < 0.05$), indicating that the anonymous on-line discussion improved the students' ability to reason from multiple perspectives.

Table 11.2 Comparisons on students' informal reasoning outcomes before and after on-line discussion task ($n = 37$)

		Mean	SD	t
Supportive argument	Before on-line discussion	1.27	0.51	0.96
	After on-line discussion	1.41	0.73	
Counter-argument	Before on-line discussion	1.35	0.72	2.25*
	After on-line discussion	1.62	0.59	
Rebuttal	Before on-line discussion	0.59	0.55	4.51**
	After on-line discussion	1.11	0.61	
Total number of arguments	Before on-line discussion	3.22	1.23	3.69*
	After on-line discussion	4.14	1.44	
Total number of reasoning modes	Before on-line discussion	2.41	0.93	2.64*
	After on-line discussion	2.89	0.84	

$^{*}p < 0.05$; $^{**}p < 0.01$

The Effect of the Use of Anonymous On-Line Discussion Forums on Students with Different Informal Reasoning Levels

In this study, the students were categorized as achieving either a "higher" reasoning level or a "lower" reasoning level according to the informal reasoning outcomes revealed in the pre-test. In the pre-test, the students who were capable of constructing not only simple claims (supportive arguments) and counter-arguments, but also rebuttals were viewed as attaining a "higher" reasoning level in this study, while those made only simple claims (supportive arguments) or counter-arguments were regarded as achieving a "lower" reasoning level. According to the students' responses in the pre-test, 21 students achieved a "higher" reasoning level (labeled as "H-R Group" in this study), while 16 students achieved a "lower" reasoning level (labeled as "L-R Group" in this study).

The effects of the use of anonymous on-line discussion forums on different reasoning level of students' informal reasoning outcomes were further explored. The results in Table 11.3 revealed that the anonymous on-line discussion task in this study only helped the students in the H-R Group generate more significant arguments ($p < 0.05$). However, the students in the H-R Group did not show any other significant difference in their informal reasoning outcomes, including supportive argumentation, counter-argument construction, rebuttal construction, and their usage of different reasoning modes, before and after the anonymous on-line discussion task ($p > 0.05$).

Table 11.4 showed that the anonymous on-line discussion task also helped the students in the L-R Group propose significantly more arguments ($p < 0.05$).

Table 11.3 Comparisons on students' informal reasoning outcomes before and after on-line discussion task in the H-R Group ($n = 21$)

		Mean	SD	t
Supportive argument	Before on-line discussion	1.29	0.56	0.94
	After on-line discussion	1.48	0.75	
Counter-argument	Before on-line discussion	1.57	0.60	1.45
	After on-line discussion	1.76	0.54	
Rebuttal	Before on-line discussion	1.05	0.22	1.45
	After on-line discussion	1.24	0.63	
Total number of arguments	Before on-line discussion	3.90	0.89	1.78*
	After on-line discussion	4.48	1.50	
Total number of reasoning modes	Before on-line discussion	2.71	0.85	0.93
	After on-line discussion	2.95	0.92	

* $p < 0.05$

Table 11.4 Comparisons on students' informal reasoning outcomes before and after on-line discussion task in the L-R Group ($n = 16$)

		Mean	SD	*t*
Supportive argument	Before on-line discussion	1.25	0.45	0.32
	After on-line discussion	1.31	0.70	
Counter-argument	Before on-line discussion	1.06	0.77	1.70
	After on-line discussion	1.44	0.63	
Rebuttal	Before on-line discussion	0.00	0	6.54**
	After on-line discussion	0.94	0.57	
Total number of arguments	Before on-line discussion	2.31	1.01	3.67**
	After on-line discussion	3.69	1.25	
Total number of reasoning modes	Before on-line discussion	2.00	0.89	3.31**
	After on-line discussion	2.81	0.75	

** $p < 0.01$

However, these students performed significantly better in their rebuttal construction and their usage of different reasoning modes after completing the on-line discussion task ($p < 0.05$), indicating that the anonymous on-line discussion task in this study did scaffold the students in the L-R Group in their rebuttal construction, which could be viewed as one of the important indicators for their reasoning quality.

Students' Argumentation in Anonymous On-Line Discussion Forums

This study further investigated the students' argumentation in anonymous on-line discussion forums. The students' argumentation in anonymous on-line discussion forums was evaluated in the three aspects: *participation*, *purpose of posts*, and *argumentation quality*.

According to Table 11.5, the students of different reasoning levels did not show any difference in "number of posts" as well as in their proposing posts for different purposes ($p > 0.05$), indicating that the students in the H-R Group study did not engage in anonymous on-line discussion more actively than their counterparts (i.e., the students achieving a higher reasoning level). Also, Table 11.5 revealed that the students in the H-R Group did not outperform their counterparts in "achieved argumentation quality" and "average argumentation quality" ($p > 0.05$). In other words, the students in the H-R Group did not show higher argumentation quality than their counterparts in anonymous on-line discussion. In summary, the students in the H-R Group did not outperform those in the L-R Group in anonymous on-line argumentation regarding a socio-scientific issue.

Table 11.5 Comparisons on students' participation, purposes for posts, and argumentation quality in on-line discussion task between different reasoning level groups ($n = 37$)

		Mean	SD	t
Participation				
Number of posts	L-R Group ($n = 16$)	2.88	2.06	−0.78
	H-R Group ($n = 21$)	2.43	1.43	
Purposes for the posts				
Proposing personal position	L-R Group ($n = 16$)	0.94	0.25	0.65
	H-R Group ($n = 21$)	1.00	0.32	
Replying	L-R Group ($n = 16$)	1.38	1.78	−0.85
	H-R Group ($n = 21$)	1.00	0.84	
Explaining or elaborating	L-R Group ($n = 16$)	1.06	1.00	−1.53
	H-R Group ($n = 21$)	0.48	1.33	
Questioning	L-R Group ($n = 16$)	0.31	0.60	0.32
	H-R Group ($n = 21$)	0.38	0.67	
Irrelevant posts	L-R Group ($n = 16$)	0.56	0.63	−1.54
	H-R Group ($n = 21$)	0.29	0.46	
Argumentation quality for each post				
Achieved argumentation quality	L-R Group ($n = 16$)	3.63	1.41	−1.46
	H-R Group ($n = 21$)	2.90	1.55	
Average argumentation quality	L-R Group ($n = 16$)	2.73	0.82	−1.21
	H-R Group ($n = 21$)	2.32	1.19	

$^{*}p < 0.05$; $^{**}p < 0.01$

Discussion and Conclusions

This study was conducted to explore the effects of anonymous on-line discussion task on students' informal reasoning outcomes regarding a socio-scientific issue. In addition, the effects of anonymous on-line discussion on students with different reasoning abilities were explored. The results of this study revealed the usefulness of anonymous on-line discussion task in students' counter-arguments construction and rebuttal construction. In other words, the students' informal reasoning quality was improved after completing the anonymous on-line discussion task.

Moreover, the effects of the use of anonymous on-line discussion forums on different reasoning level of students' informal reasoning outcomes were also investigated in this study. It was found that both the students achieving a "higher" reasoning level and those achieving a "lower" reasoning level benefited from the anonymous on-line discussion task, but in different ways. Both the students proposed significantly more arguments after on-line discussion task; but only the students achieving a "lower" reasoning level performed significantly better in their rebuttal construction and usage of different reasoning modes after the on-line discussion task. In other words, the students achieving a "lower" reasoning level benefited more from the anonymous on-line discussion task. In this study, each discussion group had four students, consisting of two students supporting genetic engineering and two students with opposite position. The treatment above may scaffold the students achieving a "lower" reasoning level to reason from multiple perspectives, generate counter-arguments, and construct rebuttals.

One may be interested in that why the students achieving a "higher" reasoning level did not perform better in their rebuttal construction or usage of different reasoning modes after completing the anonymous on-line discussion task. Their engagement in the anonymous on-line discussion forums may provide some clues for explanation. In this study, it was found that the students achieving a higher reasoning level did not show more active engagement and higher argumentation quality than their counterparts in the anonymous on-line discussion. It may due to the possibility that these students' attitude toward on-line discussion was not as positive as those with lower reasoning level. Or, the students achieving a higher reasoning level in this study were more tended to reason on a socio-scientific issue individually, rather than to argue on this issue with others. Further research is suggested to examine the aforementioned perspectives. Moreover, students' attitudes toward face-to-face and on-line anonymous socio-scientific argumentation and their preferences toward these two argumentative environments may be also important issues for further investigation.

This study is one of the initial attempts to address the effects of the use of anonymous on-line discussion forums on different reasoning level of students' informal reasoning outcomes. The findings of this study may also be helpful for science instructors. To improve students' reasoning quality regarding socio-scientific issues, science instructors may make use of anonymous on-line discussion forums. However, coupled with the use of anonymous on-line discussion forums, some teaching strategies for enhancing the engagement of learners with a higher reasoning level will be crucial.

References

American Association for the Advancement of Science. (1989). *Science for all Americans*. New York: Oxford University Press.

American Association for the Advancement of Science. (1993). *Benchmarks for science literacy.* New York: Oxford University Press.

Clark, D. B., & Sampson, V. D. (2007). Personally-seeded discussions to scaffold online argumentation. *International Journal of Science Education, 29*, 253–277.

Driver, R., Newton, P., & Osborne, J. (2000). Establishing the norms of scientific argumentation in classrooms. *Science Education, 23*, 689–698.

Erduran, S., Simon, S., & Osborne, J. (2004). Tapping into argumentation: Developments in the application of Toulmin's argument pattern for studying science discourse. *Science Education, 88*, 915–933.

Evans, J. St. B. T. (1996). Deciding before you think: Relevance and reasoning in the selection task. *British Journal of Psychology, 87*, 223–240.

Evans, J. St. B. T. (2002). Logic and human reasoning: An assessment of the deduction paradigm. *Psychological Bulletin, 128*, 978–996.

Evans, J. St. B. T. (2003). In two minds: Dual-process accounts of reasoning. *Trends in Cognitive Sciences, 7*, 454–459.

Evans, J. St. B. T., & Curtis-Holmes, J. (2005). Rapid responding increases belief bias: Evidence for the dual-process theory for reasoning. *Thinking & Reasoning, 11*, 382–389.
Jiménez-Aleixandre, M. P., Rodríguez, A. B., & Duschl, R. A. (2000). "Doing the lesson" or "doing science": Argument in high school genetics. *Science Education, 84*, 757–792.
Jonassen, D. H. (1996). *Computers in the classroom: Mindtools for critical thinking*. Columbus, OH: Merrill/Prentice-Hall.
Jonassen, D. H. (2000). *Computers as mindtools for schools*. Upper Saddle River, NJ: Prentice Hall.
Kortland, K. (1996). An STS case study about students' decision making on the waste issue. *Science Education, 80*, 673–689.
Kuhn, D. (1993). Connecting scientific and informal reasoning. *Merrill-Palmer Quarterly, 39*, 74–103.
Means, M. L., & Voss, J. F. (1996). Who reasons well? Two studies of informal reasoning among children of different grade, ability, and knowledge levels. *Cognition and Instruction, 14*, 139–178.
National Research Council. (1996). *National science education standards*. Washington, DC: National Academy.
Osborne, J., Erduran, S., & Simon, S. (2004). Enhancing the quality of argument in school science. *Journal of Research in Science Teaching, 41*, 994–1024.
Randel, J. M., Pugh, H. L., & Reed, S. K. (1996). Differences in expert and novice situation awareness in naturalistic decision making. *International Journal of Human Computer Studies, 45*, 579–597.
Sadler, T. D. (2004). Informal reasoning regarding socioscientific issues: A critical review of research. *Journal of Research in Science Teaching, 41*, 513–536.
Sadler, T. D., & Donnelly, L. A. (2006). Socioscientific argumentation: The effects of content knowledge and morality. *International Journal of Science Education, 28*, 1463–1488.
Sadler, T. D., & Fowler, S. (2006). A threshold model of content knowledge transfer for socioscientific argumentation. *Science Education, 90*, 986–1004.
Shaw, V. F. (1996). The cognitive processes in informal reasoning. *Thinking and Reasoning, 2*, 51–80.
Sloman, S. A. (1996). The empirical case for two systems of reasoning. *Psychological Bulletin, 119*, 3–22.
Weinberger, A., & Fischer, F. (2006). A framework to analyze argumentative knowledge construction in computer-supported collaborative learning. *Computers and Education, 46*, 71–95.
Wu, Y. T., & Tsai, C.-C. (2007). High school students' informal reasoning on a socio-scientific issue: Qualitative and quantitative analyses. *International Journal of Science Education, 29*, 1163–1187.
Wu, Y. T., & Tsai, C.-C. (2010). High school students' informal reasoning regarding a socio-scientific issue, with relations to scientific epistemological beliefs and cognitive structures. *International Journal of Science Education*. Available online 3 March 2010. doi: 10.1080/09500690903505661.
Zeidler, D. L., Sadler, T. D., Simmons, M. L., & Howes, E. V. (2005). Beyond STS: A research-based framework for socioscientific issues education. *Science Education, 89*, 357–377.
Zembal-Saul, C., Munford, D., Crawford, B., Friedrichsen, P., & Land, S. (2002). Scaffolding pre-service science teachers' evidence-based arguments during an investigation of natural selection. *Research in Science Education, 32*, 437–465.
Zohar, A., & Nemet, F. (2002). Fostering students' knowledge and argumentation skills through dilemmas in human genetics. *Journal of Research in Science Teaching, 39*, 35–62.

Chapter 12
The Development and Validation of the Assessment of Scientific Argumentation in the Classroom (ASAC) Observation Protocol: A Tool for Evaluating How Students Participate in Scientific Argumentation

Victor Sampson, Patrick J. Enderle, and Joi Phelps Walker

Introduction

Argumentation that is scientific in nature is often described as a form of "logical discourse whose goal is to tease out the relationship between ideas and evidence" (Duschl, Schweingruber & Shouse, 2007, p. 33) or a knowledge building and validating practice in which individuals propose, support, critique, and refine ideas in an effort to make sense of the natural world (Driver, Newton & Osborne, 2000; Kuhn, 1993; Sampson & Clark, 2011). Scientific argumentation, as a result, plays a central role in the development, evaluation, and validation of scientific knowledge and is viewed by many an important practice that makes science different from other ways of knowing (Driver et al., 2000; Duschl & Osborne, 2002). Yet, few students are given a opportunity to develop the knowledge or skills needed to participate in scientific argumentation or to learn how scientific argumentation differs from other forms of argumentation by time they graduate from high school (Duschl, et al., 2007; National Research Council, 2005, 2008) or as part of their undergraduate science education (National Research Council, 1999; National Science Foundation, 1996).

In response to this issue, several new instructional approaches and curricula have been developed over the last decade to give students more opportunities to acquire the understandings and abilities needed to participate in scientific argumentation. A continual challenge associated with this type of research, however, is the difficulties associated with documenting the nature or quality of the scientific

---

An earlier version of this chapter, entitled "Assessment of argumentation: An observation protocol," by P. Enderle, J. Walker, C. Dorgan, and V. Sampson was presented at the 2010 International Conference of the National Association for Research in Science Teaching.

V. Sampson (✉)
Florida State University, Tallahassee, FL, USA
e-mail: vsampson@fsu.edu

M.S. Khine (ed.), *Perspectives on Scientific Argumentation*,
DOI 10.1007/978-94-007-2470-9_12, © Springer Science+Business Media B.V. 2012

argumentation that takes place between students inside the classroom and tracking how students' ability to participate in scientific argumentation changes over time. Many researchers, for example, assess argumentation quality by first video or audio recording students as they engage in this complex practice, then they transcribe the discourse, and finally code or score it using a framework that focuses on the nature and function of the contributions to the conversation (Duschl, 2007; Erduran, 2007; Erduran, Simon & Osborne, 2004; Kelly, Druker & Chen, 1998; Kuhn & Reiser, 2005; Kuhn & Udell, 2003; Osborne, Erduran & Simon, 2004; Sampson & Clark, 2011). Yet, along with the various affordances that are associated with this type of approach for assessing the quality of argumentation, there are numerous constraints that stem from video taping, transcribing, and then "coding and counting" (Suthers, 2006).

First, this type of analysis is often time-consuming and expensive. Researchers, as a result, tend to study small samples of students or focus on a specific context. Second, the various aspects of a verbal argument are often difficult to identify during a discussion, which, in turn, tends to have an adverse effect on reliability (see Duschl, 2007; Erduran, 2007; Erduran, Simon et al., 2004). Another barrier to this type of approach is the nonlinear nature of scientific argumentation, which often makes it difficult for researchers to follow a line or thought through an episode of a multi-voiced argumentation or to define the boundaries of a unit of analysis. An analysis that focuses on the nature or functions of contributions to a discussion (e.g., the number of times the students support their claims or challenge the ideas of others, etc.) also limits what researchers are able to measure and forces them to disregard aspects of scientific argumentation that might be important or informative. For example, there are few studies that have examined the reasoning students employ during an episode of argumentation, the criteria they use to assess the merits of an idea, and how students interact with each other and the available materials as a way to assess quality. An assessment of argumentation quality that relies on a tabulation of the nature and function of contributions, therefore, is often limited in scope and privileges certain elements of argumentation at the expense of others.

The field therefore needs to develop new instruments that researchers can use to capture and score an episode of argumentation in a more holistic fashion, including nonverbal social interactions, and will result in a more comprehensive assessment of the overall quality of an event. Such an instrument also needs to be able to provide researchers with a reliable criterion-referenced measure of students' competency. This type of instrument is needed, as Driver et al. (2000) suggests, "to inform educational interventions designed to improve the quality of argumentation" and "to inform teachers about what to look for" (p. 295).

In this chapter, we will present a new instrument that researchers can use to measure that nature and quality of scientific argumentation. This instrument, which we call the *Assessment of Scientific Argumentation inside the Classroom* (ASAC) observation protocol, is intended to provide a criterion-referenced tool that targets the conceptual, cognitive, epistemic, and social aspects of scientific argumentation. This tool can be used to assess the nature and quality of argumentation that occurs between students inside the science classroom, to examine how students' participation in scientific argumentation changes over time in response to an intervention, or to compare the impact of different interventions on the way students participate in scientific argumentation. In the paragraphs that follow, we will first describe the method we used to develop and validate this new instrument. We will then conclude the chapter with a presentation and discussion of our findings, the limitations of work, and our recommendations about how and when to use the ASAC.

Method

Framework Used to Establish the Validity and Reliability of the ASAC

There is a large body of literature concerned with issues related to the validity and reliability of the assessment instruments that are used in educational research. In this methodological-focused literature, the legitimacy of drawing a conclusion about the knowledge or skills of people from scores on an assessment instrument is of upmost importance and shapes how new instruments are developed and validated. Many instrument developers, for example, use experts to determine if a new instrument measures what it purports to measure and estimates of internal consistency, such as Cronbach Alpha or KR-20, to evaluate the reliability of the instrument as part of the development and validation process (Burns, 1994). Although both are important, these two approaches are not sufficient. Trochim (1999), for example, suggests researchers need to focus on multiple properties of an instrument, such as the construct and criterion-related validity of an instrument, as well as its reliability in order to determine if an assessment actually measures what it is intended to measure. We therefore developed the methodological framework provided in Fig. 12.1 to guide the development and the initial validation of the ASAC.

In this framework, an instrument is deemed to possess good construct validity (i.e., the translation of a construct into an operationalization) if the theoretical construct is well defined, based on the available literature, and measures only the

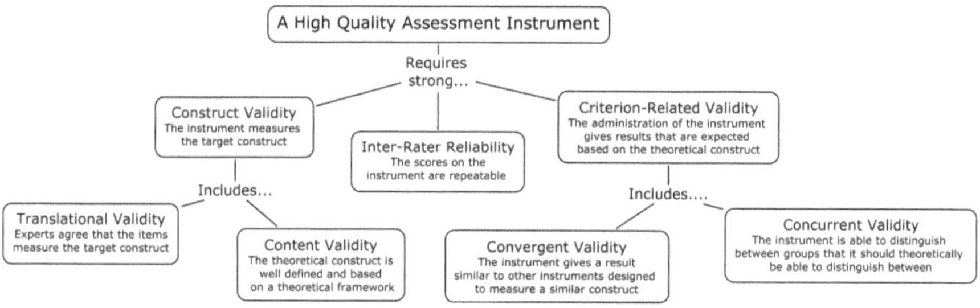

Fig. 12.1 The framework used to guide the development and initial validation of the ASAC

targeted construct (content validity). The items included in the instrument also need to be good translations of the construct based on expert opinion (translational validity). In addition to construct validity, an instrument must also have strong criterion-related validity in order to be considered credible and of high quality. Criterion-related validity considers the conclusions that can be drawn based on data generated by using the instrument. An instrument is deemed to possess strong criterion-related validity if it results in scores that are expected based on the theoretical construct (Trochim, 1999). An instrument with strong criterion-related validity, for example, should give results similar to another method that measures the same or a similar construct (convergent validity) and should be able to distinguish between groups or individuals that it is expected to distinguish between (concurrent validity). Assessment instruments, especially observation protocols, also need to have strong inter-rater reliability in order to generate scores that are consistent and repeatable.

Development of the ASAC

The method that we used to develop the ASAC began with a search of the literature in order to define the construct to be assessed and continued through item pool preparation, item refinement, and selection based upon expert review to ensure construct validity. The development then concluded with an evaluation of the instrument's criterion validity and reliability (Borg & Gall, 1989; Nunally, 1970; Rubba & Anderson, 1978). The seven-step process described below, which is based on recommendations outlined in the *Standards for Educational and Psychological Testing* (American Educational Research Association, American Psychological Association & National Council on Measurement in Education, 1999), was used to create a high-quality assessment instrument that is both valid and reliable.

Step 1: Define the construct to be measured. A clear definition of the construct that an assessment is intended to measure is needed in order to guide the development of an instrument. The construct also needs to be well defined in order to evaluate the content validity of the instrument and to determine how well an instrument measures the construct of interest. Therefore, in order to guide our work, we adopted a view of argumentation as a process where "different perspectives are being examined and the purpose is to reach agreement on acceptable claims or course of actions" (Driver et al., 2000, p. 291). This view of argumentation stresses collaboration over competition and suggests that activities that promote argumentation can provide a context where individuals are able to use each other's ideas to construct and negotiate a shared understanding of a particular phenomenon in light of past experiences and new information (Abell, Anderson & Chezem, 2000; Andriessen, Baker & Suthers, 2003; Boulter & Gilbert, 1995; deVries, Lund & Baker, 2002; Veerman, 2003). In other words, argumentation is a social and collaborative process that groups of individuals engage in "to solve problems and advance knowledge" (Duschl and Osborne, 2002, p. 41). An important distinction in this definition stems from the focus on the *process* involved in argumentation, a less researched phenomena, as opposed to arguments, the *product* of such activities that has received more attention. Given this theoretical perspective, we chose to define the construct of *scientific argumentation* as a social and collaborative process of proposing, supporting, evaluating, and refining ideas in an effort to make sense of a complex or ill-defined problem or to advance knowledge in a manner that is consistent with conceptual structures, cognitive processes, epistemological commitments, and the social norms of science (see Driver et al., 2000; Duschl, 2008; Kuhn, 1993; Sampson & Clark, 2009).

Step 2: Development of the instrument specifications (content coverage and item format). Our goal at this stage of the development process was to ensure that the instrument would measure each aspect of the target construct as defined by our theoretical framework. To accomplish this goal, we decided to focus on three aspects of scientific argumentation, which according to Duschl (2008), students need to develop in order to be able to participate in this complex practice (p. 277). First, an individual must be able to use important conceptual structures (e.g., scientific theories, models, and laws or unifying concepts) and cognitive processes valued in science when reasoning about a topic or a problem. Second, an individual must know and use the epistemic frameworks that characterize science to develop and evaluate claims. Third, and perhaps most importantly, individuals that are able to engage in scientific argumentation must understand and be able to participate in the social processes that shape how knowledge is communicated, represented, argued, and debated within scientific community. We therefore decided to develop a protocol that was divided into four sections to represent these various aspects, each with a distinct focus for assessing an episode of scientific argumentation (conceptual, cognitive, epistemological, and social).

At this stage in the development process, we also decided to include items in the protocol that are observable during an episode of argumentation regardless of context or topic of discussion. We felt that this was important decision to make at the onset of the project because some elements of scientific argumentation, although important, cannot be measured easily through direct observation (e.g., it only take place within the mind of an individual, it only occurs when someone produces a formal written argument, or it tends to only occur under certain circumstances). Finally, and perhaps most importantly, we decided to craft the items so researchers can use them to rate an element of each aspect of scientific argumentation on a scale. We decided to use a rating scale rather than a simple dichotomous option (yes/no) so researchers will be able to document the prevalence of each element of the four aspects of scientific argumentation (e.g., not at all, often, etc.), which in turn, will allow for a greater distribution of scores.

Step 3: Development of the initial pool of items. We generated an initial pool of 29 items based on important notions and issues raised in the argumentation literature. Each item contains a stem sentence describing a critical element of an aspect of scientific argumentation and a detailed description offering insight into the aim of the item. These initial stems were written based on recommendations made by Edwards (1957) to reduce item error due to ambiguity. A Likert-style scale, which ranges from 0 to 3, was also included for ranking the element based on the presence and prevalence of the observable actions described in the stem sentence (0—not at all, 1—once or twice, 2—a few times, 3—often). A few items focused on undesirable actions in regards to quality scientific argumentation and, as such, the scaling of these particular items is reversed for scoring.

Step 4: Initial expert review of the item pool. We then conducted an evaluation of the content and translational validity of the initial pool of items. To complete this evaluation, we asked a group of experts to review the items using an online survey instrument. We identified 18 experts based on their significant contributions to argumentation research and the relevant literature. We then sent the group of experts an email to explain the objective of this project, a request for their service, and a link to the online survey. We asked the reviewers to rank whether each item was an important aspect of scientific argumentation and if it should be included in the protocol on a scale of 1–5 (five being the highest). We also asked the reviewers whether the description for each item was appropriate and offer suggestions about how an item or the description of the item should be revised. The reviewers' identities were kept anonymous to allow for the utmost candor in their responses.

The online survey was kept active for a period of 2 months following the initial email that was sent in order to solicit the services of the experts. In total, eight thoughtful responses were received and used to make adjustments to the observation protocol and the initial item pool. Items that received an average ranking of 4 or higher (i.e., agree to strongly agree) were kept, the items with an average ranking between 3 and 4 (neutral to agree) were revised or combined with other items based on the reviewer's comments, and the items that had an average ranking of between 1 and 3 (i.e., strongly disagree to neutral) were discarded. This process resulted in the elimination of a total of eight items from the initial collection.

Step 5: The first field test of the instrument. The next phase of development process involved the authors attempting to use the protocol to assess the quality of several video-recorded episodes of argumentation. Our objective at this stage of the development process was to ensure that a rater could observe the element of argumentation targeted by an item and to ensure that all the items were clear and the accompanying descriptions were detailed enough to produce reliable scores across multiple raters. To accomplish this task, we viewed several videos of students engaged in a task that was designed to promote and support scientific argumentation. The students, in a small collaborative working group of three, in these videos were asked to read several different alternative explanations for why ice melts at different rates when placed different types of materials, to determine which provided explanation was the most valid or acceptable and then to craft an argument in support of their chosen explanation. These videos of high schools students engaged in an episode of argumentation were collected as part of another study (Sampson & Clark, 2009, 2011).

The authors along with another rater viewed the videos together, but did not discuss their scores on each item until the video ended and each rater completed the protocol on his or her own. Scores for each item were then compared and when significant differences among the raters emerged, the item and the description of the item were discussed, evaluated, and modified in order to reduce ambiguity. This process resulted in numerous refinements to the stems and descriptions. An additional objective of this process was to identify and remove any items from the protocol that targeted an aspect of argumentation that was too difficult to observe or assess. This iterative refinement process was repeated over several cycles until the four raters were relatively consistent in their scoring. At the end of this step of the project, the protocol was reduced to a total of 20 items.

Step 6: A second expert review of the items. The observation protocol was sent out to the same expert reviewer group for further comments and to evaluate the translational and content validity of the revised items. Reviewer comments were critically considered in making adjustments to the text of the items, with particular attention paid to the descriptions for each stem as well as the inclusion of items in one of the four broad categories used to structure the protocol. This round of reviewer input resulted in responses from seven members of the panel (although some reviewers did not rate or comment on each item). Guided by these responses, the authors made several additional adjustments.

One item was deleted from the existing protocol due to its repetitive nature, as identified by reviewers, and agreed upon by the authors. Another item was also considered by the reviewers to be too similar to another specific stem, so the authors condensed those two items into one. Another structural change suggested by the reviewers, and agreed upon by the authors, was to combine two of the categorical aspects (conceptual and cognitive) and their items into one, more cohesive grouping. The resulting instrument contains 19 items that are divided into three categories (conceptual and cognitive aspects, epistemological aspects, and social aspects of scientific argumentation). The rating of the translational and content validity items by the panel of experts, along with the literature used to develop them, can be found in the *Results*.

Step 6: Analysis of the inter-rater reliability of the instrument. At this point, our focus moved from instrument development to the process of initial validation. We used the final version of the observation protocol from step 5 to score 20 different videos of students engaged in an episode of scientific argumentation during an actual lesson. All of the episodes took place during the "argumentation session" stage of either the *Argument-Driven Inquiry* (ADI) instructional model (Sampson & Gleim, 2009; Sampson, Grooms & Walker, 2009) or the *Generate an Argument* (GaA) instructional model (Sampson & Grooms, 2010). The argumentation session that is included in both of these instructional models is designed to give small groups of students an opportunity to propose, support, critique, and revise an evidence-based argument either by using data they collected through a method of their own design (ADI) or from a corpus of data provided to them (GaA).

Two of the authors served as raters for all 20 episodes of argumentation. The raters viewed the videos at the same time and recorded observation notes in the table provided as part of the ASAC protocol (see Appendix). Then, at the completion of each video, each rater assigned a score for each item on the protocol and recorded some of their observations to justify their decision. Once the raters had completed the protocol individually, the two raters compared their scores for each item. The score assigned by each rater on each item, as well as the total score assigned to each episode by the two raters, was recorded and then compared in order to evaluate the inter-rater reliability of the ASAC observation protocol (see Results).

Step 7: Analysis of the criterion-related validity of the instrument. To assess the convergent validity of the ASAC, which, as noted earlier, is an instrument's ability to provide a similar score to other instruments that are used to measure the same construct, we used the Toulmin Argument Pattern (TAP) framework (Erduran, Osborne & Simon, 2004; Osborne et al., 2004; Simon, Erduran & Osborne, 2006) to score a subset of 12 videos from the inter-rater reliably analysis. We then compared the ASAC and TAP scores for each episode in order to determine if the two methods resulted in similar conclusions about the overall quality of the scientific argumentation (see Results). We decided to use this framework to help validate the ASAC because it places a much greater emphasis on the structural components of an argument than the ASAC does; thus we predicted that there would be a strong correlation between ASAC scores and TAP rankings but that there would also be more variation in ASAC scores due to its more holistic focus.

In order to measure the quality of an episode of argumentation using the TAP framework (see Erduran, Osborne et al., 2004), researchers must first transcribe a discussion. In this case, we used the argumentation sessions we recorded during the various classroom lessons. The argumentative operations of each conversational turn are then coded using five different categories: (a) opposing a claim, (b) advancing claims, (c) elaborating on a claim, (d) reinforcing a claim with additional data and/or warrants, and (e) adding qualifications. One of these codes is applied to each conversational turn during the discussion that takes place during an episode of argumentation. Researchers must then identify the structural components of an argument (i.e., claim, counter-claim, data, warrants, rebuttals, etc.) that are found within and across the conversational turns.

After identifying the argumentative operations of each conversational turn and the various structural components of arguments voiced by the participants in the discussion, the quality of an argumentation episode is assessed using the hierarchy outlined in Table 12.1. The hierarchy is based on two major assumptions about what makes one episode of argumentation better than another. First, higher quality

Table 12.1 TAP argumentation quality hierarchy developed by Erduran, Simon, and Osborne (2004)

Quality	Characteristics of argumentation
Level 5	Extended arguments with more than one rebuttal.
Level 4	Arguments with a claim with a clearly identifiable rebuttal. Such an argument may have several claims and counter-claims as well, but this is not necessary.
Level 3	Arguments with a series of claims or counter-claims with data, warrants, or backings with the occasional weak rebuttal.
Level 2	Arguments consisting of claims with data, warrants, or backings, but no rebuttals. Osborne advocates further distinction at this level: • Level 2B (2.5)—Arguments consisting of a claim supported by multiple pieces of data, warrants, or backings, but no rebuttals. • Level 2A (2.0)—Arguments consisting of a claim supported by a single piece of data, warrant, or backing, but no rebuttals.
Level 1	Arguments that are a simple claim versus a counter-claim or a claim versus claim.

argumentation must include arguments that consist of grounds (i.e., data, warrants, or backing) rather than unsubstantiated claims. Second, episodes of argumentation that include rebuttals (i.e., a challenge to the grounds used to support a claim) are "of better quality than those without, because oppositional episodes without rebuttals have the potential to continue forever with no change of mind or evaluation of the quality of the substance of an argument" (Erduran, Osborne, et al., 2004, p. 927).

Next, we used 20 videos from the inter-rater reliability analysis to evaluate the concurrent validity of the ASAC. Concurrent validity, as discussed earlier, concerns that ability of an instrument to discern between theoretically different groups. Therefore, if the ASAC has strong concurrent validity, we would expect an expert group of individuals (e.g., graduate students who understand the epistemological commitments, cognitive processes, and social norms that govern how people participate in scientific argumentation) to score higher on the ASAC than a group of novices (e.g., high school students who have little or no experience participating in scientific argumentation). We would also expect students to score higher on the ASAC after having numerous opportunities to participate in scientific argumentation.

The 20 videos were therefore divided into four theoretically distinct groups. The first group consists of high school students without any genuine experience with scientific argumentation ($n = 3$). The second group consists of undergraduate students who were video recorded as they participated in lab activity designed using the ADI instructional model for the first or second time at the beginning of a general chemistry lab course ($n = 7$). The third group also consists of undergraduate students; however, these videos were recorded at the end of a semester of general chemistry lab after the students had a chance to participate in four different ADI labs ($n = 7$). The final group consists of science education graduate students ($n = 3$). We then compared the average ASAC score of these four groups in order to determine if the ASAC could be used to distinguish between them as we predicted (see Results).

Results

In this section, we describe the results of our analysis of the reliability and validity of the ASAC in light of the methodological framework outlined in Fig. 12.1.

Construct validity. We evaluated the construct validity of the ASAC in two ways. During the construction of the first iteration of the ASAC protocol, the authors drew upon the argumentation literature to identify many common elements of what researchers and science educators considered to be characteristics of quality argumentation. To further strengthen the theoretical foundation of these items and their content validity, the authors relied on three important aspects of argumentation which, according to Duschl (2008), are fundamental to the process and make scientific argumentation different from the argumentation that takes place between individuals in other contexts. These aspects, as noted earlier, include: (1) the conceptual structures and cognitive processes used when reasoning scientifically; (2) the epistemic frameworks used when developing and evaluating scientific knowledge; and, (3) the social processes and contexts that shape how knowledge is communicated, represented, argued, and debated (p. 277).

We also evaluated the translational validity of the items and content validity of the instrument through two rounds of expert review. The comments and suggestions generated during this process assisted in shaping the wording and structure of the items, which, in turn, enhanced the translational validity of the instrument. The first iteration of the ASAC protocol underwent several major structural changes, resulting in an instrument comprised of 19 items measuring various aspects of quality argumentation, organized into three broad categories related to the theoretical framework. The reviewers' rating of the content and translational validity of each item from the second round of review as well as the empirical or theoretical foundation of each item is provided in Table 12.2. As illustrated in this table, the pool of expert reviewers agreed that each item is an important aspect of scientific argumentation and should be included in the instrument (min = 4.14/5, max = 5/5). The expert reviewers also agreed that the items, with the exception of items 8 and 15 (which were moved to a different section based on their feedback), were placed in the appropriate category and the corresponding section of the protocol (min = 4.17/5, max = 5/5).

The *Conceptual and Cognitive Aspects of Scientific Argumentation* section of the ASAC consists of seven items. These items allow a researcher to evaluate important elements of scientific argumentation, such as how much the participants focus on problem solving or advancing knowledge, how often individuals evaluate alternative claims, the participants' willingness to attend to anomalous data, the participants' level of skepticism, and the participants' use of appropriate or inappropriate reasoning strategies. The *Epistemic Aspects of Scientific Argumentation* section contains seven items. These items focus on important elements of scientific argumentation such as the participants' use of evidence, their evaluation of the evidence, the extent to which the participants use scientific theories, laws or models during the discussion, and how often the participants use the language of science to communicate their ideas. Finally, the *Social Aspects of Scientific Argumentation* section contains five items, which provides a means for assessing how the participants communicate and interact with each other. These items assess important elements of argumentation such as the participant's ability to be reflective about what they say, their respect for each other, their willingness to discuss ideas introduced by others, and their willingness to solicit ideas from others.

Inter-rater reliability. We used the scores from 20 different episodes of argumentation that were generated by two different raters to evaluate the inter-rater reliability of the ASAC. We first calculated a correlation coefficient between the two sets of total scores. The results of the analysis indicate that there was a significant and strong correlation between the scores of the two raters, $r(20) = 0.99$, $p < 0.001$. Figure 12.2 provides a scatter plot of the data points along with the equation for the best fitting line and the proportion of variance accounted for by that line. This estimate of the total score inter-rater reliability, $R^2 = 0.97$, is high for an observational protocol.

Table 12.2 The empirical and theoretical foundation of the items included in the ASAC along with the experts ratings of the content and translational validity of the items

Items	Empirical or theoretical foundation	Important element[a]		Inclusion in the category[a]	
		M	*SD*	*M*	*SD*
Conceptual and cognitive aspects					
• The work and talk of the group focused on the generation or validation of knowledge.[b]	Abell et al. (2000), Berland and Reiser (2009), Sandoval and Reiser (2004)	5.00	0.00	4.50	0.84
• The participants sought out and discussed alternative claims or explanations.	Driver et al. (2000), Duschl and Osborne (2002), Sampson and Clark (2011)	4.83	0.41	5.00	0.00
• The participants modified their claim or explanation when they noticed an inconsistency or discovered anomalous information.	Berland and Reiser (2009), Duschl and Osborne (2002), Vellom and Anderson (1999)	5.00	0.00	4.80	0.45
• The participants were skeptical of ideas and information.	Driver et al. (2000), Osborne et al. (2004)	4.71	0.49	4.57	0.79
• The participants provided reasons when supporting or challenging an idea.	Erduran and Jimenez-Aleixandre (2007)	5.00	0.00	4.71	0.76
• The participants based their decisions or ideas on inappropriate reasoning strategies.[c]	Duschl (2008), Vellom and Anderson (1999), Zeidler (1997)	4.57	0.79	4.33	0.98
• The participants attempted to evaluate the merits of each alternative explanation or claim in a systematic manner.	Duschl (2008), Osborne et al. (2004)	4.86	0.38	4.43	0.79
Epistemological aspects					
• The participants relied on the "tools of rhetoric" to support or challenge ideas.[c,d]	Kuhn (1993), Zeidler (1997)	4.14	0.90	3.29	1.50
• The participants used evidence to support and challenge ideas or to make sense of the phenomenon under investigation.	Berland & Reiser (2009), Duschl (2008), Erduran and Jimenez-Aleixandre (2007)	5.00	0.00	4.43	0.79
• The participants examined the relevance, coherence, and sufficiency of the evidence.	Driver et al. (2000), Duschl (2007)	5.00	0.00	4.17	1.33
• The participants evaluated how the available data was interpreted or the method used to gather the data.	Duschl (2007), Duschl and Osborne (2002)	4.71	0.76	4.57	0.79

Table 12.2 (continued)

Items	Empirical or theoretical foundation	Important element[a]		Inclusion in the category[a]	
		M	*SD*	*M*	*SD*
• The participants used scientific theories, laws, or models to support and challenge ideas or to help make sense of the phenomenon under investigation.	Driver et al. (2000), Duschl (2008), Sandoval (2003)	5.00	0.00	4.43	0.98
• The participants made distinctions and connections between inferences and observations explicit to others.	Driver et al. (2000), Erduran and Jimenez-Aleixandre (2007), Sandoval (2003)	4.57	0.54	4.43	0.98
• The participants used the language of science to communicate ideas.	Carlsen (2007), Erduran and Jimenez-Aleixandre (2007)	4.57	0.54	4.00	1.16
Social aspects					
• The participants were reflective about what they know and how they know.[e]	Alexopoulou and Driver (1996), Erduran and Jimenez-Aleixandre (2007)	4.50	0.55	3.83	0.98
• The participants respected what each other had to say.	Boulter and Gilbert (1995), Richmond and Striley (1996)	4.43	1.13	4.71	0.76
• The participants discussed an idea when it was introduced into the conversation.	Berland and Reiser (2009), Sampson & Clark (2011)	4.86	0.38	4.86	0.38
• The participants encouraged or invited others to share or critique ideas.	Boulter and Gilbert (1995), Sampson and Clark (2011)	4.71	0.76	5.00	0.00
• The participants restated or summarized comments and asked each other to clarify or elaborate on their comments.	Alexopoulou and Driver (1996), Boulter and Gilbert (1995), Richmond and Striley (1996)	5.00	0.00	5.00	0.00

[a] The items were scored using the following scale: 1—strongly disagree, 2—disagree, 3—neutral, 4—agree, 5—strongly Agree.
[b] Item was reworded to read, "The conversation focused on the generation or validation of claims or explanations" based on reviewer feedback.
[c] Item represents an undesirable element of scientific argumentation and is therefore scored in reverse.
[d] Item was moved from the conceptual and cognitive aspects section of the protocol to the epistemological aspects section based on expert feedback.
[e] Item was moved from the conceptual and cognitive aspects section of the protocol to the social aspects section based on expert feedback.

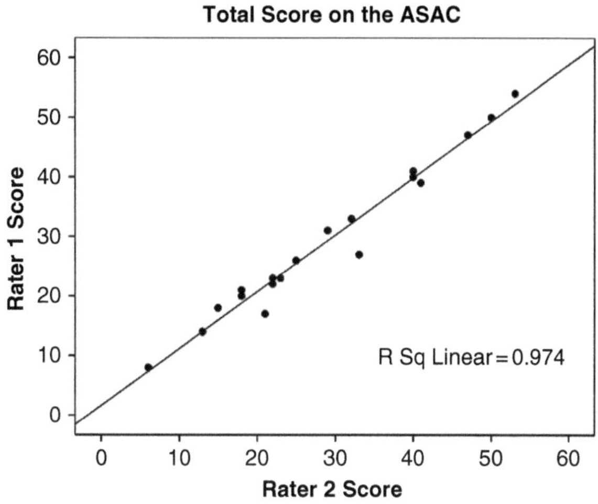

Fig. 12.2 A scatter plot of the total ASAC score generated by two different raters

Next, we calculated a correlation coefficient between the two sets of scores for the items. The results of this analysis, once again, indicate that there was a significant and strong correlation between the scores of the two raters, $r(380) = 0.89$, $p < 0.001$. Figure 12.3 provides a scatter plot of the 380 data points along with the equation for the best fitting line and the proportion of variance accounted for by that line ($R^2 = 0.795$). This estimate of the inter-rater reliability for the items is also high (although not as high as for the total score). The scatter plot in Fig. 12.3 also illustrates that most of the observed discrepancies between the scores produced by the two raters for the various items were with a single point. This indicates that even

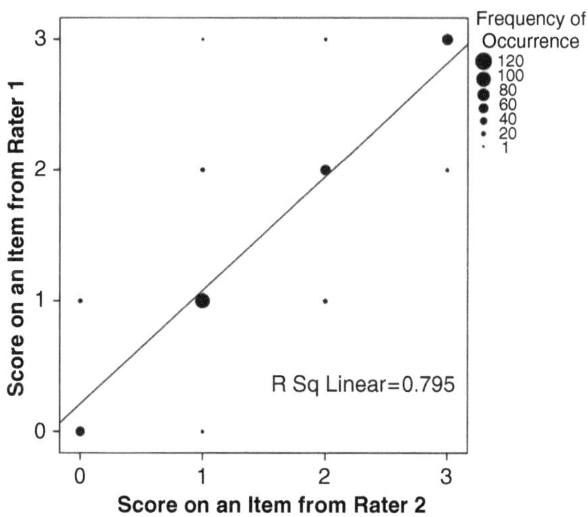

Fig. 12.3 A scatter plot of the scores on each item generated by two different raters

when the raters did not assign the same score on an item their evaluation was at least similar. It is important to note, however, that this type of analysis does not take into account chance agreement between the two raters. The influence of chance agreement on estimates of inter-rater reliability is important to consider when a scale is used because a scale limits choice and can artificially inflate the measure.

We therefore also used this corpus of data to calculate a Cohen's Kappa value as an additional estimate of inter-rater reliability. Cohen's Kappa, which is often used to measure the consistency and repeatability of scores between two raters, takes into account chance agreement unlike correlations or measures of percent agreement. The maximum value of Kappa is 1.0, which indicates perfect agreement, and a value of 0.0 indicates that the observed agreement is the same as that expected by chance. Landis and Koch (1977) suggest that values of Kappa above 0.60 indicate "good to excellent" agreement between scores of two raters, and values of 0.40 or less show "fair to poor" agreement. The inter-rater reliability of the ASAC, as measured with Cohen's Kappa, was 0.69. Thus, this analysis indicates that two different raters can use the ASAC to score an episode of argumentation in a manner that is consistent and repeatable.

Criterion-related validity. As a first test of this important aspect of validity, we compared ASAC scores from 12 episodes of argumentation to the scores we generated using the TAP framework. We predicted, as discussed earlier, that these two approaches would result in similar conclusions because they both are designed to measure the same construct. We therefore calculated a correlation coefficient between the two sets of total scores. The results of the analysis indicate that there was a significant and strong correlation between the two sets of scores, $r(12) = 0.96$, $p < 0.001$. Figure 12.4 provides a scatter plot of the 12 episodes of argumentation that we scored using both measures. As illustrated in this figure, the videos of the high school students engaged in argumentation received a total ASAC score in the range of 8–22 (out of a possible 57) and a TAP ranking of Level 2B (2.5 out of a possible 5). The videos of the undergraduate students were given a total ASAC score in the range of 7–33 and TAP rankings between 2.5 and 4. The videos of the graduate students, in contrast with the other two groups, received total ASAC scores in the range of 47–53.5 and all earned a Level 5 TAP ranking. Overall, these findings suggest that the ASAC and the TAP framework measure the same underlying construct and the ASAC, as a result, has adequate convergent validity.

We then examined how well the instrument is able to distinguish between groups that it should be able to distinguish between based on our theoretical framework (i.e., students with different levels of knowledge and skills needed to participate in scientific argumentation). A Kruskal–Wallis test (a nonparametric alternative to an ANOVA) was conducted to evaluate differences among the four groups (high school students, undergraduate students at the beginning of the semester, undergraduate students at the end of the semester, and graduate students) on the median ASAC score. The test, which was corrected for tied ranks, was significant, $\chi^2(3, N = 20) = 10.88$, $p = 0.01$. Follow-up tests were conducted to evaluate pairwise difference among the four theoretical groups, controlling for type I error across the tests by using the Holm's sequential Bonferroni approach. The results of these tests indicate

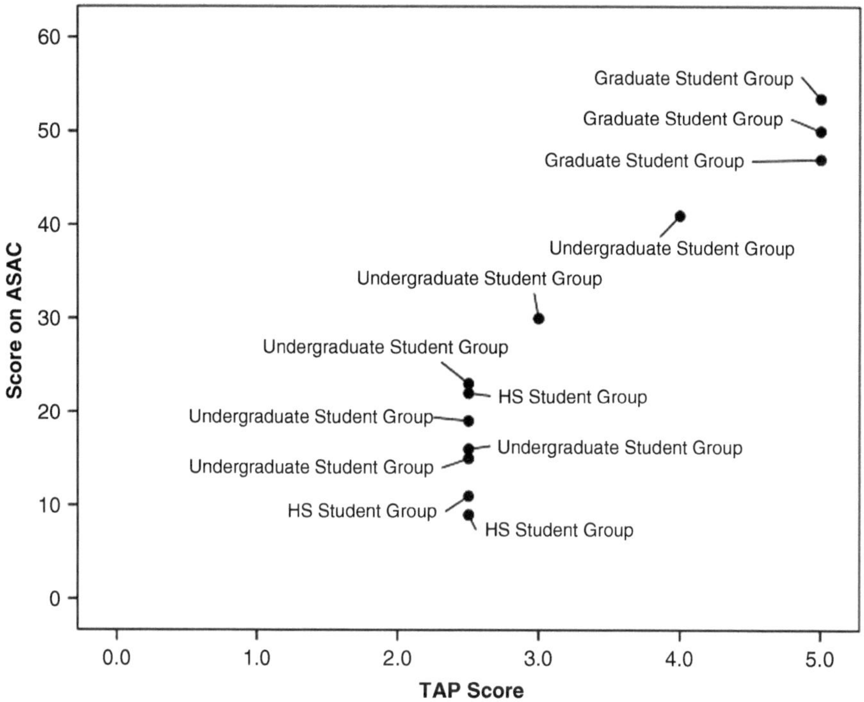

Fig. 12.4 Comparison of ASAC scores and TAP rankings

a significant difference in the median ASAC scores between the high school students and the graduate students, the undergraduate student group at the beginning of the semester and the graduate students, and the undergraduate students at the end of the semester and the graduate students. Figure 12.5 provides a boxplot of the distribution in ASAC scores for the four different groups. Overall, this analysis indicates that the instrument can distinguish between groups of students who should have different argumentation skills as expected and therefore has adequate concurrent validity.

Discussion and Limitations

The development of an ASAC observation protocol serves as another contribution to the expanding field of research that examines argumentation in science education. As the importance of developing students' ability to engage in productive scientific argumentation continues to grow, researchers and educators must have methods for assessing how students participate in this complex practice. Several frameworks have been developed in recent years that focus on the structure of the arguments produced by students during an episode of argumentation in order to help fulfill

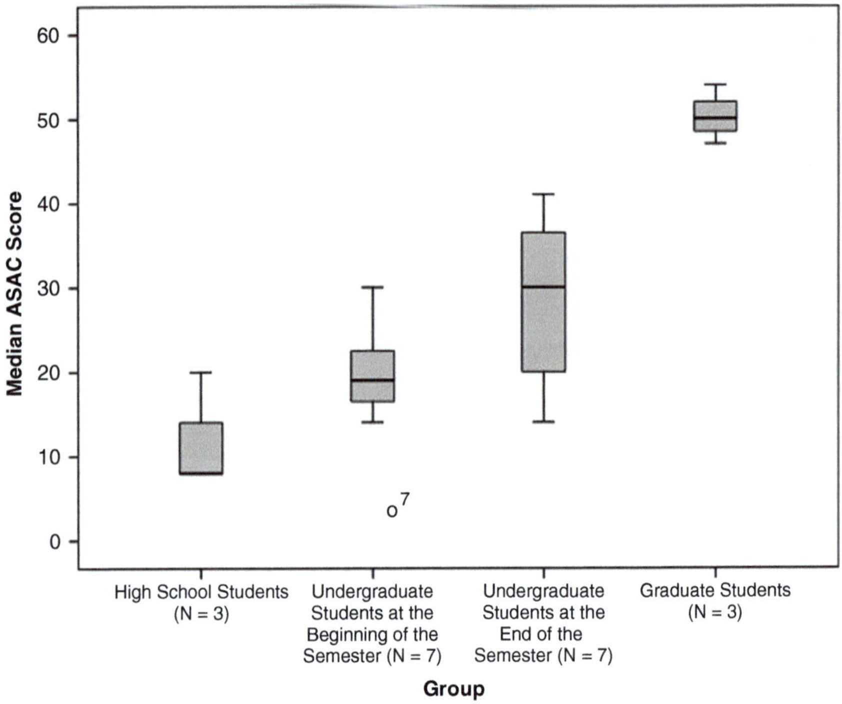

Fig. 12.5 Distribution of ASAC scores for different groups of students

this need (Berland & Reiser, 2009; Erduran, Osborne et al., 2004; McNeill, Lizotte, Krajcik & Marx, 2006; Sampson & Clark, 2008; Sandoval, 2003). These approaches have aptly focused on the general structure of arguments and their justifications, the content validity of these products, and the epistemological appropriateness of them. However, there has been less work that emphasizes and investigates another critical component, the social process and activities involved in generating those arguments.

The ASAC observation protocol, as developed and initially validated through this study, should serve as a useful tool to incorporate all of the aforementioned elements as well as bringing the social activities involved in the process into consideration. The research presented here helps to establish the construct validity of the instrument by sharing the literature that was used to define the construct of scientific argumentation and develop the items used to measure it. The construct validity was further supported through the guidance and critique provided during instrument construction by a panel of several expert reviewers comprised from science education researchers with a research record in this topic. These stages of validation support the assertion that the ASAC observation protocol does measure the construct of *scientific argumentation*.

Likewise, the initial criterion-related validity of the ASAC, which is the ability of the instrument to provide scores expected on theoretical grounds, was also established through the several different approaches. The ASAC, when used to assess a collection of videos of students engaged in structured activities, resulted in scores that varied substantially between groups of students that theoretically should differ in argumentation skills. The authors also used the TAP framework to help demonstrate that the ASAC and TAP both measure argumentation quality. Finally, inter-rater reliability, an assessment of the possibility for researchers to produce similar and repeatable scores, was demonstrated through several quantitative procedures. The satisfactory Cohen's Kappa value obtained along with the strong and significant correlations between two raters' sets of scores provide evidence that two raters can produce a similar, if not identical, assessment of an episode of argumentation when using the ASAC.

The authors, however, readily acknowledge that the ASAC protocol presented here and the validation measures described represent a first attempt to develop a research instrument that can measure conceptual, cognitive, epistemological, and social aspects of argumentation that emerge during classroom activities. Although the data set used during this research provides sufficient evidence for the validity and reliability of the ASAC protocol, additional data should be collected to further strengthen the quantitative data set and related measures. Increasing the amount of scores from further video reviews offers the potential to demonstrate further inter-rater reliability and an increase in Kappa and correlation values. Another benefit from an increased data set would be the capability to analyze each item in the instrument using Kappa calculations. At this stage of our research, the data set was not large enough to allow for these more detailed assessments. Increased data collection using the ASAC protocol could also increase the variety of contexts to which the instrument is applied, beyond the structured "argumentation session" that were used as a source of data in this study.

It is also important to note that the videos of the activities used in the data set reported here represent a particular type of approach for promoting and supporting student engagement in scientific argumentation, one that is structured to encourage students to share, critique, and refine evidence-based arguments that provide an answer to a research question. The authors concede that using this type of activity does provide a minimal amount of forced argumentation (because the argumentation sessions are design to foster it), excluding the validity of the instrument for more naturally emerging instantiations of classroom-based argumentation. However, we feel that this limitation should be considered to be an invitation to extend the use of the ASAC protocol to a variety of contexts, particularly in light of its unique and contextually bound social component.

The context of an episode of argumentation, especially the structure of the activity used to engage students in this practice, will influence the magnitude of the score on the ASAC (or the score on any instrument for that matter). For example, any classroom activity that does not allow students to collaborate with each other would not be able to score as well on the instrument, and a classroom culture that discourages verbal interactions between students will also result in lower scores. Another contextual aspect of the activity that might influence ASAC scores is an opportunity for students to generate their own data and explanations. We noticed during our review of the videos that the groups that were provided with a collection of data and a list of explanation to select from were not as likely to engage in extended episode of argumentation. Instead, these students seemed to select an explanation and then search for confirming evidence. In activities where students generated their own data, students had to make sense of it and develop their explanations, offering more opportunities to engage in scientific argumentation. Thus, the nature of the activity must be deeply considered when raters use the ASAC protocol to assess argumentation quality or to make comparisons across groups. However, we feel that this issue is a potential strength of the instrument; it will allow researchers to examine how the structure of activity influences the way students engage in scientific argumentation.

Another implication for using the ASAC protocol that can be noted from this study involves the necessity of familiarization and training the raters that will use the instrument. The authors of this chapter, who also served as raters, were involved with every step of the instrument development; therefore they were rather familiar with the content of the protocol. However, even in light of knowledge of the instrument, the two raters still had to refresh their understandings and align their interpretation of the items before beginning a scoring session. We found that watching a few practice videos, scoring them, and then discussing the discrepancies in scores were all that was needed to "calibrate" the raters. Raters should complete at least two "trail runs" by watching and scoring videos of an activity similar to the one that will be assessed before scoring the actual data set in order to help ensure the highest possible inter-rater reliability.

In conclusion, the ASAC observation protocol should serve educational researchers well, as investigations into the benefits and effectiveness of argumentation in science classrooms continues to grow. This growth has potential to move into other areas of concern, such as gender and cultural dynamics that can influence the process and the product. The ability for researchers to measure quality over periods of time within specific groups or environments can be facilitated through this instrument. This use can allow researchers and teachers alike to measure the progress of change in students' abilities to engage in productive scientific argumentation and enhancement of scientific argumentation skills. Thus, the ASAC instrument, although nascent in its development, offers a much-needed tool to help researchers understand and science education realize some of the visions of reform and literacy underpinning many current efforts.

Recommendations for Using the ASAC

The ASAC, as noted earlier, is criterion-referenced assessment of the quality of an episode of argumentation. The observers' judgments should therefore not reflect a comparison with any other instructional setting or event. The instrument contains 19 items. Seventeen items are rated on a scale from 0 (not at all) to 3 (often) and two items, which target undesirable aspects of scientific argumentation, are rated in reverse (0—often, 3—not at all). Possible scores range from 0 to 57 points, with higher scores reflecting higher quality scientific argumentation. The ASAC can be used in a wide range of educational contexts and levels including middle schools, high schools, and universities. It can be used to score a live event (e.g., to score an episode of argumentation as it unfolds in the classroom, to score the nature of argumentation inside a classroom where permission to video record students has not been granted, etc.) or a video recording of a past event.

An observer should adhere to the following procedure when using the ASAC (assuming that the observer has been trained about how to use the instrument and has completed several trail runs with another rater to ensure that his or her interpretations of the items are aligned with the content of the item descriptions). First, the observer should turn to the *record of events* section (see Appendix) and take observational notes in the provided table while watching the entire episode of argumentation. After the episode is complete, the observer should then turn the section of the protocol with the 19 items and score them. The observer should also include observations they made in the space for *comments* under each item description in order to support his or her ranking of an item. Finally, the rater should return to the sections called *observational information*, *group characteristics*, and *activity design* and fill in all the necessary background information based on his or her observations of the episode of argumentation and, if necessary, ask the classroom teacher to provide any additional information that is needed.

We, however, recommend that only trained observers use the ASAC. Although the protocol includes a detailed description of the aim of each item, raters need to participate in a formal training program. This training program, at a minimum, should include an opportunity for the trainees to examine the content and aim of each item, observe videotapes of episodes of argumentation or an actual instance of argumentation occurring in a classroom, score them using the ASAC, and discuss their interpretations of the items with others. As part of this process, raters should be encouraged to review videos together and discuss discrepancies in order to bring their personal interpretations of the items into alignment with the actual content of the item descriptions (e.g., raters tend to disregard an aspect of a description or interpret the content of an item description in unintended ways). We also recommend, as noted earlier, that raters watch several trail videos, score them, and discuss any discrepancies in order to "calibrate" with the instrument and each other before beginning a scoring session associated with a research study.

The ASAC can be used for several different research purposes. First, it can be used in longitudinal studies to examine how students' ability to participate in scientific argumentation changes over time. Researchers, for example, can use the ASAC to assess the quality of a series of subsequent events that provide students with an opportunity to engage in an episode of argumentation in order to determine how much students improve as a result of a new curriculum or new instructional method over the course of a semester or a school year. Second, the instrument can be used in comparison studies to examine the efficacy of a new curriculum or instructional strategy as a way to improve scientific argumentation skills. It can also be used to compare various designs of a new instructional method or ways of organizing the structure of an activity as part of development project. Researchers, for example, might be interested in comparing the nature of the quality of the argumentation that takes place between students when they are required to generate and make sense of their own data as part of the activity versus being supplied with an existing data (which we discussed earlier). Researcher can also use the ASAC to determine gains in argumentation skills in an experimental or quasi-experimental research study that uses a performance task as a pre–post intervention assessment.

Science teacher educators can also use the ASAC for professional development purposes because science teachers often do not know "what to look for and how to guide their students' arguments" (Driver et al., 2000, p. 295) or how to monitor their students' progress as they learn how to participate in scientific argumentation. We think the ASAC will help teachers with this difficult task. Science teacher educators, for example, can train science teachers to use the observational protocol which, in turn, would help science teachers develop a better understanding of what counts as high-quality scientific argumentation (i.e., increase their understanding of scientific argumentation). Once trained, these teachers could then use the ASAC in their own classrooms to assess how well their students participate in argumentation. These teachers could then use the information they gathered using the ASAC to guide their own classroom practice and to plan future lessons. Thus, the ASAC should provide science teachers, science teacher educators, and science education researchers with a valid and reliable way to assess the quality of argumentation, so better curricular and instructional decision can be made about what works and what needs to be fixed.

Acknowledgements The development and validation of the instrument reported here was supported, in part, by the Institute of Education Science, U.S. Department of Education through grant #R305A100909 to Florida State University. The opinions expressed are those of the authors and do not represent the views of the Institute or the U.S. Department of Education.

Appendix

ASSESSMENT OF SCIENTIFIC ARGUMENTATION IN THE CLASSROOM OBSERVATION PROTOCOL

OBSERVATION INFORMATION

Teacher:	______________	School:	______________
Subject:	______________	Grade:	______________
Observer:	______________	Date:	______________

Duration of the episode: ______________________________

GROUP CHARACTERISTICS

Size:	☐ 2 ☐ 3 ☐ 4 ☐ 5 ☐ 6 or More ☐ Whole Class	Number of times that these students have been placed into this same group before:	☐ Never ☐ 1 ☐ 2 ☐ 3 ☐ 4 or more ☐ Unknown
Assignment to the Group:	☐ Random ☐ Self-Selected ☐ Achievement – Mixed ☐ Achievement – High ☐ Achievement – Low ☐ Teacher choice – Other ☐ Unknown	Gender Composition:	☐ All Male ☐ All Female ☐ # of Males > # of Females ☐ # of Females > # of Males ☐ # of Females = # Males
Racial/Ethnic Composition:	______________	Native Language Composition:	______________

ACTIVITY DESIGN

Provide a brief description of (a) the way the activity or lesson was designed in an effort to promote and support argumentation and (b) the way the teacher encouraged students to engage in argumentation.

RECORD OF EVENTS

In the space provided keep a running record of the events that occurred as the participants interacted with each other, the materials, and ideas.

Time	Description of Event

CONCEPTUAL AND COGNITIVE ASPECTS OF SCIENTIFIC ARGUMENTATION

How the group attempts to negotiate meaning or develop a better understanding

1. The conversation focused on the generation or validation of claims or explanations.	0	1	2	3
	Not at all	Once or Twice	A few times	Often

Description: The emphasis on the generation or validation of claims or explanations indicates that there were some significant claims or explanations at the heart of discussion. Groups that score high on this item maintain the focus of their talk and efforts on understanding or solving the problem rather than the best way to finish their work quickly or with the least amount of effort. *Note:* Groups that stay on topic but never go engage in an in-depth discussion about what is happening should be scored low on this item.

Comments:

2. The participants sought out and discussed alternative claims or explanations.	0	1	2	3
	Not at all	Once or Twice	A few times	Often

Description: Divergent thinking is an important part of scientific argumentation. A group that meets this criterion would talk about more than one claim, explanation, or solution. Individuals that valued alternative modes of thinking would respect and actively solicit new or alternative claims, explanations, or solutions from the other participants. *Note:* Groups that discuss multiple types of grounds or support for a claim, explanation, or solution but only one claim, explanation, or solution should be scored low on this item.

Comments:

3. The participants modified their claim or explanation when they noticed an inconsistency or discovered anomalous information.	0	1	2	3
	Not at all	Once or Twice	A few times	Often

Description: Inconsistencies between claims or explanation and the phenomenon under investigation are common in science. A group that modified their claim or explanation when they noticed inconsistencies or anomalies would not ignore "things that do not fit" or attempt to discount them once they are noticed by one of the participants. Groups that score high on this item try to modify their claim or explanation (not just their reasons) in order to account for an inconsistency or an anomaly rather than attempting to "explain them away" or simply deciding that something "doesn't matter."

Comments:

4. The participants were skeptical of ideas and information.	0	1	2	3
	Not at all	Once or Twice	A few times	Often

Description: During scientific argumentation, allowing a variety of ideas to be presented, but insisting that challenge and negotiation also occur would indicate that group members were skeptical. Accepting ideas without accompanying reasons would result in a low score because it is a sign of credulous thinking. In other words, students must be willing to ask, "how do you know?" or "Are you sure?" Groups that respond to the ideas of others with comments such as "ok", "that sounds good to me", or "whatever you think is right" would score low on this item.

<table>
<tr><td colspan="5">Comments:</td></tr>
<tr><td rowspan="2">5. The participants provided reasons when supporting or challenging an idea.</td><td>0</td><td>1</td><td>2</td><td>3</td></tr>
<tr><td>Not at all</td><td>Once or Twice</td><td>A few times</td><td>Often</td></tr>
<tr><td colspan="5">Description: Providing reasons to support or challenge a claim, conclusion, or explanation is a crucial characteristic of argumentation. Claims must have some support provided for them beyond simply restating the claim itself. Making claims with out support would result in a low score on this item and including any reason like “that’s what I think”, “it doesn’t make sense”, “the data suggests…” or “but that doesn’t fit with…” would result in a higher score. Note: Personal or past experiences count as a reason for this item.</td></tr>
<tr><td colspan="5">Comments:</td></tr>
<tr><td rowspan="2">6. The participants based their decisions or ideas on inappropriate reasoning strategies.</td><td>3</td><td>2</td><td>1</td><td>0</td></tr>
<tr><td>Not at all</td><td>Once or Twice</td><td>A few times</td><td>Often</td></tr>
<tr><td colspan="5">Description: When people are trying to support ideas they often: (a) jump to hasty generalizations, (b) attribute causality to random events, (c) insist that a correlation is evidence of causality, and (d) exhibit a confirmation bias (for example saying, “now we need some data to prove this”). Groups that avoid inappropriate reasoning strategies or recognize them when they occur would score high on this item. Groups where these types of reasoning strategies are common would score low on this item.</td></tr>
<tr><td colspan="5">Comments:</td></tr>
<tr><td rowspan="2">7. The participants attempted to evaluate the merits of each alternative explanation or claim in a systematic manner.</td><td>0</td><td>1</td><td>2</td><td>3</td></tr>
<tr><td>Not at all</td><td>Once or Twice</td><td>A few times</td><td>Often</td></tr>
<tr><td colspan="5">Description: This addresses the tentative or responsive nature of science. The idea that there is often more than one way to interpret data or evidence and that only through careful analysis can an idea be accepted or eliminated. This gets at the “gut” response factor. Conclusions are not based on opinion or inference.</td></tr>
<tr><td colspan="5">Comments:</td></tr>
</table>

EPISTEMIC ASPECTS OF SCIENTIFIC ARGUMENTATION

How consistent the process is with the culture and norms of science

8. The participants relied on the "tools of rhetoric" to support or challenge ideas.	**3**	**2**	**1**	**0**
	Not at all	Once or Twice	A few times	Often

Description: "Tools of rhetoric" refer to tricks or strategies used to win a debate. Tool of rhetoric include: (a) claiming that if someone cannot disprove a claim it must be true, (b) using emotive words and false analogies, (c) directing the focus of the discussion from thinking about a claim or an explanation to thinking about the person holding or proposing a claim or an explanation, (d) over-relying on authorities, (e) dichotomizing issues so that if you discredit one position, then the observer is forced to accept the other view, and (f) making claims that are a simple restatement of one of the premises. Groups that avoided using the tools of rhetoric would score high on this item. *Note:* This item focuses on how the content of a discussion is presented or supported (i.e., how they are saying it) rather than the content of the discussion (i.e., what they are saying).

Comments:

9. The participants used evidence to support and challenge ideas or to make sense of the phenomenon under investigation.	**0**	**1**	**2**	**3**
	Not at all	Once or Twice	A few times	Often

Description: A goal of scientific argumentation is the use of data as evidence to defend a claim, conclusion, or explanation. This item implies that students were attempting to use evidence in their arguments. This should more than an opinion; they must include data. Statements like "that's what I think" or "it doesn't make sense" would result in a low score. Statements like "the data we found suggests that ..." or "our evidence indicates..." would result in a higher score.

Comments:

10. The participants examined the relevance, coherence, and sufficiency of the evidence.	**0**	**1**	**2**	**3**
	Not at all	Once or Twice	A few times	Often

Description: This item draws attention to the amount and kinds of evidence used to support a claim or explanation. Groups that attempt to (a) determine the value of a piece of evidence (e.g., "does that matter?"), (b) look at links or the relationship between multiple pieces of evidence (e.g., "This supports X and Y but this only supports X"), or (c) attempt to determine if there is enough evidence to support an idea (e.g., "We do not have any evidence to support that") would score higher on this item.

Comments:

11. The participants evaluated how the available data was interpreted or the method used to gather the data.	**0** Not at all	**1** Once or Twice	**2** A few times	**3** Often
Description: The evidence provided for a claim or explanation should be evaluated based on how well the data was gathered and interpreted. A question such as "Why is that evidence included?" or "How did they gather their data?" or "Where did that data come from?" indicates that the participants are assessing methods or an interpretation of data and would result in a higher score.				
Comments:				

12. The participants used scientific theories, laws, or models to support and challenge ideas or to help make sense of the phenomenon under investigation.	**0** Not at all	**1** Once or Twice	**2** A few times	**3** Often
Description: Science is theory-laden. In other words, scientists rely on broad, well-supported organizing ideas to frame their arguments and claims. Students should also employ these paradigmatic ideas in providing warrants for the evidence and claims they make or use to refute others' claims. Explicit reference to these "big ideas" will result in a higher score on this item.				
Comments:				

13. The participants made distinctions and connections between inferences and observations explicit to others.	**0** Not at all	**1** Once or Twice	**2** A few times	**3** Often
Description: The structure of scientific arguments includes evidence involving both empirical (such as quantitative measurements and systematic observations) and inferential (noting of trends and logical connections among observations) aspects. Making these distinctions and their connections explicit to others enhances the quality of the argumentation and thus results in a higher score.				
Comments:				

14. The participants used the language of science to communicate ideas.	**0** Not at all	**1** Once or Twice	**2** A few times	**3** Often
Description: This item stresses the importance of the accurate use of scientific language by students. The adoption and use appropriate terms (e.g., condensation, force, etc), phrases (e.g., "it supports" rather than "it proves") or ways of describing information is a characteristic of argumentation that is scientific. *Note:* Ideas may be explicated before being labeled with the correct terminology.				
Comments:				

SOCIAL ASPECTS OF SCIENTIFIC ARGUMENTATION

How the participants interact with each other

15. The participants were reflective about what they know and how they know.	**0** Not at all	**1** Once or Twice	**2** A few times	**3** Often
Description: It is important for members of the group to agree on what they know and to be specific about how they know. Statements such as, "do we all agree?" or "is there anything else we need to figure out?" or "can we be sure?" indicate that participants are monitoring their progress and have an end goal in mind.				
Comments:				

16. The participants respected what each other had to say.	**0** Not at all	**1** Once or Twice	**2** A few times	**3** Often
Description: Respecting what others have to say is more than listening politely or giving tacit agreement. Respect also indicates that what others had to say was actually heard and considered (e.g., "that is a good point", interesting idea", or "I hadn't thought of that"). A group that scored high on this would allow everyone to present their ideas and express their opinions without censure or ridicule.				
Comments:				

17. The participants discussed an idea when it was introduced into the conversation.	**0** Not at all	**1** Once or Twice	**2** A few times	**3** Often
Description: To be a participating and contributing member of the group, it is important to feel valued. Ideas and opinions need to be critically acknowledged. This means they are considered and given weight by the group. Groups that ignore ideas when they are proposed (results in the same idea being mentioned over and over) would earn a low score on this item.				
Comments:				

18. The participants encouraged or invited others to share or critique ideas.	**0** Not at all	**1** Once or Twice	**2** A few times	**3** Often
Description: Good argumentation comes from considering and comparing competing ideas from multiple individuals to construct the most robust explanation of the phenomenon under study. Groups that consist of individuals that invite others to share (e.g., "what do you think"), critique (e.g., "do you agree" or "it is ok to disagree with me"), or discuss an idea (e.g., "let's talk about this some more") would score higher that a group with an alienating leader that dominates the conversation and the work of the group.				

Comments:				
19. The participants restated or summarized comments and asked each other to clarify or elaborate on their comments.	**0**	**1**	**2**	**3**
	Not at all	Once or Twice	A few times	Often
Description: The depth of discussion will be enhanced by not making implicit judgments or assumptions about another person's ideas or views, and it demonstrates that their point of view is valued and is furthering the discussion. Communication provides students with opportunities to identify the strengths and weaknesses of their understanding.				
Comments:				

Total: /57

References

Abell, S. K., Anderson, G., & Chezem, J. (2000). Science as argument and explanation: Exploring concepts of sound in third grade. In J. Minstrell & E. H. Van Zee (Eds.), *Inquiry into inquiry learning and teaching in science* (pp. 100–119). Washington, DC: American Association for the Advancement of Science.

Alexopoulou, E., & Driver, R. (1996). Small-group discussion in physics: Peer interaction modes in pairs and fours. *Journal of Research in Science Teaching, 33*(10), 1099–1114.

American Educational Research Association, American Psychological Association, & National Council on Measurement in Education. (1999). *Standards for educational and psychological testing*. Washington, DC: American Psychological Association.

Andriessen, J., Baker, M., & Suthers, D. (2003). Argumentation, computer support, and the educational contexts of confronting cognitions. In J. Andriessen, M. Baker, & D. Suthers (Eds.), *Arguing to learn: Confronting cognitions in computer-supported collaborative learning environments* (pp. 1–25). Dordrecht: Kluwer.

Berland, L., & Reiser, B. (2009). Making sense of argumentation and explanation. *Science Education, 93*(1), 26–55.

Borg, W. R., & Gall, M. D. (1989). *Educational research: An introduction* (5th ed.). White Plains: Longman.

Boulter, C. J., & Gilbert, J. K. (1995). Argument and science education. In P. J. M. Costello & S. Mitchell (Eds.), *Competing and consensual voices: The theory and practices of argument* (pp. 84–98). Clevedon: Multilingual Matters.

Burns, R. B. (1994). *Introduction to research methods in education*. Chesire: Longman.

Carlsen, W. S. (2007). Language and science learning. In S. K. Abell & N. Lederman (Eds.), *Handbook of research on science education* (pp. 57–74). Mahwah, NJ: Lawrence Erlbaum Associates.

deVries, E., Lund, K., & Baker, M. (2002). Computer-mediated epistemic dialogue: Explanation and argumentation as vehicles for understanding scientific notions. *Journal of the Learning Sciences, 11*(1), 63–103.
Driver, R., Newton, P., & Osborne, J. (2000). Establishing the norms of scientific argumentation in classrooms. *Science Education, 84*(3), 287–313.
Duschl, R. (2007). Quality argumentation and epistemic criteria. In S. Erduran & M. Jimenez-Aleixandre (Eds.), *Argumentation in science education: Perspectives from classroom-based research* (pp. 159–175). Dordrecht: Springer Academic.
Duschl, R. (2008). Science education in three-part harmony: Balancing conceptual, epistemic, and social learning goals. *Review of Research in Education, 32*, 268–291.
Duschl, R., Schweingruber, H., & Shouse, A. (Eds.). (2007). *Taking science to school: Learning and teaching science in grades K-8*. Washington, DC: National Academies Press.
Duschl, R. A., & Osborne, J. (2002). Supporting and promoting argumentation discourse in science education. *Studies in Science Education, 38*, 39–72.
Edwards, A. (1957). *Techniques for attitude scale construction*. New York: Appleton-Century-Crofts.
Erduran, S. (2007). Methodological foundations in the study of argumentation in science classrooms. In S. Erduran & M. Jimenez-Aleixandre (Eds.), *Argumentation in science education: Perspectives from classroom-based research.* (pp. 47–69). Dordrecht: Springer Academic.
Erduran, S., & Jimenez-Aleixandre, M. (Eds.). (2007). *Argumentation in science education: Perspectives from classroom-based research*. Dordrecht: Springer Academic.
Erduran, S., Osborne, J., & Simon, S. (2004). The role of argument in developing scientific literacy. In K. Boersma, O. deJong, H. Eijkelhof, & M. Goedhart (Eds.), *Research and the quality of science education* (pp. 381–394). Dordrecht: Kluwer.
Erduran, S., Simon, S., & Osborne, J. (2004). Tapping into argumentation: Developments in the application of Toulmin's argument pattern for studying science discourse. *Science Education, 88*, 915–933.
Kelly, G. J., Druker, S., & Chen, C. (1998). Students' reasoning about electricity: Combining performance assessments with argumentation analysis. *International Journal of Science Education, 20*(7), 849–871.
Kuhn, D. (1993). Science as argument: Implications for teaching and learning scientific thinking. *Science Education, 77*(3), 319–337.
Kuhn, D., & Udell, W. (2003). The development of argument skills. *Child Development, 74*(5), 1245–1260.
Kuhn, L., & Reiser, B. (2005). *Students constructing and defending evidence-based scientific explanations.* Paper presented at the annual meeting of the National Association for Research in Science Teaching, Dallas, TX.
Landis, J., & Koch, G. (1977). The measurement of observer agreement for categorical data. *Biometrics, 33*, 159–174.
McNeill, K. L., Lizotte, D. J., Krajcik, J., & Marx, R. W. (2006). Supporting students' construction of scientific explanations by fading scaffolds in instructional materials. *Journal of the Learning Sciences, 15*(2), 153–191.
National Research Council. (1999). *Transforming undergraduate education in science, mathematics, engineering, and technology*. Washington, DC: National Academies Press.
National Research Council. (2005). *America's lab report: Investigations in high school science*. Washington, DC: National Academies Press.
National Research Council. (2008). *Ready, set, science: Putting research to work in K-8 science classrooms*. Washington, DC: National Academies Press.
National Science Foundation. (1996). *Shaping the future: Strategies for revitalizing undergraduate education.* Paper presented at the National Working Conference, Washington, DC.
Nunally, J. C. (1970). *Introduction to psychological measurement*. New York: McGraw-Hill.
Osborne, J., Erduran, S., & Simon, S. (2004). Enhancing the quality of argumentation in science classrooms. *Journal of Research in Science Teaching, 41*(10), 994–1020.

Richmond, G., & Striley, J. (1996). Making meaning in the classroom: Social processes in small-group discourse and scientific knowledge building. *Journal of Research in Science Teaching, 33*(8), 839–858.

Rubba, P. A., & Anderson, H. O. (1978). Development of an instrument to assess secondary school students: Understanding of the nature of scientific knowledge. *Science Education, 62*(4), 449–458.

Sampson, V., & Clark, D. (2008). Assessment of the ways students generate arguments in science education: Current perspectives and recommendations for future directions. *Science Education, 92*(3), 447–472.

Sampson, V., & Clark, D. (2009). The effect of collaboration on the outcomes of argumentation. *Science Education, 93*(3), 448–484.

Sampson, V., & Clark, D. (2011). A comparison of the collaborative scientific argumentation practices of two high and two low performing groups. *Research in Science Education, 41*(1), 63–97.

Sampson, V., & Gleim, L. (2009). Argument-driven inquiry to promote the understanding of important concepts and practices in biology. *American Biology Teacher, 71*(8), 471–477.

Sampson, V., & Grooms, J. (2010). Generate an argument: An instructional model. *The Science Teacher, 77*(5), 33–37.

Sampson, V., Grooms, J., & Walker, J. (2009). Argument-driven inquiry: A way to promote learning during laboratory activities. *The Science Teacher, 76*(7), 42–47.

Sandoval, W. A. (2003). Conceptual and epistemic aspects of students' scientific explanations. *Journal of the Learning Sciences, 12*(1), 5–51.

Sandoval, W. A., & Reiser, B. J. (2004). Explanation driven inquiry: Integrating conceptual and epistemic scaffolds for scientific inquiry. *Science Education, 88*(3), 345–372.

Simon, S., Erduran, S., & Osborne, J. (2006). Learning to teach argumentation: Research and development in the science classroom. *International Journal of Science Education, 28*(2&3), 235–260.

Suthers, D. (2006). Technology affordances for intersubjective meaning making: A research agenda for CSCL. *International Journal of Computer Supported Collaborative Learning, 1*(3), 315–337.

Trochim, W. M. (1999). *The research methods knowledge base* (2nd ed.). Cincinnati, OH: Atomic Dog.

Veerman, A. (2003). Constructive discussions through electronic dialogue. In J. Andriessen, M. Baker, & D. Suthers (Eds.), *Arguing to learn: Confronting cognitions in computer-supported collaborative learning environments* (pp. 117–143). The Netherlands: Kluwer.

Vellom, R. P., & Anderson, C. W. (1999). Reasoning about data in middle school science. *Journal of Research in Science teaching, 36*(2), 179–199.

Zeidler, D., L. (1997). The central role of fallacious thinking in science education. *Science Education, 81*, 483–496.

Chapter 13
Beyond Argumentation: Sense-Making Discourse in the Science Classroom

Scott P. McDonald and Gregory J. Kelly

Introduction

Science classrooms are complex ecosystems of norms and practices that develop over the life of a classroom community. Since the late 1980s, with the emergence of learning theories focused on communities and their practices, there has been an increasing emphasis on creating activities in science classrooms that better reflect the practices of the science community. The idea that elementary and secondary science students should be acculturated into the practices of science as part of learning the content of science is now *de rigueur* in our field. However, the specific way this is instantiated or characterized differs from approach to approach. For example, inquiry as characterized in the National Research Council (2000) is one way that processes and practices of science are described. The essential features of inquiry (Table 2.6, p. 29) include attention to evidence, ability to communicate scientific ideas, and engagement with scientific questions. The emergence of the nature of science as a seminal part of science instruction—either explicitly or implicitly—is another form of characterizing the community practices of scientists. One component that these different perspectives agree upon is that evidence and argument are central to the practices of science. Over the past decade, a strong vein of research in science education has emerged focused on the scientific argumentation practices of students. Initially building on the work of philosophers, in particular Toulmin (1958) and his description of argument, there has been a sharp increase in attempts to both characterize the nature of argument constructed by students as well as develop curricular and pedagogical supports for students' scientific argumentation.

In this chapter, we suggest that—while important—an increasingly specific focus on argumentation in student discourse has significant limitations in terms of supporting student learning, developing students' understandings of the way scientists practice within their community, and supporting the development of productive

S.P. McDonald (✉)
The Pennsylvania State University, University Park, PA, USA
e-mail: smcdonald@psu.edu

M.S. Khine (ed.), *Perspectives on Scientific Argumentation*,
DOI 10.1007/978-94-007-2470-9_13, © Springer Science+Business Media B.V. 2012

norms and practices in communities of science learning. Research in this area shows signs of increasingly specified and calcified definitions of argument and how discourse in science classrooms is analyzed. There are both social justice and equity concerns around a narrow pedagogical focus on one type of student discourse and significant analytical limitations in terms of understanding the quality and productivity of students' classroom science discourse. We begin by providing examples of classroom science discourse as a context for our comments. Following that, we briefly characterize the trends in science education research, including some of the work on productive discourse done outside the specific line of argumentation research. We close by suggesting scientific sense making as a broader perspective on science discourse practices that would be more productive both to support science teaching and learning as well as science education research.

Sense Making in Classroom Discourse, Part 1

The following short episode illustrates not only the complexity of classroom discourse—showing the potential of viewing reasoning from an argumentation point of view—but also the confusing nature of real-time interpretation of discourse in a sense-making situation. In this case, a chemistry class is discussing mole ratios and is attempting to make sense of the reaction of zinc and sulfur. Prior to this conversation, students had been working on an extended analogy over multiple days where they were making cookies using different ingredients to get a sense of ratios and recipes. This recipe making is then connected to the mole concept as a unit of measure that describes the recipes of chemistry. The students have been given an amount of zinc (6.54 g) and are asked to determine how much sulfur they need for a reaction. Different groups have come up with different solutions and reasoning and the groups are debriefing as a class to determine a solution to test. This selection of transcript comes at the point the students are working in groups evaluating the merits of at least two possible solutions that have been suggested by the class. As the students discuss the two solutions, the teacher enters to point out aspects of the argument in an effort to guide the students to her preferred response—or at least direct them toward certain evidence.

1. Teacher: What do you think? [asking students in one group about another group's idea]
2. Alex: I don't think it [the idea presented] will work.
3. Teacher: No? Who . . . who's do you like better—this one or the other one?
4. Colin [pointing with pen over his shoulder at group A]: That one.
5. Alex: Theirs [also indicating group A] made more sense.
6. Bonnie: Like the one thing bothering me was like the one box over there [referring to sulfur on the Periodic Table] . . . like, when you move to the right . . .

7. Teacher: Wait! Wait! What? Wha- . . . what? Danielle . . .[Danielle, a member of group A, turns around]. Question here . . . for you. [To Bonnie]: Ask [Danielle] that.

At this point in the discussion, the teacher recognized that for one student group to be able to understand a key aspect of reasoning in chemistry, they would have to understand relevant knowledge already established by another student. The teacher sought to enter this into the student groups' conversation by directing Bonnie to ask Danielle the question about the relative position of elements in the periodic table (turn 7), knowing that this would provide key information about the relative reactivity of the elements.

8. Bonnie: Oh. Why'd you move 16 blocks over? What made you decide to do that?
9. Danielle: Because sulfur [points left index finger toward Periodic Table] is more reactive [makes gesture where she moves right index finger slightly up then to the right]. They're more reactive as you go over. . . and up.
10. Bonnie: Oh. Okay. [Said with some uncertainty]
11. Danielle: [Turns around to engage with her group]
12. [Bonnie along with her group, Alex and Earl look at the Periodic Table] Although Bonnie's question is "answered" by Danielle, Bonnie showed some uncertainty and it is unclear whether this exchange has forwarded her thinking. Nevertheless, she focused on her group, only to have Earl again seek clarification from Danielle:
13. Earl: Wait. "Over and up"?
14. Alex: Yeah—what do you mean by "over and up"?
15. Bonnie: Like, sulfur's more reactive than zinc.

Bonnie translated Danielle's "up and over" interpretation in terms of chemical reactivity—knowledge that is potentially relevant to solving the problem at hand. Alex was not satisfied with this explanation and again sought Danielle:

16. Alex: What do you mean by "over and up"?
17. Danielle: Like, reactivity increases to the right and up.
18. Alex: Yeah.
19. Fiona: Francium to fluorine [points to the Periodic Table].
20. Alex: Yeah, so did you go from, like, where potassium is over to zinc and then past zinc to aluminum and then to sulfur?
21. Danielle: Over and up.
22. Alex: Okay.

This ended the second exchange regarding how to interpret the periodic table. The talk suggest that the two students, Alex and Danielle, were seeking ways of understanding each other and Danielle's initial interpretation of how translations on the periodic table relate to chemical reactivity.

23. [Both Alex and Danielle turn back toward respective groups]
24. Earl [talking to Bonnie]: I don't even . . . I don't know why they did that.
25. Bonnie: I don't . . . yeah.
26. Alex [after looking at the Periodic Table]: Because when they were explaining it, they said they went twelve over [to Zn] and then, that was it. And then they went from sulfur . . .
27. Bonnie: Yeah [expression of not believing it].
28. Alex: I don't understand it.
29. Colin: I don't understand . . . I don't think they understand it either.

The group that includes Earl, Bonnie, Alex, and Colin do not reach the conclusion suggested and seemingly understood by the other group, as suggested by Danielle's use of the periodic table. The argument was incomplete and not persuasive for this group. There are a number of interesting aspects to looking at this transcript from the point of view of argumentation. First, we see that the nature of actual talk rarely resembles even the most informal arguments—people while engaged in conversational cooperation (Gumperz, Cook-Gumperz, & Szymanski, 1999) rarely make explicit all the details needed for a tight philosophical argument (Kelly, Druker, & Chen, 1998). Furthermore, in this case, the speakers are uncertain and still in the process of attempting sense making. Second, while we could identify aspects of the argument that are missing, there is more going on than just making and understanding an argument or set of arguments. There are two possible solutions put to the groups. The groups have differing interpretations about the respective merits of the solutions when the teacher directs one member of one group to pose a question to another. Group members have a history, with recognition and reputations for knowledge and problem-solving ability. Furthermore, as the teacher marked one student as possessing relevant knowledge, this differentiated the students' interpretation and added status to the knowledge of Danielle and her problem-solving path. Third, the teacher is doing discursive work to set up the arguments in particular kinds of ways. She set the two solutions (although they were generated by the students), sent the student groups to work, and favored certain ideas as they emerged and marked them as significant. Some arguments are favored and achieve status by the ways the teacher positions students in the class. The respective merits of the substantive argument, both in terms of the quality of its reasoning and evidence and also its normative correctness, is only part of the story; such arguments occur in the flux of everyday life where substantive evidence does not stand alone, but is rather talked into being by how such evidence is accomplished and made significant to the speakers through interaction. It is critical to understand that seen through the lens of argumentation there is very little quality discourse occurring here, but viewed through the lens of scientific sense making there is a great deal of interesting intellectual work being done.

Still Making Sense? Classroom Discourse, Part 2

The following section of the transcript continues as the teacher is bringing the class to closure and is attempting to connect a number of student ideas together to create an understanding of the core mole concept. She begins by returning to Danielle (and her groups') solution to the initial question of how much sulfur they need to react with zinc. The teacher begins with the ratio of one-to-one that, at this point, is not focused on atoms, but is rather attending to the mass of the macro elements. The teacher is focused both on the mathematics of the ratio concept, in terms of determining if one-tenth to one-tenth is the same as one-to-one, as well as this issue of mass or atoms being the object of the ratio in chemical reactions.

30. Teacher: Okay, the key came from the atomic mass—just like Danielle said. One-tenth the mass of zinc reacted with one-tenth the mass of sulfur. Did you guys think [points to group B] that anyone else's plan tied in with yours?
31. Anthony: {Inaudible} [shrugs shoulders].
32. Bella: [Laughs at Anthony]
33. Teacher: Or could?
34. Bella: Theirs [nodding head toward group A].
35. Teacher: Whose?
36. Bella: [Tilting her head toward group A]
37. Teacher: Theirs [pointing to group A]?
38. Bella: Yeah.
39. Cassandra: [Shakes head in agreement]

In this example, the teacher was seeking to generate a conversation where multiple ideas can be considered by the students in the class (turn 30)—i.e., a dialogic conversation following Mortimer & Scott (2003). The teacher has made choices to set up a comparison across the groups' solutions. This may lead to an explicit comparison of evidence, and thus a form of persuasive discourse or argument. The discourse form in this case does not readily match well with analytics for considering argumentation. The conversation is heavily reliant on gesture and indexicality (turns 30–39). The conversation is complicated by the choice made by the students to choose Group A for a comparison of respective plans—from a normative science point of view another group's idea was closer to theirs both in terms of the numerical value and the nature of the explanation, a point the teacher may have realized in situ.

40. Teacher [to group A]: What was your idea?
41. Devon: One-to-one.
42. Teacher: One-to-one.
43. Earnest: Balanced equation.
44. Teacher: Okay, one-to-one. Wait, wait—what's that [on the white board] say?
45. Devon [pointing to white board]: Which was {inaudible, but does look at group A}. One-tenth to one-tenth.

46. Teacher: One-tenth to one-tenth. Is that the same as one-to-one [holding both hands up with palms facing]?
47. Devon: Mmhh.
48. Martha: Yeah.
    The teacher elicited (line 40) and received a seemingly approved response (line 41), yet chose to reflect the question in a different form back to the students.
49. Teacher: Yeah it is, right. Interesting. Okay, so. . . they're say::ing . . . oh, here's [gestures toward white board] my question. So, why is it one-tenth for one-tenth? And they [group A] have an idea, right. Hmmm. Why one-tenth for one-tenth? Anthony, what's their [group A] idea?
50. Anthony: One-to-one ratio.
51. Teacher: From where?
52. Anthony [2 sec. elapse]: It . . . {inaudible} . . .[Request for re-explanation]
53. Teacher: From where. . .Fanny?
54. Fanny: What was the question? I'm sorry I {inaudible}.
55. Teacher: Your idea—where'd that come from—the one-to-one idea?
56. Fanny [shaking head]: Yeah.
57. Teacher: Yeah, where'd that come from?
58. Fanny: The chemical equations.

In this sequence, the teacher seeks responses from the students regarding the idea of ratio, and especially the chemical equation (lines 58). The students had already identified the one-tenth to one-tenth ratio, but had not used the specific word "ratio." In this section of discourse, the teacher is also trying to build to the idea that this particular ratio can be derived from the chemical equation.

59. Teacher: Equation. So, in the chemical equation for the "Question of the Day"—which you have written down, you can look right at it—it says one-for-one-for-one, right? Yeah [shakes her head]—you know what I mean? Let's look at it. So, it says, "zinc plus sulfur give zinc sulfide" [writes "$Zn + S \rightarrow ZnS$"] so, my ratio here is one-to-one-to-one, because my coefficients out here, right, are one [pointing to Zn], one [pointing to S], one [pointing to ZnS]. See how both those ideas can tie together. So, if your coefficients weren't one-to-one would it still be one-tenth for one-tenth? [1 sec. pause] What do you think? Cassandra [points to her], what do you think?
60. Cassandra: {No audible comment for a couple of seconds, may have been gesture of uncertainty}
61. Teacher: Not sure. Cassandra's saying, "No." Why not?
62. Cassandra [after 1 sec. pause]: I don't know.
63. [A couple of students laugh]
64. Teacher: Not sure? You're right, though, it wouldn't be. Okay, if you had a different ratio in the balanced equation, like one-to-two, you would get one-tenths for two-tenths. Okay. So, both groups actually had really good ideas that tied together. Also, the idea of reactivity isn't totally off, right, because [points to A] does theirs make sense?

The class proceeds toward finalization of the solution; nevertheless, this section of discourse is adequate for the purposes of our discussion. The teacher is attempting to help students with the process of sense making as well as build a consensus by combining ideas and showing how they can complement and build on each other. She is attempting to build connections across a set of concepts (mathematic ratios, chemical equations, nature of how elements react), but at this point has omitted the difference between the mass the students can physically measure and how the respective masses relate to the mole ratios. A great deal of the discourse here is focused on the mathematical ratios, while the science concept that it underlies is in the background as the students discuss the nature of ratios. The students are also grappling with connecting multiple representations, including the equation for the reaction on the smart board, the molar masses on the periodic table, and the discourses including gestures from the teacher and their peers. In all this complexity, there is little that can be directly pointed to as argumentation, especially final form of scientific argumentation (claims, evidence, warrants, etc.); however, again there is a great deal of productive science talks occurring. Given these two examples—as context for our discussion—we now turn to the contrast between sense-making argumentation as ways of framing discourse in science classrooms.

Science Discourse and Practice

Everyday life in classrooms is accomplished through language and associated social processes. As members of a group affiliate over time and build ways of being talking, and acting, common norms and expectations are constructed, contested, and redefined. Groups make sense of their reality through communication and social actions. Thus, ways of aligning evidence in argument emerge from broader ways of being and sense making in a group that are constructed through social interaction (Kelly & Green, 1998). Importantly, the norms created in classrooms provide intellectual space, and potential academic identities, for members of the classroom to draw upon as they engage and participate in group actions. Similarly, in science contexts, opportunities to engage in research and other relevant practices are constructed through social interaction. The cultural aspects of scientific practice have been well documented through the empirical study of scientific communities in such fields as sociology, anthropology, and rhetoric of science (Kelly & Chen, 1999).

Argumentation in science fields is the product of both genre conventions, constructed over extensive time periods (Bazerman, 1988), and one of a set of everyday actions constructed in the moment in particular contexts (Collins, 1985). A scientific argument in a professional journal or presented to colleagues at a research conference represents only one of the many genres and ways of speaking and writing in science. Drawing from the social studies of scientific practice, particularly the anthropology of science (Knorr-Cetina, 1999; Latour, 1987; Traweek, 1988), we argue that the final form argumentation, with explicitly stated evidence tailored to a professional audience, is just one discourse of science and is a particularly structured and formal genre of science discourse. Thus, there are many types of discourse and ways of using language needed to accomplish the work of science. Such discourses vary across audiences (e.g., at the laboratory bench, among colleagues during a discovery, to editors of a professional journal, for a press release regarding new findings), purposes (e.g., thinking aloud to solve a problem, persuading a colleague, defending published data), and venues (e.g., PowerPoint to laboratory group, email to collaborator, discussion with a student). Thus, while educational reform has called for emulation of scientific practice in educational settings, there has not been careful analysis of the range and typicality of such practices. Often, educational reform is based on a set of assumptions about scientific practice with little empirical evidence to substantiate the assumed normative goals. The pattern of thinking of science as the scientific method and teaching it as a final form process is an example that was later problematized by the social studies of science. Argumentation provides an example of how bringing scientific practices to education offers potential for new forms of learning, but such practices need to be considered in a broader context of discourse and practices if they are not to become another formulaic and largely empty characterization of science practices.

The rationale for the focus on argumentation as a pedagogical goal in science classrooms, and often as a measure of students' abilities to engage in science, is grounded in three central premises. First, argumentation offers opportunities to engage students in seemingly authentic scientific practices. The focus on engaging students in discourse practices is tied to research identifying the importance of students "talking science" and learning the genres of such discourse through participation (Kelly, 2010; Kelly & Crawford, 1997; Lemke, 1990; Roth, 2005). Second, argumentation may offer ways for students to learn the knowledge of a given discipline in a more thorough and deeply conceptual way. Student learning of scientific concepts poses a challenge for educators, as often even after instruction, central scientific concepts are not well understood. Argumentation is seen as a pattern of science discourse that leads to sense making and thus deeper conceptual understanding. Third, argumentation is often seen as a means to teach about the nature of science as a discipline. Engaging students in making evidence-based claims may foster such understandings (Kelly, 2008). We now discuss the problems embedded in each of these three premises and then discuss the limitation of a focus on argumentation both pedagogically and analytically.

Nature of Authenticity

Argumentation is a practice in science and therefore engaging science students in argumentative practices may be viewed as a reasonable method for engaging students in authentic practices of science. We generally agree with the fact that argumentation offers ways to engage students in authentic science; however, we do so with a number of caveats. What counts as an authentic process raises questions both about the actual scientific practice in question and also the educational wisdom of authenticity from a pedagogical point of view. As we have argued, scientists engage in a range of discourses to accomplish scientific work. Some of these discourses include marshaling evidence. Nevertheless, there are other ways of communicating, building affiliation, and interacting where explicit statements of claims supported by evidence are not part of the discourse. Focusing on the more formal types on discourse of a community can lead to the discourse being reduced to a list or heuristic as happened with five paragraph essay in English (another attempt to formally structure a dynamic form of argument) or the steps of the scientific method.

The other imbedded assumption in this premise that warrants examination concerns the nature of authenticity in classroom practice. It is not necessarily the case that making the discourse practices of a science classroom more like the discourse of practicing scientists leads to better science learning environments as the two sets of practices have different purposes and contextual constraints (Kelly & Brown, 2003). Authentic learning contexts may require ways of speaking, listening, writing, and so forth, that are central for learning, but not related to the work of producing new knowledge beyond a limited and local audience (McDonald & Songer, 2008).

Argumentation, spoken or written, occurs with and by real people, in situated contexts, with real and intended audiences. Thus, while attention to the substantive aspects of evidence use gives us some insights into the uses of discourse in science contexts, there will be other dimensions of language use, for other purposes, including, but not limited to, taking social positions, building alliances, saving face, and so forth. Ryu and Sandoval (2008) indicated that the amount of normative argumentation students in small groups engaged in varied across groups based on student ability. Students spent discourse time sorting through the tasks of group work or finishing the task without disagreement and had few substantive disagreements about the science where normative arguments occurred. The fact that the focus of much of the discourse was not argumentation does not mean it was not pedagogically valuable; it points to the fact that argumentation occurs with and by real people, in situated contexts, with the real and intended audiences. People are living, thinking beings, with multiple goals, many of which have nothing to do with the cognitive aspects of creating an argument, even if they need to go through the motions to accomplish tasks in a classroom.

The complexity of goals that occur in real contexts means that part of the conversation about authenticity in science classrooms must take into account that school science talk is not the same as science talk. There have been many discussions of the idea that students should be apprenticed in or acculturated to a community of science practitioners. However, this is problematic, as the classroom teacher is not an authentic participant in the community of science practitioners nor are schools context designed to produce new scientific knowledge. Students are being acculturated into classroom science learning practices. This means that it is not only actually impossible to construct authentic science classroom practices in this way, but also that it is likely not desirable, as the practices that best help students to understand science are not identical to the practices that help scientist develop new scientific understandings.

Arguing to Learn

Learning how to align evidence through verbal, written, and symbolic representation may support student learning of scientific concepts. Science learning clearly needs to include more than the "final form" science of known theory and facts (Duschl, 1990). Learning science concepts should include understanding the evidentiary basis for how concepts were derived and how and why they are used to understand the natural world. For example, knowing there is a theory of plate tectonics, and even some of the key characteristics of this theory, does not necessarily entail understanding. As noted by Duschl (2008), a thorough understanding of the theory includes knowing the conceptual, epistemic, and social dimensions of the theory. Knowing the evidentiary bases for the theory, and ways that it can be applied in a variety of context, includes understanding what counts as a good argument in the relevant field. Thus, argumentation may serve a role in this type of learning, but nevertheless other learning goals and means are needed to scaffold student learning.

Argumentation can be part of a discourse-rich learning environment supporting student understanding. There is some evidence that students' conceptual understandings in science can be deepened and enriched via argumentation, although some level of experience and knowledge of the content seems critical (von Aufschnaiter et al., 2008). Much of the research around the impact of argument as a pedagogical tool has occurred in contexts where argumentation was an explicit structure of the activity. Sampson and Clark (2011), for example, asked students to evaluate different explanations of a discrepant event and then examined the quality of their written arguments. They found differences in the way that students argued based on their ability groupings and indicated more about the nature of different types of argumentation, rather than the degree to which it develops content understanding. There is also some evidence that the role of questions between peers in the context of argumentative discourse is critical to learning (Chin & Osbourne, 2010). The explicit inclusion of other discourse practices (questioning) as part of the claims/evidence/reasoning notion argumentation literature indicates there is complexity to the pedagogical enactment of argumentation as a support for learning. Such complexity is contingent on a large number of contextual factors, such as the participants' view of the purpose of the tasks, the participants' personal and interpersonal goals, the group dynamics (for the case of small group discourse), the real and intended audience, the extant knowledge drawn into the conversation, and the established norms for speaking, listening, and interacting. It seems to follow then that arguing to learn can only happen in the larger context of science sense-making discourse that occurs around and within the arguments.

Nature of Science

One more explicit description of the understandings of the practices and norms of science as a community has become characterized and studied as the nature of science. A number of scholars have argued that understanding the nature of science needs to include some experiential components (Duschl, 2008; Kelly, 2008; Sandoval & Millwood, 2005). Argumentation may be a means for students to understand the practical, conceptual, and epistemic nature of scientific practice (Duschl, 2008). Such engagement in one of the discourses of science (i.e., argumentation) will not be sufficient to develop students' views of the complexity of the nature of science. Furthermore, not all aspects of science are made evident in argumentative practices. For example, learning how to observe particular features within an observable field of vision is an important aspect to participation in science (Kelly, 2010). Coming to observe an instance of a phenomenon as seeing as a particular feature often requires more knowing others with relevant knowledge and experience making the phenomenon witnessable and recognizable to a novice observer (Goodwin, 1994).

While such forms of learning to observe require discourse, the form may not involve argumentation. Rather, rendering the phenomenon witnessable requires other forms of discourse. Through such participation, novices or students may learn how to observe from a disciplinary point of view and thus learn both aspects of the nature of science and how to establish facts to be subsequently used in arguments. It is clear that the pedagogical work of the teachers' discourse involves orders of complexity of fostering epistemic practices related to argumentation. Jiménez-Aleixandre and Reigosa (2006) indicated that the epistemic operations of a teacher include three distinct referential levels: specific examples, a class of referents, and abstract referents and exemplifies the complexity of using discourse with students as a method for developing their understanding of the nature of science, which is itself a form of discourse. Thus, to view learning the nature of science, not as stipulative definitions, but as engagement in scientific practices, entails understanding the ways that argument and other discourse forms contribute to such participation. For example, learning what counts as a "good" or "acceptable" observation may entail sense making through gesture and other representational forms among members of a epistemic community.

We now return to the examples from the chemistry classroom to consider how argumentation may contribute to our understanding of the events, and if taken as the primary lens for viewing classroom discourse, impose constraints on the ways sense is made for participants.

Return to Sense-Making Examples

Returning to the sections of classroom discourse above, we want to examine the discourse from the point of view of the argumentation that occurs. The chemical reaction and its equation are as simple as it can be: $Zn + S \rightarrow ZnS$. Yet, reviewing the transcript, the discourse cannot be easily reconstructed into an argumentative pattern. There is much discussed, but the teacher's central claim must be discerned through the clutter of naturalistic talk and action—both for us as analysts and presumably by the students in the class. Furthermore, the evidence in this case is not empirical; indeed, the teacher, drawing from a cooking analogy, is trying to set up a thought experiment so that the students first predict the most efficient ratios for the reactants, and from this lead them to devise an experiment, consider the consequences, and subsequently redesign the experiment to adjust the reactants' respective masses to adhere to the inferred chemical principle of mole ratios.

At this point, the best we can do to reconstruct the main argument is as follows:

Claim: Zn and S react in an atomic ratio of one-to-one.
Evidentiary support: the chemical equation $Zn + S \rightarrow ZnS$
Warrant: chemical equations are expressions of chemical reaction in molar ratios

Yet, in the flux of the actual discourse, and with the absence of a clear distinction between the mass and molar mass (at least at this point in the conversation), the sought conclusion is not at all obvious to the student, Casssandra, and perhaps others. The reading of the chemical equation 1 atom plus 1 atom yields one molecule and the mathematical identity of the one-tenth-to-one-tenth ratio to the one-to-one ratio seem to be the ostensive goals of the extended discussion. The teacher sought participation, asked students to explain, set up a comparison across groups, and reiterated some conclusions, and yet the pedagogical goal remained elusive. The confusion on the student's part is not an indictment of argumentation as a process—the teacher could have framed the argument more effectively and so forth. Rather, the case shows that the elaborate plan, complete with the cooking analogy, the potential for multiple experiments, and revisiting of the main ideas, is constructed both (a) in ways more complex than can be readily captured through argumentation analysis and (b) through discursive work that includes social and expressive functions of language that cannot be characterized through argumentation analysis.

It is impossible to tell if the discourse in class would have been more productive if the teacher had focused more explicitly on argument components such as claims, evidence, and warrants. The students were proposing solutions to the overall questions, the amount of sulfur needed for the reaction. These proposed solutions could be taken as claims and the teacher could have pressed for evidence and reasoning around those claims. However, the discourse that is present in these excerpts shows productive sense-making activity where both the teacher and the students are engaged with connecting their proposals with each other and with the target concepts of the phenomenon. Both pedagogically and analytically there are significant advantages to viewing this classroom activity as an engagement in science sense making rather than in argumentation.

Conclusion

By contextualizing our ideas in examples of classroom practice, we hoped to show that viewing science classrooms from an argumentation point of view can be both beneficial (in seeking a method to consider how participants use evidence) and a limitation for understanding the propositional, social, and expressive functions of language use (Cazden, 2001). While much can be said about the uses of argumentation to consider how evidence gets talked and written in science learning environments, we focused on some of the limitations of an argumentation framework, limitations that are particularly acute when argumentation is not considered in the broader contexts of everyday discourse. We conclude this chapter by discussing the analytic, pedagogic, and equity limitations present in an argumentation framework.

First, there are analytical limitations to any argumentation framework. Argumentation, as a presupposed normal goal, does not readily occur in many perfectly successful conversations, even conversations around science ideas. Conversation cooperation often entails assumptions about common understandings that go unstated—this is an efficient way to speak—and has been shown empirically in studies of classroom discourse. We will not always see argumentation in everyday life, even in science contexts of various sorts, even when evidence is readily available (Kelly et al. 1998). Even when argumentation, or at least evidence use, is employed by speakers, the methodological challenges include understanding the norms for interaction—typically constructed outside the substance of evidence use itself—within the community in question. We have argued that science includes many discourses, including importantly, sense-making conversation where persuasion is not the goal of the interaction.

Another way to consider the idea of sense making discourse is in terms of Wittgenstein's (1969) notion of language games or in terms of Gee's (2010) notion of D/discourse. While argumentation is one of the language games of science, it is not the only one, and is not even the most common one, it is simply the one that is most analytically accessible as we have given it a formal structure in terms of claim, evidence, warrant, and rebuttal (and associated uses with the rules of the language games). The discourses that construct the practices of science and learning science are many fold. Narrowing the focus to one aspect of the discourse, in large part due to its analytical accessibility, can lead to missing the forest for the trees. When arguments are used, the norms and expectations for conversation need to be considered and recognized so that attempts at persuasion are not viewed merely from the substantive content of the argument, as questions about what counts as evidence, explanation, a reasonable expectation for inferences from the audience, and other genre conventions frame how arguments can be understood by interlocutors and analysts alike.

Second, there are important pedagogic limitations of an argumentation approach. The field of science education has moved from using argumentation analysis as a research method to assess evidence to using argumentation both as a tool for analysis as well as a tool to support students' uses of evidence through instruction. This is an important development and many interesting and innovative techniques are being developed (see e.g., Erduran & Jiménez-Aleixandre, 2008). While we recognize these developments as productive with much potential, we caution that the views of classroom discourse do not become too narrow. Instruction includes more than marshaling evidence for certain concepts. Much like concept change theory that became routinized to the chagrin of some of the founders (Strike & Posner,

1992), pedagogy drawing on argumentation should be wary about becoming too mechanized.

The multifarious using of language and other symbolic systems in science alluded to earlier provide a model of open, diverse uses of language to support the multiple goals of the relevant social group. As the field develops better instructional design and research analytics, we look for the development considerations of how norms for evidence use and interpersonal communication developed over time. Argumentation may provide excellent ways to achieve cognitive, epistemic, and communicative goals called for in science education reform (Duschl, 2008). We recognize that the trifold cognitive, epistemic, and communicative goals can move instruction from a focus on achieving only the normative conceptual understanding to broader understandings of knowledge and practices of science. Argumentation is one tool that can advance pedagogy in this manner, but researchers need to examine both the supports and constraints imposed by argumentation.

One final pedagogical concern is that argumentation will become calcified in an effort to turn it into a tool for support student learning. Just as happened with the scientific method or the five-paragraph essay, there is a risk that turning analytical descriptions of argument such as claim, evidence, or warrant turns them into an empty form. Science educators have spent decades railing against the idea that something as complex and nuanced as the development of new knowledge in science could be characterized in a linear stepwise process of moving from question to conclusion. In fact, the focus on argumentation is largely the result of a focus on inquiry and other attempts to make the norms and practices of science more authentic. It would be ironic to have argumentation become the snake eating its own tail by turning classroom science discourse into a linear stepwise process of building an argument.

Finally, there are equity concerns derived from pedagogic and analytic uses of argumentation. We suggest three possible equity concerns. First, research regarding language and student identity has shown that the ways of using language in science is potentially alienating for at least some students (Brown, 2004; Carlone, 2004). Students may have ways of talking at home and in other contexts that offer opportunities to make a case, but do not adhere to the narrow requirements of what might count as a good argument in science in certain contexts. While argumentation poses the potential to expand the students' repertoire of ways of speaking and listening, it may also limit participation or differentially favor students whose everyday discourse align more closely with that of science or science teaching.

Second, there may be important gender differences about the ways students choose to engage in evidence use and assessment. Argument has a vernacular meaning suggesting disagreement and possibly consternation. Furthermore, framing evidence as a contest of better arguments may enter competition that leads less to the best analysis of facts and theory and more toward producing winners and losers. Students—particularly some female students—may find such competition less attractive than the science itself, and thus lose interest that could be otherwise fostered. In such a case, the school science practice of introducing argumentation itself could be alienating to students with interests in science.

Third, argumentation has been formulated in particular sorts of ways in science education. These forms of argumentation are framed around substantive, but relatively formal, ways of aligning evidence. There may be other ways of making sense of evidence, such as through informal reasoning and everyday reasoning that are not being considered due to the focus on argumentation. Thus, our current forms of analysis of evidence may have implicit biases derived extant argumentation theory. Maintaining an interest and analytic focus on uses of evidence broadly construed, and across multiple contexts in learners' lives, would help identify the many ways that everyday reasoning contributes to understanding.

Final Thoughts

Our argument has been that a focus on argumentation offers some potentially new and exciting ways to engage students in scientific practices. We have suggested that normative goals for science education, such as understanding concepts and developing the ability to use and assess evidence, can be enhanced through the lens of argumentation. We have made this argument with the caveat that such argumentation must be understood as one of a range of plausibly useful science discourses, and one of the many discourses of school science. What counts as evidence is often determined only after the many heterogeneous, confused, and incomplete conversations around a topic. The final form science appearing in known theories is often the work of many people over many years, occurring in ways adhering to the genres and social practices similarly constructed over time. Furthermore, learning to communicate in a highly technical genre of this sort is difficult work in any context where questions about what counts as evidence, theory, explanation, and so forth are as much as stake as the putative claim in question. These questions about what counts offer opportunities for learning and need to be part of science instruction.

References

Bazerman, C. (1988). *Shaping written knowledge: The genre and activity of the experimental article in science.* Madison, WI: University of Wisconsin Press.

Brown, B. A. (2004), Discursive identity: Assimilation into the culture of science and its implications for minority students. *Journal of Research in Science Teaching, 41*, 810–834.

Carlone, H. B. (2004), The cultural production of science in reform-based physics: Girls' access, participation, and resistance. *Journal of Research in Science Teaching, 41*, 392–414.

Cazden, C. B. (2001). *Classroom discourse: The language of teaching and learning* (2nd ed.). Portsmouth, NH: Heinemann.

Chin, C., & Osborne, J. (2010). Supporting argumentation through students' questions: Case studies in science classrooms. *Journal of the Learning Sciences, 19*(2), 230–284. doi:10.1080/10508400903530036

Collins, H. M. (1985). *Changing order: Replication and induction in scientific practice*. London: Sage.

Duschl, R. A. (1990). *Restructuring science education: The importance of theories and their development.* New York: Teacher's College Press.

Duschl, R. A. (2008). Science education in three-part harmony: Balancing conceptual, epistemic, and social learning goals. *Review of Research in Education, 32*, 268–291.

Erduran, S., & Jiménez-Aleixandre, M. P. (Eds.). (2008). *Argumentation in science education: Recent developments and future directions*. New York: Springer.

Gee, J. P. (2010). *An introduction to discourse analysis: Theory and method.* Routledge, New York, NY: Psychology Press.

Goodwin. (1994). Professional vision. *American Anthropologist, 96*(3), 606–663.

Gumperz, J. J., Cook-Gumperz, J., & Szymanski, M. H. (1999). *Collaborative practices in bilingual cooperative learning classrooms.* Santa Cruz, CA: Center for Research on Education, Diversity, and Excellence.

Jiménez-Aleixandre, M. P., & Reigosa, C. (2006). Contextualizing practices across epistemic levels in the chemistry laboratory. *Science Education, 90*, 707–733.

Kelly, G. J. (2008). Inquiry, activity, and epistemic practice. In R. Duschl & R. Grandy (Eds.), *Teaching scientific inquiry: Recommendations for research and implementation* (pp. 99–117; 288–291). Rotterdam: Sense Publishers.

Kelly, G. J. (2010). Scientific literacy, discourse, and epistemic practices. In C. Linder, L. Östman, D. A. Roberts, P. Wickman, G. Erikson, & A. McKinnon (Eds.), *Exploring the landscape of scientific literacy* (pp. 61–73). New York: Routledge.

Kelly, G. J., & Brown, C. M. (2003). Communicative demands of learning science through technological design: Third grade students' construction of solar energy devices. *Linguistics & Education, 13*(4), 483–532.

Kelly, G. J., & Chen, C. (1999). The sound of music: Constructing science as sociocultural practices through oral and written discourse. *Journal of Research in Science Teaching, 36*, 883–915.

Kelly, G. J., & Crawford, T. (1997). An ethnographic investigation of the discourse processes of school science. *Science Education, 81*(5), 533–559.

Kelly, G. J., Druker, S., & Chen, C. (1998). Students' reasoning about electricity: Combining performance assessments with argumentation analysis. *International Journal of Science Education*, 20(7), 849–871.

Kelly, G. J., & Green, J. (1998). The social nature of knowing: Toward a sociocultural perspective on conceptual change and knowledge construction. In B. Guzzetti & C. Hynd (Eds.), *Perspectives on conceptual change: Multiple ways to understand knowing and learning in a complex world* (pp. 145–181). Mahwah, NJ: Lawrence Erlbaum Associates.

Knorr-Cetina, K. (1999). *Epistemic cultures: how the sciences make knowledge*. Cambridge, MA: Harvard University Press.

Latour, B. (1987). *Science in action: How to follow scientists and engineers through society.* Cambridge, MA: Harvard University Press.

Lemke, J. L. (1990). *Talking science: Language, learning, and values.* Norwood, NJ: Ablex.

McDonald, S., & Songer, N. (2008). Enacting classroom inquiry: Theorizing teachers' conceptions of science teaching. *Science Education, 92*, 973–993.

Mortimer, E. F., & Scott, P. (2003). *Meaning making in secondary science classrooms*. Berkshire, England: Open University Press.

National Research Council. (2000). *Inquiry and the national science education standards.* Washington, DC: National Research Council.

Roth, W. M. (2005). *Talking science. Language and learning in science classrooms*. Oxford, UK. Rowman & Littlefield.

Ryu, S. R., & Sandoval, W. S. (2008). *Interpersonal influences on collaborative argument during scientific inquiry.* Paper presented at the annual meeting of the American Educational Research Association in New York.

Sampson, V., & Clark, D. B. (2011). A comparison of the collaborative scientific argumentation practices of two high and two low performing groups. *Research in Science Education, 41*(1), 63–97.
Sandoval, W. A., & Millwood, K. A. (2005). The quality of students' use of evidence in written scientific explanations. *Cognition and Instruction, 23*(1), 23–55.
Strike, K. A., & Posner, G. J. (1992). A revisionist theory of conceptual change. In R. Duschl & R. Hamilton (Eds.), *Philosophy of science, cognitive psychology, and educational theory and practice* (pp. 147–176). Albany, NY: SUNY.
Toulmin, S. (1958). *The uses of argumentation.* London: Cambridge University Press.
Traweek, S. (1988). *Beamtimes and lifetimes: The world of high energy physicists.* Cambridge, MA: Harvard University Press.
von Aufschnaiter, C., Erduran, S., Osborne, J., & Simon, S. (2008). Arguing to learn and learning to argue: Case studies of how students' argumentation relates to their scientific knowledge. *Journal of Research in Science Teaching, 45*(1), 101–131. doi:10.1002/tea.20213
Wittgenstein, L. W. (1969). *On certainty.* New York: Harper Torchbooks.

Chapter 14
Development of Argumentative Knowledge in Science Education

Myint Swe Khine

Introduction

In recent years, argumentation has emerged as one of the major topics of discussion among science educators and researchers. There is a firm belief that fostering argument in learning activities can develop students' critical thinking and reasoning skills. In addition, argumentation can help students in knowledge integration and complex meaning making. In order to engage students in scientific argumentation, students need to get involved in dialogic and collaborative inquiries. The consensus among educators is that student engagement in scientific argumentation needs to play a critical role in the education process. Recent analysis of research trends in science education indicates that the research topic of argumentation is prevalent in the literature.

Bottcher and Meisert (2011) quoting Erduran and Jiménez-Aleixandre (2008) state that argumentation can support the following aspects in science education:

- The access to the cognitive and metacognitive processes characterising expert performance and enabling modelling for students.
- The development of communicative competences and particularly critical thinking.
- The achievement of scientific literacy and empowering of students to talk and to write the languages of science.
- The enculturation into the practices of the scientific culture and the development of epistemic criteria for knowledge evaluation.
- The development of reasoning, particularly the choice of theories or positions based on rational criteria.

M.S. Khine (✉)
Bahrain Teachers College, University of Bahrain, Manama, Kingdom of Bahrain
e-mail: mskhine@btc.uob.bh; Dr.MKhine@gmail.com

M.S. Khine (ed.), *Perspectives on Scientific Argumentation*,
DOI 10.1007/978-94-007-2470-9_14,

This book attempts to consolidate contemporary thinking and research efforts in the role of scientific argumentation in education. The importance of language, discursive practice, social interactions and culture in the classrooms was investigated. The book brings together prominent scholars in the field to share their contemporary knowledge about the place of scientific argumentation in teaching and learning. The book is organized into three parts. The first part covers the theoretical premises of the study of argumentation, the second part presents practice perspectives in argumentation and the last part includes recent research studies on argumentation in science education.

Theoretical Premises of the Study of Argumentation

In Chapter 1, Osborne and colleagues introduce the importance of argumentation not only in science education but also in mathematics education, language, arts and history teaching. Their extensive review of scientific argumentation research covers the last 40 years of work by the academic community. Particularly, the research has grown exponentially in the last decade, but much of the role and value of augmentation are yet to be discovered. They point to the need to discern more about how argumentation functions, how students' processes of argumentation might be assessed and how the use of argumentation can become a common practice among science teachers.

In Chapter 2, Nussbaum and his team describe that socio-scientific discussions informed by basic science are more meaningful than other types of argumentation. They illustrate a case study on using argumentation in climate change education in a seventh-grade classroom and analyse the discussion in terms of the effect of asking critical questions. The study uses 'Losing the Lake' computer game to specifically look into the quality and depth of students' deliberative argumentation. The authors argue that for students to understand different argumentation schemes, they should be asked appropriate questions that address the schemes. They conclude that the more the students learn about the nature of argumentation, the more they can think critically about science and hence a broader view of argumentation can be encouraged.

Cavagnetto and Hand in Chapter 3 reiterate the fact that argumentation is a critical element to science instruction for building a student's conceptual understanding. They explore how scientific argument is fostered in science classrooms using questions such as 'What is the role of language in science? And what is the relationship between data and evidence?' They have developed the Science Writing Heuristic (SWH) approach that consists of a framework to guide activities in science lessons. The chapter reports the role of language in argument-based interventions in scientific argumentation.

In Chapter 4, Szu and Osborne explore scientific reasoning and argumentation from a Bayesian perspective. They describe that classroom instructions are mostly dominated by explanation rather than argument and recall rather than reasoning.

They note the three perspectives on the nature of scientific reasoning, namely the psychological, the philosophical and the sociological. While the psychological perspective is related to the work of Piaget, the philosophical account is related to a number of thinkers including Bacon, Popper, Khun, Feyerabend, Toulmin and others. The authors introduce the Bayesian perspective that describes the certainty of knowledge that reflects on possibilities assigned to a given hypothesis or event. The authors concluded that Bayesian inference will provide an alternative lens to explain science education and educational psychology research.

In Chapter 5, Berland and Hammer discuss the literature on framing and see this as both psychological constructs that can be distinguished and as a dynamic process. They give a brief overview of research on framing and discuss the significance for researchers and educators who are interested in studying and fostering scientific argumentation in the classroom. The authors draw attention to the fact that research has progressed from focusing on students' argumentation skills to whether they recognize the 'point of argument'.

Practice Perspectives in Argumentation

Part II of this book includes chapters from practice perspectives in argumentation. This part begins with the chapter by Simon, Richardson and Amos from the University of London. Chapter 6 covers how the authors develop a pedagogical practice framework that consists of three stages, namely activity design, enactment and transfer. In the first stage argumentation, activities are purposely designed with specific learning aims for argumentation and the enactment of an activity is described to show how teachers can use an activity to design and implement the instruction. Finally, the transfer of activities between curriculum designers and teachers are studied as fidelity or adaptation from the original design. By using this framework, the authors illustrate an example on design for the development of students' knowledge and reasoning on the topic of Olympic Park. The authors conclude that teachers need careful guidance in argumentation activities to ensure that they can implement the lesson effectively.

In Chapter 7, Bricker and Bell synthesize their works on argumentation for school science learning environments. They describe design-based research that is needed to investigate students' appropriation of scientific argumentation using their everyday argumentative practices. The research is based on the design of a computer program called SenseMaker that provides a argument-mapping environment. The software is used to study conceptual change in middle school science students. The chapter reports findings from a series of experiments conducted by both the researchers. They suggest that teachers need to provide opportunities for students to argue scientifically.

In order to help students engage in higher-order thinking associated with argumentation, Bulgren and Ellis develop a set of instructional tools that would support science teachers to improve the scientific argumentation among students. In

Chapter 8, the authors focus on the development of Argumentation and Evaluation Intervention (AEI) and the graphic organizer Argumentation and Evaluation Guide (AEG) for use in middle and secondary science classrooms. The authors present a detailed description of AEG that is framed according to the components of argumentation by Toulmin (1958). The AEG graphic organizer provides a space for each step and an associated question of the argumentation process. They provide the information from teachers about teaching students at different grade levels, components of argumentation that were particularly challenging and required different levels of support and feedback from students. In conclusion, the authors pose a question on whether AEI would enhance learning in different content areas with challenges in higher thinking.

Researching Argumentation in Science Education

In Part III, a number of researchers share their recent findings from their projects. Chapter 9 is the work carried out by Clark and his team on critique and argumentation from the Technology Enhanced Learning in Science Center (TELS). The TELS is organized and conducted in the Web-based Inquiry Environment (WISE).

Climate change is one of the widely argumented debates in recent years. The questions about how climate change impacts on society and to what extent it affects the way of life are still unanswered. In Chapter 10, Corner describes how people evaluate arguments about climate change and outlines some possible answers to this question. With references to popular climate change narratives, empirical data on climate change argument evaluation and first-hand experiences of climate change experts, the author examines the way that people evaluate arguments about climate change.

In Chapter 11, Wu and Tsai present their findings on university students' argumentation on socio-scientific issues via on-line discussion. As many as 37 university students from a national university in Taiwan took part in this study. The students took a course on 'Natural, Technology and Society' that covers aspects of socio-scientific issues such as nuclear power usage, global warming and genetic engineering. The authors explore the effects of anonymous on-line discussion task on students' informal reasoning outcomes with regards to socio-scientific issues.

Sampson and his colleagues share their experience in the development and validation of the Assessment of Scientific Argumentation in the Classroom (ASAC) observation protocol in Chapter 12. They use the seven-step process to create a high-quality assessment instrument that is both valid and reliable. The chapter details the results of the item-by-item analysis together with the expert ratings of the content and translational validity of the items. The development of this observation protocol has made a contribution to the growing field of scientific argumentation research in science education. The ASAC serves different purposes for science education research, including to access the quality of the argumentation and to examine the efficacy of a new curriculum or instructional strategy to improve the arugmentation. The ASAC can also be used for professional development for teachers.

In Chapter 13, McDonald and Kelly describe the sense-making discourse in the science classroom. They begin the chapter with the transcript of an exchange between teachers and students that shows the potential of viewing reasoning from an argumentation point of view as well as the confusing nature of interpretation of discourse in a sense-making situation. The second example depicts the discourse in which the teacher brings the class to closure and attempts to connect the students' ideas to create an understanding of the concept taught during the lesson. By using these two examples, the authors discuss the rationale for the focus on argumentation as pedagogical goal in science classrooms. The three premises to measure students' ability to engage in science are identified.

Conclusion

This book explores scientific argumentation as a means of addressing and solving problems in conceptual change, reasoning, knowledge building and promoting science literacy. The chapters in this book inform theoretical frameworks, new directions and changing practices from intervention studies, discourse analyses, classroom-based experiments, and design-based research. It is hoped that the book will be a critical and specialized source that describes perspectives on scientific argumentation and implications for science education.

References

Bottcher, F., & Meisert, A. (2011). Argumentation in science education: A model-based framework. *Science and Education, 20*, 103–140.

Erduran, S., & Jiménez-Aleixandre, M. P. (2008). *Argumentation in science education*. Dordrecht, The Netherlands: Springer.

Toulmin, S. (1958). *The uses of argument*. Cambridge: Cambridge University Press.

Index

M.S. Khine (ed.), *Perspectives on Scientific Argumentation*,
DOI 10.1007/978-94-007-2470-9, © Springer Science+Business Media B.V. 2012

MIX
Papier aus verantwortungsvollen Quellen
Paper from responsible sources
FSC® C105338

If you have any concerns about our products,
you can contact us on
ProductSafety@springernature.com

In case Publisher is established outside the EU,
the EU authorized representative is:
Springer Nature Customer Service Center GmbH
Europaplatz 3, 69115 Heidelberg, Germany

Printed by Libri Plureos GmbH
in Hamburg, Germany